Computational Earthquake Science Part I

Edited by
Andrea Donnellan
Peter Mora
Mitsuhiro Matsu'ura
Xiang-chu Yin

Springer Basel AG

Reprint from Pure and Applied Geophysics
(PAGEOPH), Volume 161 (2004), No. 9/10

Editors:

Andrea Donnellan
Earth and Space Sciences Division
Jet Propulsion Laboratory
California Institute of Technology
Pasadena, CA 91109-8099
USA
e-mail: donnellan@jpl.nasa.gov

Peter Mora
QUAKES, Earth Systems Science
Computational Centre
(ESSCC), Department of Earth Sciences
The University of Queensland
4072 Brisbane, Qld
Australia
e-mail: morap@quakes.uq.edu.au

Mitsuhiro Matsu'ura
Department of Earth and Planetary Science
The University of Tokyo
Bunkyo-ku
Tokyo 113-0033
Japan
e-mail: matsuura@eps.s.u-tokyo.ac.jp

Xiang-chu Yin
Center for Analysis and Prediction
CSB & Laboratory of Nonlinear Mechanics
Institute of Mechanics
China Academy of Sciences
Beijing 100080
China
e-mail: xcyin@public.bta.net.cn

A CIP catalogue record for this book is available from the Library of Congress, Washington D.C., USA

Bibliographic information published by Die Deutsche Bibliothek:
Die Deutsche Bibliothek lists this publication in the Deutsche Nationalbibliografie;
detailed bibliographic data is available in the internet at <http://dnb.ddb.de>

ISBN 978-3-7643-7142-5   ISBN 978-3-0348-7873-9 (eBook)
DOI 10.1007/978-3-0348-7873-9

Printed on acid-free paper produced from chlorine-free pulp

ISBN 978-3-7643-7142-5

9 8 7 6 5 4 3 2 1   www.birkhauser-science.com

Contents

Pure appl. geophys. 161 (2004) 1823–1825
0033–4553/04/101823–3
DOI 10.1007/s00024-004-2533-2

❙ Pure and Applied Geophysics

Computational Earthquake Science
Part I

ANDREA DONNELLAN[1], PETER MORA[2], MITSUHIRO MATSU'URA[3],
and XIANG-CHU YIN[4]

Key words: Earthquake physics, numerical modeling, finite element models, faults, rheology, computational science.

1. Introduction

During the last decade earthquake science has benefited from new observations, improved computational technologies, and improved modeling capabilities. Combining approaches in computational science, data assimilation, and information technology are improving our understanding of earthquake physics and dynamics. The scientific method relies on development of a theoretical framework or simulation model describing nature. While no such model exists for the complete earthquake generation process, conceptual developments in understanding earthquake physics, numerical simulation methodology and advances in advanced computing offer the possibility to develop such models. Development of simulation models represents a grand scientific challenge due to of the complexity of phenomena and range of scales involved from microscopic to global. Such models are providing powerful new tools for studying earthquake precursory phenomena and the earthquake cycle. They will have direct application to earthquake hazard studies and earthquake engineering, and the potential to yield spin-offs in sectors such as mining, geophysical exploration, high performance computing, material science, and geotechnical engineering.

---

[1] Earth and Space Sciences Division, Jet Propulsion Laboratory, California Institute of Technology, Pasadena, CA 91109-8099, U.S.A. E-mail: donnellan@jpl.nasa.gov

[2] QUAKES, Earth Systems Science Computational Centre (ESSCC), Department of Earth Sciences, The University of Queensland, 4072 Brisbane, Qld, Australia. E-mail: morap@quakes.uq.edu.au

[3] Department of Earth and Planetary Science, The University of Tokyo, Bunkyo-ku, Tokyo 113-0033, Japan. E-mail: matsuura@eps.s.u-tokyo.ac.jp

[4] Centre for Analysis and Prediction, CSB and Laboratory of Nonlinear Mechanics, Institute of Mechanics, China Academy of Sciences, Beijing 100080, China. E-mail: xcyin@public.bta.net.cn

To understand a nonlinear earthquake fault system necessarily implies that predictions about the future behavior and dynamics of the system can be made whose accuracy can be tested by future observations. This procedure is the true essence of the scientific method. Predictive models and simulations that capture the essential physics and dynamics of the system covered by earthquake observations can be tested by future observations. The construction of models is necessary, since earthquake observations can only be taken at the boundary (surface) of the earth, or at most in a small number of selected internal locations. Moreover, most of the significant nonlinear dynamical processes within earthquake fault systems operate over a vast range of spatial and temporal scales, from scales much smaller than human experience (tiny fractions of seconds and meters), to scales far larger (thousands of kilometers and many millions of years). Linkage of the processes over these scales means that understanding the physics at one set of scales cannot, in principle, be achieved without consideration of many other scales. Since our ability to make observations will always be limited by practical considerations, simulations are needed to interpolate between, and extrapolate beyond, the scales of resolution at which we can observe.

Modern developments in computational science and information technology have fundamentally altered the means by which knowledge is acquired, stored, manipulated, represented, and used during the modeling process. Specifically, the advent of the World Wide Web and the development of computational grids enabled by object definitions, middleware, and multi-tiered information architectures allow data and models to be manipulated by symbolic, and far more intuitive procedures. Thus new modes of scientific collaboration, discovery, and advance emerge as the people, databases and web pages, simulations and their results, sensors and their filtered data interact.

During the week of May 5–10, 2002 the United States hosted the Third International Workshop of the ACES (APEC Cooperation for Earthquake Simulations) in Maui, Hawaii. The workshop consisted of five days of technical discussions with no parallel sessions. The sessions focused on microscopic simulations, scaling physics, macro-scale simulations on earthquake generation and cycles and on dynamic rupture and wave propagation, computational environment and algorithms, data assimilation and understanding, and model applications.

The inaugural Workshop of ACES was held in 1999 in Brisbane and Noosa, Australia, during which time five topical working groups were formed and initial working group goals were identified. At a subsequent Working Group Meeting held in Tokyo in January 2000, two new working groups were added. This two-part volume represents articles from the seven working groups. Approximately 70 people attended the inaugural meeting in Brisbane, Australia in 1999, 130 people attended the meeting in Hakone, Japan in 2000, and 150 people attended the third meeting in Maui in 2002.

ACES aims to develop realistic supercomputer simulation models for the complete earthquake generation process, thus providing a "virtual laboratory" to probe earthquake behavior. This capability will provide a powerful means to study the earthquake cycle, and hence, offers a new opportunity to gain an understanding of the earthquake nucleation process and precursory phenomena. The project represents a grand scientific challenge due to of the complexity of phenomena and range of scales from microscopic to global involved in the earthquake generation process. It is a coordinated international effort linking complementary nationally based programs, centers, and research teams.

This issue is divided into two parts. The first part incorporates microscopic simulations, scaling physics, and earthquake generation and cycles. The second part encompasses dynamic rupture and wave propagation, computational environment and algorithms, data assimilation and understanding, and model applications.

Articles in Part I address constitutive properties of faults, scaling properties, and statistical properties of fault behavior. It also focuses on plate processes and earthquake generation from a macroscopic standpoint.

Part II addresses dynamic properties of earthquakes and the applications of models to earthquakes. It also contains articles on the computational approaches and challenges of constructing earthquake models. Data assimilation is critical to improving our understanding of earthquake processes, and papers addressing it are found in Part II.

We thank all of the participants of the 3rd ACES workshop and the contributors to these special issues. We particularly thank the Secretary General, John McRaney, for all of his undertakings working with the sponsors and implementing the workshop. We are grateful to Teresa Baker who served as the assistant editor for this volume, and to Ziping Fang who oversaw the web pages, and submittals and revisions of the papers. Without the efforts of these three people the workshop and publication of this volume would not have been possible. We also thank our sponsors including NASA[1], NSF[2], the USGS[3], DEST[4], ARC[5], QUAKES[6], UQ[7], RIST[8], NSFC[9], MOST[10], and CSB[11]. Portions of this work were carried out at the Jet Propulsion Laboratory, California Institute of Technology under contract with NASA.

---

[1] National Aeronautics and Space Administration

[2] National Science Foundation

[3] United States Geological Survey

[4] Department of Education, Science and Training, Commonwealth of Australia

[5] Australian Research Council

[6] Queensland University Advanced Centre for Earthquake Studies

[7] The University of Queensland

[8] Research Organisation for Information Science and Technology, Japan

[9] National Natural Science Foundation of China

[10] Ministry of Science and Technology of the People's Republic of China

[11] China Seismological Bureau

A. Microscopic Simulation

Pure appl. geophys. 161 (2004) 1829–1839
0033–4553/04/101829–11
DOI 10.1007/s00024-004-2534-1

| Pure and Applied Geophysics

Statistical Tests of Load-Unload Response Ratio Signals by Lattice Solid Model: Implication to Tidal Triggering and Earthquake Prediction

YUCANG WANG[1,2], PETER MORA[1], CAN YIN[1], and DAVID PLACE[1]

Abstract—Statistical tests of Load-Unload Response Ratio (LURR) signals are carried in order to verify statistical robustness of the previous studies using the Lattice Solid Model (MORA *et al.*, 2002b). In each case 24 groups of samples with the same macroscopic parameters (tidal perturbation amplitude A, period T and tectonic loading rate k) but different particle arrangements are employed. Results of uni-axial compression experiments show that before the normalized time of catastrophic failure, the ensemble average LURR value rises significantly, in agreement with the observations of high LURR prior to the large earthquakes. In shearing tests, two parameters are found to control the correlation between earthquake occurrence and tidal stress. One is, $A/(kT)$ controlling the phase shift between the peak seismicity rate and the peak amplitude of the perturbation stress. With an increase of this parameter, the phase shift is found to decrease. Another parameter, AT/k, controls the height of the probability density function (Pdf) of modeled seismicity. As this parameter increases, the Pdf becomes sharper and narrower, indicating a strong triggering. Statistical studies of LURR signals in shearing tests also suggest that except in strong triggering cases, where LURR cannot be calculated due to poor data in unloading cycles, the larger events are more likely to occur in higher LURR periods than the smaller ones, supporting the LURR hypothesis.

Key words: Load-Unload Response Ratio (LURR), Lattice Solid Model (LSM), numerical simulation, tidal trigger, earthquake prediction.

1. Introduction

Tidal trigger of earthquakes or possible correlation between tidal forces and earthquakes aroused interest for several decades (CATTON, 1922; TSURUOKA *et al.*, 1995; EMTER, 1997). It seems plausible that tidal forces should trigger earthquakes considering the fact that the peak rate of tidal forces may be significantly higher than average rates of tectonic stress buildup. Although there are studies reporting a positive correlation between earthquake occurrence and the earth tide (TSURUOKA

[1] QUAKES, Department of Earth Sciences, The University of Queensland, Qld 4072, Brisbane, Australia. E-mails: wangyc@quakes.uq.edu.au, mora@quakes.uq.edu.au, canyon@quakes.uq.edu.au, place@quakes.uq.edu.au
[2] LNM, Institute of Mechanics, Chinese Academy of Sciences, Beijing, 100080, China. E-mail: yin@lnm.imech.ac.cn

et al., 1995; WILLIAM and WILCOCK, 2001; TOLSTOY *et al.*, 2002), negative results have also been published (HEATON, 1982; RYDELEK *et al.*, 1992; VIDALE *et al.*, 1998). This discrepancy at least suggests that no significant strong correlation has been widely observed.

The Load-Unload Response Ratio hypothesis (LURR) views this problem from a different perspective. The basic assumption of LURR is that at an earlier stage of damage a system is stable, and not sensitive to minor external disturbances, whereas when approaching the macroscopic failure, the system will have a quite different response to minor external loading and unloading. At this stage it may be possible for tidal correlation to be observed. The reason why no significant tidal triggering has been widely observed may be that the instantaneous tidal loads are not large enough to always trigger earthquakes and are not the only factor effecting earthquakes. In retrospective studies, high LURR values have been observed months to years prior to most events and intermediate-term earthquake predictions have been made (YIN *et al.*, 2000, 2002). Recent study indicates that optimal LURR region sizes scale with magnitudes of forthcoming events, similar to Accelerating Moment Release (AMR) observations (BOWMAN *et al.*, 1998), which suggest that both LURR and AMR may be used as a measure of the proximity to a critical state, and may have the same underlying physical mechanism (YIN *et al.*, 2002).

As a first step towards understanding the underlying physical mechanism for the LURR observations, numerical studies are conducted using the Lattice Solid Model (LSM) (MORA *et al.*, 2002b). The preliminary simulations reproduce signals similar to those observed in earthquake prediction practice with a high LURR value followed by a sudden drop prior to macroscopic failure of the sample, suggesting that LURR provides a good predictor for catastrophic failure in elastic-brittle systems. Since random-sized particles are involved in the model, different random configurations and parameters are needed to test statistical robust and possible dependence on the parameters.

In this paper, we first present the statistical features of LURR under uni-axial compression using the Lattice Solid Model. Since shearing simulations generate more events, in the second part, we investigate the effects of parameters on earthquake occurrence and statistical features of LURR under shearing loading.

2. Statistical Tests of LURR under Uni-axial Compression Using Different Random Configurations

As in the previous study (MORA *et al.*, 2002b), the model is initialized as a heterogeneous 2-D block made up of random-sized particles with diameters ranging from 0.2 to 1 model units. The system is subjected to uni-axial compression from rigid driving plates on the upper and lower edges of the model. Experiments are conducted using both strain and stress control to load the plates. The detailed calibration and fracture orientation can be found in the previous papers (MORA et al., 2002a, 2002b; PLACE et al., 2002). A sinusoidal stress perturbation is added to the constant loading rate to simulate loading and unloading cycles induced by tidal forces

$$\Delta\sigma_{zz} = A\cos(2\pi t/T) \tag{1}$$

where T and A are period and amplitude of tidal stress. Thus total stress is $\sigma_{zz} = \Delta\sigma_{zz} + kt$, k represents loading rate of tectonic stress σ_{zz}. LURR is calculated using

$$LURR = E^+/E^- \tag{2}$$

where E^+ and E^- respectively denote the cumulative seismic energy release during loading and unloading within a given time window. The cumulative seismic energy release is obtained by summing total kinetic energy released during all loading or unloading cycles within the specified time window, where we define loading when $d\sigma_{zz}/dt \geqslant 0$, and unloading when $d\sigma_{zz}/dt < 0$. The total kinetic energy release at any given instant t is the sum of the kinetic energy within the system and the energy lost to the artificial viscosity energy prior to time t (MORA et al., 2002b).

In this test, 24 groups of simulations with the same parameters of T, A and k but different configurations of random particles are conducted. LURR values are calculated for each sample in the same way as previous work (MORA et al., 2002b). Due to sample specificity (XIA et al., 1996; BAI et al., 2000), the fracture pattern and catastrophic failure times are different for each sample. Therefore failure times are normalized to unit for comparison and average. Figure 1 shows plots of the averaged LURR values from 24 groups versus normalized time for the same parameters used in the previous study. One can see that the averaged LURR begins to rise from $t = 0.5$, and reaches its peak at $t = 0.9$ and then drops just before the main event at

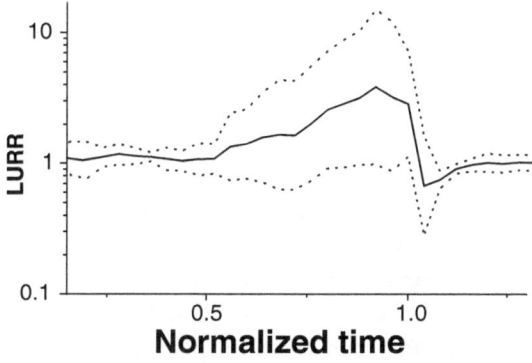

Figure 1
Evolution of ensemble average LURR computed from 24 samples. The dotted line marks the standard deviations. Parameters are: $k = 30$ MPa/100,000 time steps, $T = 4000$ time steps and $A = 0.96$ MPa.

$t = 1.0$. After $t = 1.0$, although there are many events, it is flat. It is also noted that the standard deviations of LURR increase before the main event.

Due to the limitation of computer power, it is unfeasible to scan all the parameter range, simulations are conducted using values T, A and k with a factor of five larger and smaller. Similar results suggest that our previous results are statistically robust.

Figure 2 shows the evolution of averaged cumulative energy release before the catastrophic failure using the same 24 groups of samples as in Figure 1. The dotted line is the fitted time-to-failure function according to $\Omega(t) = A + B(t_f - t)^m$ (BUFE *et al.*, 1993), here t_f is set to 1.0. Apparent accelerated energy release is observed, suggesting that both high LURR and AMR appear in these tests just before the main fracture.

There are still very few events before the catastrophic failure (generally 20–30 events) under uni-axial compression. In order to generate more events before the larger earthquakes, shearing tests are used in the next section.

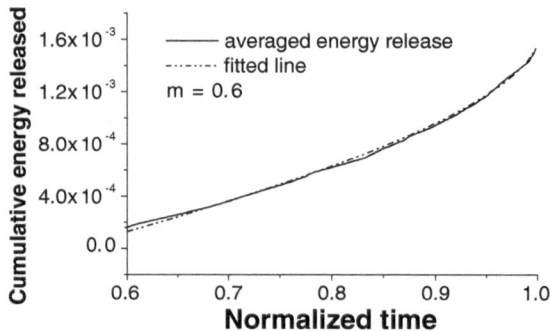

Figure 2

Evolution of ensemble average cumulative energy release . The dotted line is fitted time-to-failure function according to $\Omega(t) = A + B(t_f - t)^m$ (BUFE *et al.*, 1993), t_f is set to 1.0. Parameters are the same as above.

3. Statistical Test of LURR under Shearing Tests

3.1 Model Description

In this test the model consists of a central fault sandwiched between elastic regions which are attached to rigid driving plates at their outer boundaries (Fig. 3). The fault region is composed of random-sized particles. The surface roughness of the fault is controlled by two parameters: a fractal dimension D and height of asperities H (0.5 and 1.6 in this simulation). Between the fault particles, simple frictional force with magnitude proportional to the normal force is employed (PLACE *et al.*, 2001; MORA *et al.*, 2002a). Boundary condition is circular in a horizontal direction. The shearing is achieved by moving the driving plates at a moderate and constant rate which

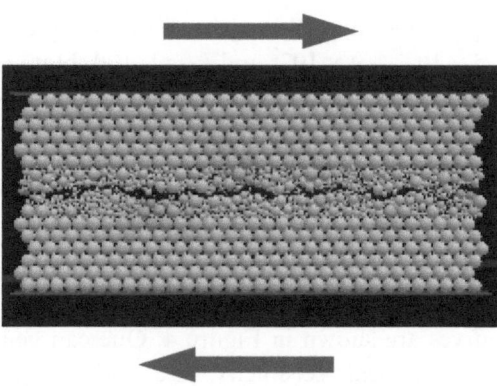

Figure 3
Snapshot of a shearing test. Boundary conditions are circular in the horizontal direction. A constant normal stress of 150 MPa is applied at the upper and lower rigid driving plates.

models the slow tectonic loading while maintaining a constant normal stress of 150 MPa. Similarly, a slight sinusoidal perturbation is added to the slow tectonic loading, simulating tidal stress. The model allows rupture to occur along the internal surface within the fault region, modeling the slip and stick process (and even breaking of asperities). Thus this model can be considered as a simplified model for an interacting fault system containing a long single fault.

3.2 Effects of the Parameters on Earthquake Occurrence

To simulate the effects of tidal stress on earthquake occurrence, the parameters cannot be chosen arbitrarily, but according to the following conditions

$$|\Delta\sigma_{zz}| \ll \sigma_{zz} \tag{3}$$

$$\frac{d|\Delta\sigma_{zz}|}{dt} \gg k \tag{4}$$

$$T_e \ll T \ll T_L. \tag{5}$$

where T_e is the earthquake rupture duration and T_L is the average time interval between large earthquakes. Two dimensionless parameters are introduced to describe the conditions above, $k_1 = \frac{|\Delta\sigma_{zz}|}{\sigma_{zz}}$, $k_2 = \frac{d|\Delta\sigma_{zz}|}{dt}/k = \frac{2\pi A}{Tk}$ Equations (3) and (4) require $k_2 >> 1$, and $k_1 \ll 1$, or $Tk/2\pi \ll A \ll kt$, where t is the tectonic loading time.

In nature, tidal stresses can be up to 0.001–0.004 MPa, with its rate of about 0.001 MPa/h and typical period T of 12 hours. The tectonic stress should be in the order of 10–100 MPa, with long-term rate of approximately 10^{-6}–10^{-4} MPa/h. so the approximate ranges of two parameters are: $k_1 = 10^{-5}$–10^{-4} and $k_2 = 10^{1}$–10^{3}.

Due to limited computer power, it is impossible to use the observed parameters in our numerical experiments, therefore we use a higher loading rate. In this simulation, the parameters are: $A = 0.06$–4.8 MPa, $T = 2000$–8000 time steps and

$k = 10^{-4}$–10^{-3} MPa per time step. Strength of the sample is found to be about 120 MPa. Thus $k_1 = 5 \times 10^{-4} - 4 \times 10^{-2}$ and $k_2 = 1$–100. Note that $k_2 < 1.0$ means a constant positive $d\sigma_{zz}/dt$, and a violation of Eq. (4), therefore LURR is not calculated in this case, nonetheless the probability density functions (see below) are still carried out to compare parameter effects.

To study the effects of parameters such as A, T and k on earthquake occurrence, we compared the probability density function (Pdf) of modeled events versus phase angle for different parameters. Here phase angle $\theta \in [-\pi, \pi]$, and $\theta + 2\pi n = \frac{2\pi}{T} t$, n is the integer number of cycles and t is event time.

The typical Pdf curves are shown in Figure 4. One can generally see a peak and phase shift ϕ between the peak seismicity rate and the peak amplitude of the perturbation stress. Detailed studies indicate that the Pdf curve is mainly controlled by two parameters, rather than A, T and k individually. One is k_2, controlling phase shift ϕ. This is clearly seen in Figure 4, where three curves have different A, T and k , but the same k_2, the Pdf peaks are found to appear in a similar place.

When k_2 is increased, the phase shifts seemingly decrease from 2π to 0 (Fig. 5), indicating that the peak seismicity occurrence rate shifts from the position of peak perturbation rate towards that of peak perturbation.

Another parameter, $k_3 \propto AT/k$ controls the peak height of the Pdf curves. As this parameter increases, the Pdf becomes sharper and narrower, indicating strong correlation (Fig. 6 and Fig. 4). When k_3 is too small (e.g., 0.08), however, Pdf is almost flat, indicating no apparent tidal correlation.

Therefore we can conclude that in such an elastic-brittle system, a larger amplitude of perturbation and smaller loading rate contribute stronger correlation

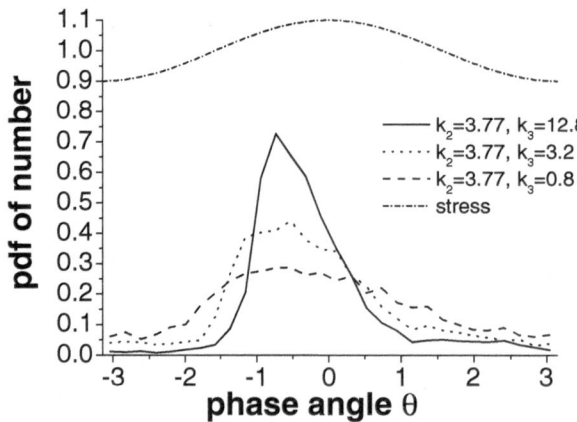

Figure 4

Plots of Pdf of events versus phase angle for different A/k and T although the same k_2 The perturbation stress is plotted for comparison. Parameters are: $k = 5$ Mpa/10000 time step, $T = 8{,}000$ time steps, $A = 2.4$ MPa ($k_3 = 12.8$); $k = 5$ MPa/10,000 time step, $T = 4{,}000$ time steps, $A = 1.2$ MPa ($k_3 = 3.2$); $k = 5$ MPa/10,000 time step, $T = 2{,}000$ time steps, $A = 0.6$ MPa ($k_3 = 0.8$).

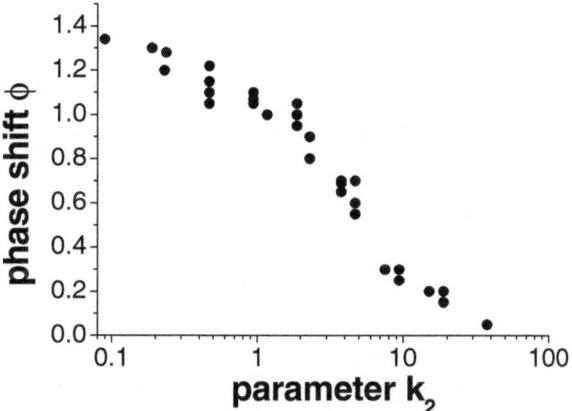

Figure 5

Dependence of phase shift ϕ on k_2. When k_2 increases, ϕ decreases, indicating a shift of the peak seismicity rate towards the position of peak tidal stress.

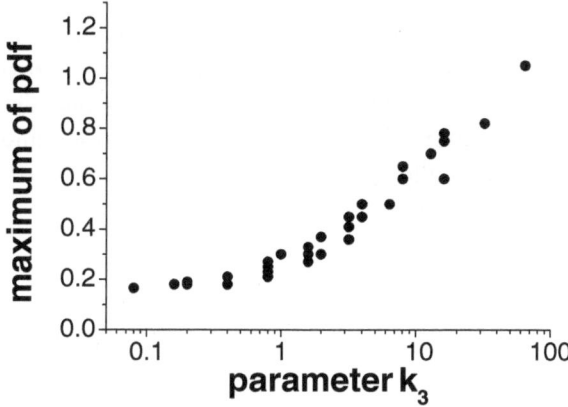

Figure 6

Relation between the peak height of the Pdf curves and parameter k_3 The larger k_3 corresponds to the higher and narrower Pdf curves, indicating stronger triggering.

between tidal stress and seismicity. Besides A and k, the period of tidal stress T also plays an important role. On one hand, decreasing T makes the peak seismicity rate approach the peak of the perturbation stress. On the other hand, smaller T causes the low peak height of Pdf curve, lessening the tidal effects.

3.3 Statistics of LURR in Shearing Tests

Since a phase shift is observed between the peak seismicity rate and the peak of the perturbation stress, loading and unloading are redefined here. In this study

$\cos(2\pi t/T + \phi) > 0$ is defined as loading, otherwise as unloading. Meanwhile, the averaged triggering effect should be deducted in order to keep averaged LURR around 1. A parameter is defined to roughly describe the average triggering effect.

$$\alpha = \frac{\text{total event number in loaing cycles}}{\text{total event number in unloading cycles}}. \tag{7}$$

Then LURR values are calculated using

$$LURR = \frac{\sum E^+ / \sum E^-}{\alpha} \tag{8}$$

where E^+ and E^- are kinetic energy released during modeled earthquake events in loading and unloading cycles respectively. Time window of $3\,T$ and energy cut-off of 3e–5 (events with energy less than 3e–5 are employed in calculations) are used to avoid large fluctuation. Unlike the compression test, no predominant large events are found in this simulation, a different statistical method is adopted instead of normalizing the failure time into unit. For 8 groups of simulation with the same parameters, we divide the total events into two groups: the 10% largest events and the rest of the smaller ones. Then we compare Pdf versus LURR values for the two groups. Dot line in Figure 7 shows the percentage of smaller events which occur in a certain pin of LURR; we can see that Pdf is nearly random distributed. However the apparent right shift of Pdf for the 10% largest events (solid lines) means that the larger events may occur more likely in high LURR periods and less likely in low LURR periods than the smaller ones.

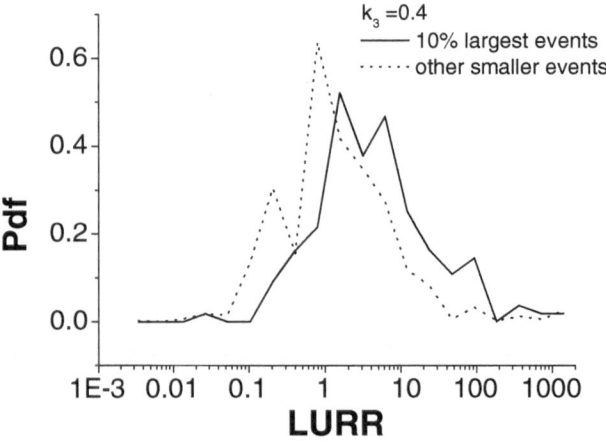

Figure 7

Pdf versus LURR for the largest 10% of events (solid) and other smaller events (dotted). $k = 5$ MPa/ 10,000 time step, $T = 2,000$ time steps, $A = 0.6$ MPa and $k_3 = 0.4$, $\alpha = 1.31$.

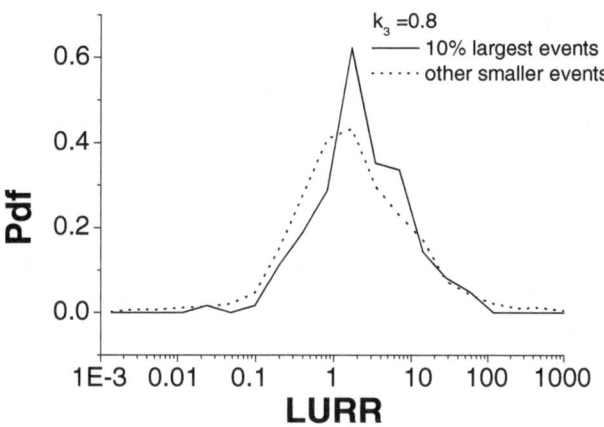

Figure 8

Pdf versus LURR for the largest 10% of events (solid) and other smaller events (dotted). $k = 5$ MPa/
10,000 time step, $T = 2,000$ time steps, $A = 0.3$ MPa and $k_3 = 0.8$, $\alpha = 1.45$.

Since in a strong triggering case very few events would occur in unloading cycles, making it impossible to calculate LURR. LURR is calculated only in weak triggering cases. Our statistical results show that except for strong triggering cases with large k_3 (e.g. $k_3 > 0.8$), similar results are observed in all other cases (Fig. 8). These results suggest that although no apparent overall tidal correlations of earthquake are widely observed, it is still possible that the large events are more likely to occur in high LURR periods.

4. Discussions and Conclusions

The statistical studies presented in this paper verified statistical robustness of previous results (MORE et al., 2002b). In the weak triggering cases of shearing tests, larger events are found more likely to occur when LURR is high than the smaller events, indicating that although no overall tidal triggering is observed, it is still possible to observe some correlation prior to strong earthquakes.

Within our parameter ranges, a seemingly intrinsic phase shift between the peak seismicity rate and the peak amplitude of the perturbation stress, decided by $A/(kT)$, is observed in this simple elastic-brittle system. This phenomenon is also supported by experimental results (LOCKNER and BEELER 1999) and simulation (TSURUOKA et al., 2002). The height of the probability density function of seismicity, or degree of tidal triggering, is controlled by another parameter, AT/k. The physical mechanisms of these two parameters remain unclear, they deserve further study, both observationally and numerically.

The results in this paper present another possible reason to explain why tidal triggering has not been widely observed. According to Vidale (VIDALE et al., 1998), the preseismic stress rate is much higher (1,000 times) than the long-term tectonic rate, subsequently the effect of tides is lessened. Our simulations do not contradict this idea, for the high preseismic stress rate means low k_3 in our study, therefore low correlation. A mechanism for delayed failure has been suggested to explain the observations (LOCKNER and BEELER 1999) and such a mechanism should be incorporated into the simulation model in future work. However, our studies suggest that even in elastic-brittle system without time-delayed failure, smaller period of perturbation T can also contribute to weak correlation between tidal stress and seismicity. This also implied that when testing tidal triggering of earthquakes, it is necessary to devote attention to the longer period components of tidal stress instead of only the shortest one.

Acknowledgments

This study is supported by the Australia-China special Fund for Scientific and Technological Cooperation, UQ New Staff Research Start-up Fund Chinese NSF Fund for International Exchange and Cooperation and Chinese NSF (grant number 40004002). The authors thank two anonymous reviewers for their valuable advice enhancing this paper.

REFERENCES

BAI, Y. L. WEI, Y. J., XIA, M. F., and KE, F. J. (2000), *Weibull Modulus for Diverse Strength due to Sample-specificity*, Theoret. Appl. Fracture Mech. *34* (3), 211–216.

BOWMAN, D. D., OUILLON, G., SAMMIS, C. G., SORNETTE, A., and SORNETTE, D. (1998), *An Observational Test of the Critical Earthquake Concept*, J. Geophys. Res. *103*, 24359–24372

CATTON, L. A. (1922), *Earthquake Frequency, with Special Reference to Tidal Stress in the Lithosphere*, Bull. Seismol. Soc. Am. *12*, 47–198.

EMTER, D. (1997), *Tidal Triggering of Earthquakes and Volcanic events, in (Wilhelm, H., et al. ed,)* Tidal Phenomena, Springer-Verlag (1997), pp 293–309.

HEATON, T. H. (1982), *Tidal Triggering of Earthquakes*, Bull. Seismol. Soc. Am. *72*, 2181–2200.

LOCKER, D. A. and BEELER, N. M. (1999), *Premonitory Slip and Tidal Triggering of Earthquakes*, J. Geophys. Res. *104*, 20,133–20,151.

MORA, P. and PLACE, D. (2002a), *Stress Correlation Function Evolution in Lattice Solid Elasto-dynamic Models of Shear and Fracture Zones and Earthquake Prediction*, Pure Appl. Geophys. *159*, 2413–2427.

MORA, P. WANG, Y. C YIN, C. PLACE, D., and YIN, X. C. (2002b), *Simulation of the Load-unload Response Ratio and critical Sensitivity in the Lattice Solid model*, Pure Appl Geophys. *159*, 2525–2536.

PLACE, D. and MORA, P., (2001), *A random lattice solid model for simulation of fault zone dynamics and fracture processes*. In Bifurcation and Localisation Theory for Soils and Rocks'99, (eds. Mühlhaus, H-B., DYSKIN, A. V. and PASTERNAK, E.) (AA Balkema, Rotterdam/Brookfield 2001).

PLACE, D., LOMBARD, F., MORA, P., and ABE, S., 2002, *Simulation of the Micro-physics of Rocks Using LSmearth*, Pure Appl Geophys., *159*, 1933–1950.

RYDELEK, P. A., SACKS, I. S., and SCARPA, R. (1992), *On Tidal Triggering of Earthquake Swarms at Campi Flegrei*, Geophys. J. Int. *109*, 125–137.

TOLSTOY, M., VERNON, F. L., ORCUTT, J. A. and WYATT, F. K., (2002), *Breathing of the Seafloor: Tidal Correlations of Seismicity at Axial Volcano*, Geology *30*, 503–506.

TSURUOKA, H. and OHTAKE, M., 2002, *Effects of the Earth Tide on Earthquake Occurrence: An Approach by Numerical Simulation*, J. of Geography *111*, 256–267. (Japanese with English abstract).

TSURUOKA, H., OHTAKE, M., SATO, H. (1995), *Statistical Test of the Tidal Triggering of Earthquakes: Contribution of the Ocean Tide Loading Effect*, Geophys. J. Int. *122*, 183–194.

VIDALE, J. E., AGNEW, D. C., JOHNSTON, M. J. S., and OPPENHEIMER, D.H.1998, *Absence of Earthquakes Correlation with Earth Tides: An Indication of High Preseismic Fault Stress rate*, J. Geophys. Res. *103*, 24,567–24,572.

WILLIAM, S. and WILCOCK, D, 2001, *Tidal Triggering of Microearthquakes on the Juan de Fuca Ridge*, Geophys. Res. Lett. *28*, 3999–4002.

XIA, M. F., SONG, Z. Q., XU, J. B., ZHAO, K. H., and BAI, Y.L. (1996), *Sample-specific Behavior in Failure Models of Disordered Media*, Communications in Theoretical Physics, *25*(1), 49–54.

YIN, X. C., CHEN, X. Z., SONG, Z. P., and YIN, C. (1995), *A New Approach to Earthquake Prediction: the Load\Unload Response Ratio (LURR) Theory*, Pure Appl. Geophys. *145*, 701–715.

YIN, X. C., WANG, Y. C., PENG, K. Y., BAI, Y. L. (2000), *Development of a New Approach to Earthquake Prediction: Load/Unload Response Ratio (LURR) Theory*, Pure Appl. Geophys. *157*, 2365–2383.

YIN, X. C., MORA, P., PENG, K. Y., WANG, Y. C., and WEATHERLEY, D. (2002), *Load-unload Response Ratio and Accelerating Moment/Energy Release critical Region Scaling and Earthquake Prediction*, Pure Appl. Geophys. *159*, 2511–2523.

(Received September 27, 2002, revised January 27, 2003, accepted February 10, 2003)

 To access this journal online:
http://www.birkhauser.ch

Pure appl. geophys. 161 (2004) 1841–1852
0033–4553/04/101841–12
DOI 10.1007/s00024-004-2535-0

┃Pure and Applied Geophysics

Long-range Stress Redistribution Resulting from Damage in Heterogeneous Media

YILONG BAI[1], ZHAOKE JIA[1], XIAOHUI ZHANG[1], FUJIU KE[1,2],
and MENGFEN XIA[1,3]

Abstract—It has been shown in CA simulations and data analysis of earthquakes that declustered or characteristic large earthquakes may occur with long-range stress redistribution. In order to understand long-range stress redistribution, we propose a linear-elastic but heterogeneous-brittle model. The stress redistribution in the heterogeneous-brittle medium implies a longer-range interaction than that in an elastic medium. Therefore, it is surmised that the longer-range stress redistribution resulting from damage in heterogeneous media may be a plausible mechanism governing main shocks.

Key words: Long-range stress redistribution, damage, heterogeneous media.

1. Introduction

Recently, the significance of long-range stress redistribution in understanding the earthquake mechanism has drawn considerable attention (HILL *et al.*, 1993; KLEIN *et al.*, 2000; WEATHERLEY *et al.*, 2000; RUNDLE, J.B. 1995; KNOPOFF, 2000). Various models of cellular automata (CA) with long-range stress redistribution for earthquake faults were widely used in these studies. Nonetheless, determination of the nature and significance of the long-range interaction is by no means an easy problem. Different research groups have different understandings. For instance, KLEIN *et al.* (2000) stated that "linear elasticity yields long-range stress tensors for a variety of geological applications" and "for a two-dimensional dislocation in a three-dimensional homogeneous elastic medium, the magnitude of the stress tensor goes as $\sim 1/r^3$." They noticed "while geophysicists do not know the actual stress tensors for real faults, they expect that long-range stress tensors, which are similar to the $\sim 1/r^3$ interaction, apply to faults." Moreover, they stressed that "it is suspected that microcracks in a fault, as well as other "defects" such as water, screen the

[1] State Key Laboratory of Nonlinear Mechanics, Institute of Mechanics, Chinese Academy of Sciences, Beijing 100080, China. E-mail: baiyl@lnm.imech.ac.cn
[2] Department of Applied Physics, Beijing University of Aeronautics and Astronautics, Beijing 100083, China. E-mail: kefj@lnm.imech.ac.cn
[3] Department of Physics, Peking University, Beijing 100871, China. E-mail: xiam@lnm.imech.ac.cn

$\sim 1/r^3$ interaction, leading to a proposed $\sim e^{-\alpha r}/r^3$ interaction, where $\alpha << 1$, implying a slow decay to the long-range interaction over the fault's extent."

On the contrary, WEATHERLEY *et al.* (2000) pointed out in their cellular automaton model that "the interaction exponent (p in $\sim 1/r^p$) determines the effective range for strain redistribution in the model. The effective range decreases rapidly as the exponent (p) increases. The event-size distributions illustrate three different populations of events in the dissipative healing models (two-dimensional models):

- Characteristic large events ($p < 1.5$),
- Power-law scaling events ($1.5 < p \leq 2.0$),
- Overdamped, no large events ($p > 2.0$).

They concluded that the models display a smooth transition from characteristic large events preceded by strain correlation evolution and accelerating energy release, to a power-law distribution of events preceded by linear energy release, as the effective range of interactions decreases. Given that the stress redistributions in three and two-dimensional homogeneous linear elastic media are $\sim 1/r^3$ and $\sim 1/r^2$ respectively, the difficulty in understanding long-range stress redistribution is obvious.

Physically, the existence of cracks and other "defects" like water may have two opposite effects on stress redistribution. One is to screen stress and then lead to a shorter-range redistribution, as claimed by KLEIN *et al.* (2000). On the other hand, the stress balance requires a compensational increase of the stress beyond the "defects," implying a longer-range redistribution of stress.

Recently, KNOPOFF (2000) investigated the magnitude distribution of declustered earthquakes in Southern California. He concluded that the characteristic length of 3 km in the magnitude distribution is a crossover between two different mechanisms in the physics of earthquake occurrence.

All of the results remind us that a declustered or characteristic large earthquake may occur relevant to some intrinsic length scales and with a longer-range stress redistribution. Then, instead of the commonly used homogeneous linear elastic theory, can we find possible alternative models with longer-range stress redistribution? This is the aim of this paper.

2. Possible Long-range Stress Redistribution in Heterogeneous Media

It is well known that the main terms of stress in a homogeneous linear elastic medium are $\sim 1/r^3$ in a three-dimension model with a spherical void and $\sim 1/r^2$ in a two-dimensional model with a cylindrical hole. Now, let us examine the stress redistribution owing to a void in a linear-elastic but heterogeneous-brittle medium to determine the effect of microdamage resulting from the heterogeneity on stress redistribution. In particular, we wish to determine whether microdamage is a "screen" leading to a shorter-range redistribution or a compensation implying a longer-range redistribution of stress.

It is assumed in the model that every mesoscopic element has the same elastic moduli, like Young's modulus E and Poisson ratio v, but various breaking strengths σ_c or strain threshold ε_c. Moreover, the strain threshold of the element follows a distribution function (Fig. 1),

$$h(\varepsilon_c) = \begin{cases} 0; & \text{when} \quad \varepsilon_c < \varepsilon_c^* \\ \frac{q-1}{\varepsilon_c^*}\left(\frac{\varepsilon_c}{\varepsilon_c^*}\right)^{-q}, q > 1; & \text{when} \quad \varepsilon_c \geq \varepsilon_c^* \end{cases} \tag{1}$$

where ε_c^* is the minimum of strain threshold, i.e., the strain thresholds of all elements are larger than ε_c^*, and the number of elements with strain threshold higher than ε_c^* decreases as a power law. Then, for strain less than ε_c^*, all mesoscopic elements remain solid, namely the system is elastic; however, as the strain is higher than ε_c^*, some mesoscopic elements will break and damage occurs. Also, the parameter q in the heterogeneous-brittle model should remain greater than unit and the greater the parameter q is, the stronger the heterogeneity relevant damage is, see Figure 1.

We must confess that we do not know the actual distribution of strength in geological media. However, the above simple distribution looks qualitatively reasonable and makes it easy to perform some analysis to study stress redistribution in heterogeneous media. In fact, the heterogeneity must imply some intrinsic length scales relevant to structures of geological media. Nonetheless, as first-order approximation, we use local mean field to deal with the problem. In this way, the intrinsic length scales are eliminated in the approximation and the stress-strain relation in uniaxial stress state is (Fig. 2),

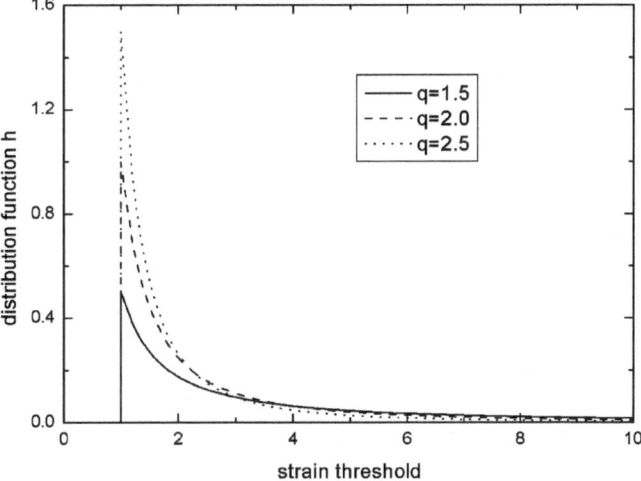

Figure 1

The distribution function of strength of mesoscopically heterogeneous elements, $h(\varepsilon_c)$. It shows that the greater the value of q is, the smaller the mesoscopic strength scatter becomes.

$$\sigma = \begin{cases} E\varepsilon; & \text{when} \quad \varepsilon_c < \varepsilon_c^* \\ E\varepsilon_c^* \left(\frac{\varepsilon}{\varepsilon_c^*}\right)^{2-q} & \text{when} \quad \varepsilon_c \geq \varepsilon_c^* \end{cases} \tag{2}$$

$$D = \int_0^\varepsilon h(\varepsilon_c)\,d\varepsilon_c = \begin{cases} 0; & \text{when} \quad \varepsilon_c < \varepsilon_c^* \\ 1 - \left(\frac{\varepsilon}{\varepsilon_c^*}\right)^{1-q}; & \text{when} \quad \varepsilon_c \geq \varepsilon_c^* \end{cases} \tag{3}$$

where D is damage. In accord with damage mechanics, the effect of damage can be described by the reduced modulus, such as

$$E' = E(1 - D) = E\bar{\varepsilon}^{1-q}, \tag{4}$$

where $\bar{\varepsilon} = \left(\frac{\varepsilon}{\varepsilon_c^*}\right)$.

In multi-axial stress state, it is presumed that the damage or the reduced moduli are governed by maximum strain, i.e., the circumferential strain ε_θ. We call this the θ-model. Then, when $\bar{\varepsilon}_\theta > 1$, the elastic-brittle constitutive relation in spherical configuration (3-D) becomes

$$\bar{\sigma}_r = [(1 - v)\bar{\varepsilon}_r + 2v\bar{\varepsilon}_\theta]\bar{\varepsilon}_\theta^{1-q} \tag{5}$$

$$\bar{\sigma}_\theta = [v\bar{\varepsilon}_r + \bar{\varepsilon}_\theta]\bar{\varepsilon}_\theta^{1-q} \tag{6}$$

where v is Poisson ratio and $\bar{\sigma} = (1 - 2v)(1 + v)\sigma/E\varepsilon_c^*$. In the following we will ignore the bar above all dimensionless variables. The stress balance equation in spherical configuration (3-D) is

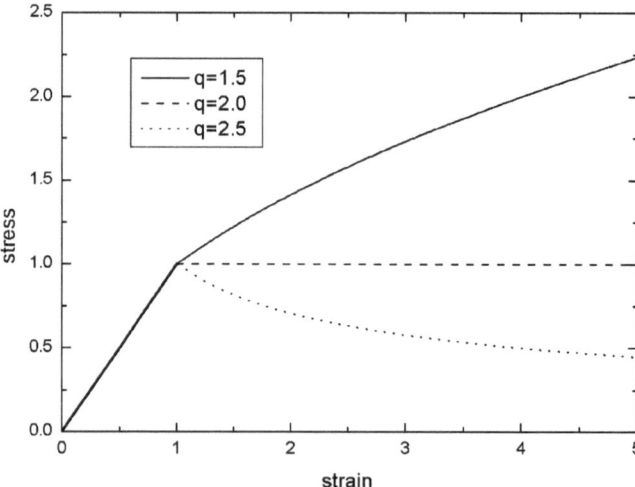

Figure 2

The one-dimensional stress and strain relation of the linear-elastic but heterogeneous-brittle model with different q values. It shows that the greater the mesoscopic strength scatter is, the softer the model becomes.

$$\frac{d\sigma}{dr} + 2\frac{\sigma_r - \sigma_\theta}{r} = 0. \qquad (7)$$

The strains can be expressed by dimensionless displacement u and radius r (all are nondimensionalized by the inner radius of the spherical void, Figure 3),

$$\varepsilon_r = \frac{du}{dr}, \qquad (8)$$

$$\varepsilon_\theta = \frac{u}{r}, \qquad (9)$$

Substitution of the strain definition (8–9) and the elastic-brittle relation (5–6) into the balance equation (7) leads to a nonlinear ordinary differential equation (Lambert equation),

$$(1 - v)u'' + (1 - v)(1 - q)\frac{u'^2}{u} + (1 + q + v - 3vq)\frac{u'}{r} + 2(vq - 1)\frac{u}{r^2} = 0. \qquad (10)$$

Equation (10) works for the presumed θ–model with constitutive relation (5 and 6). It is worth noting that Eq. (10) reduces to the linear elastic version, when $q = 1$. However, when q is greater than unity, the nonlinear second term in Eq. (10) plays a significant role. We use the following nonlinear transformation to simplify equation (10). Let

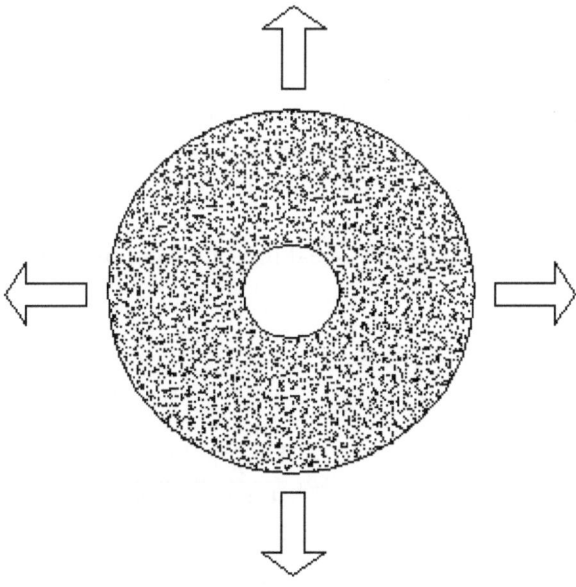

Figure 3

The configuration for the discussion of stress redistribution in heterogeneous media. The central void represents initial damage. The grayness indicates heterogeneity, and the big circle shows a two-dimensional axisymmetric configuration of the model.

$$u = U^{\alpha}, \tag{11}$$

where α is an undetermined parameter. After substituting (11) into Eq. (10) and taking $\alpha = 1/(2 - q)$, one can find that the equation for variable U becomes a linear one,

$$(1 - v)\alpha U'' + (1 + q + v - 3vq)\alpha \frac{U'}{r} + 2(vq - 1)\frac{U}{r^2} = 0. \tag{12}$$

There is a power-law solution to Eq. (12) of the form,

$$U = Ar^{-\beta}, \tag{13}$$

where A is an arbitrary constant and β is an undetermined exponent. Substitution of the solution (13) into equation (12) gives the following dependence of the exponent β on q and v,

$$\beta_1 = q - 2 \quad \text{and} \quad \beta_2 = \frac{2(1 - vq)}{1 - v}. \tag{14}$$

Hence, the stresses, either circumferential or radial, will be in the form

$$\sigma = \left[O(1) + O\left(\frac{\varepsilon_r}{\varepsilon_\theta}\right) \right] \varepsilon_\theta^{2-q} = \left[O(1) + O\left(\frac{\varepsilon_r}{\varepsilon_\theta}\right) \right] [A_1 + A_2 r^{-p}], \tag{15}$$

where A_1 and A_2 are two arbitrary constants, and

$$p = \beta + 2 - q = 2\frac{2 - v}{1 - v} - \frac{1 + v}{1 - v}q. \tag{16}$$

One can verify that the ratio of strains in the expression of stress (15),

$$O\left(\frac{\varepsilon_r}{\varepsilon_\theta}\right) \approx O(1). \tag{17}$$

The reason is as follows. Generally speaking, the solution of the variable U can be written as

$$U = A_1 r^{-\beta_1} + A_2 r^{-\beta_2}. \tag{18}$$

Then, the term

$$\left(\frac{\varepsilon_r}{\varepsilon_\theta}\right) = -\alpha\beta_1 \frac{1 + \frac{A_2\beta_2}{A_1\beta_1} r^{-\beta_2+\beta_1}}{1 + \frac{A_2}{A_1} r^{-\beta_2+\beta_1}}. \tag{19}$$

Provided $\beta_2 > \beta_1$,

$$O\left(\frac{\varepsilon_r}{\varepsilon_\theta}\right) \approx O(1) + O(r^{-\beta_2+\beta_1}). \tag{20}$$

So,

$$\sigma \approx A_1' + A_2' r^{-2\frac{2-v}{1-v}-\frac{1+v}{1-v}q}. \tag{21}$$

Notably, the power-law exponent p in stress redistribution, see Eq. (16) or (21), approaches 3 when q tends to 1, as linear homogeneous elasticity gives in textbook, and p decreases with increasing q. That is to say, stress redistribution in a heterogeneous elastic-brittle medium has longer interaction range with stronger heterogeneity relevant damage.

Similarly, we have derived the stresses for the two-dimensional (cylindrical) configuration under the same assumptions of heterogeneity and reduced modulus. The corresponding versions in the two-dimensional plane stress case (2-D) are as follows:

The elastic-brittle constitutive relation when $\bar{\varepsilon}_\theta > 1$ becomes

$$\bar{\sigma}_r = [\bar{\varepsilon}_r + v\bar{\varepsilon}_\theta]\bar{\varepsilon}_\theta^{1-q}, \tag{5a}$$

$$\bar{\sigma}_\theta = [v\bar{\varepsilon}_r + \bar{\varepsilon}_\theta]\bar{\varepsilon}_\theta^{1-q}, \tag{6a}$$

where $\bar{\sigma} = (1 - v^2)\sigma/E\varepsilon_c^*$. Later we ignore the bar above dimensionless variables again. The stress balance equation in cylindrical configuration (2-D) is

$$\frac{d\sigma}{dr} + \frac{\sigma_r - \sigma_\theta}{r} = 0. \tag{7a}$$

The non-linear ordinary differential equation of displacement u is.

$$u'' + (1 - q)\frac{u'^2}{u} + (q + v - vq)\frac{u'}{r} + (vq - 1 - v)\frac{u}{r^2} = 0. \tag{10a}$$

The corresponding power-law exponent p is

$$p = \beta + 2 - q = \begin{cases} 0 \\ 3 + v - (1 + v)q \end{cases}. \tag{22}$$

Similarly, the power-law exponent p in stress redistribution, see Eq. (22), approaches 2 when q tends to 1 and p decreases with increasing q also, although at a slower rate than that in three-dimensions (Table 1).

3. Results, Finite Element Computation and Discussions

Before discussing concrete calculated results, certain remarks on the stress redistribution in heterogeneous media obtained in the previous section should be emphasized.

Noticeably, the obtained power distribution of stresses in the model should be testified to be at least a proper approximation of real geological media. However,

Table 1

The formula and values of power exponent p in approximate power law, $\sigma \approx r^{-p}$ in three- and two-dimensional configurations

q	p (3-D) $p = 2\frac{2-v}{1-nu} - \frac{1+v}{1-v}q$	p (2-D) $p = 3 + v - (1+v)q$ when $v = \frac{1}{4}$
1	3	2
1.2	2.66	1.75
1.5	2.16	1.37
1.75	1.75	1.06

geophysicists do not know the actual stress tensors for real faults. Therefore we have to consider all possible models and resort to numerical simulations. The two major assumptions made in the θ - model are: Poisson ratio v remains invariant and the circumferential strain ε_θ governs the reduced moduli. In order to check the significance of the second assumption in the θ-model, we also calculate an alternative model—the mixed model, termed the M-model, which consists of elastic and damaged deformations. In the three-dimensional configuration,

$$\sigma_r = [(1 - v)\varepsilon_r + 2v\varepsilon_\theta] \tag{23}$$

$$\sigma_\theta = [v\varepsilon_\theta + \varepsilon_\theta]\epsilon_\theta^{1-q}, \tag{24}$$

and in the two-dimensional case,

$$\sigma_r = [\varepsilon_r + v\varepsilon_\theta] \tag{23a}$$

$$\sigma_\theta = [v\varepsilon_r + \varepsilon_\theta]\varepsilon_\theta^{1-q}. \tag{24a}$$

Figures 4 and 5 show the comparisons of circumferential (Fig 4) and radial (Fig 5) stresses with radial distance for $q = 1.2$ in the θ-model, mixed model and the elastic one, respectively. It is clear that the two damage models present slower attenuation than that of the elastic one. As another comparison, Figure 6 gives circumferential stress with radial distance for $q = 1.5$, in the three models, respectively. One can notice that with increasing q value, i.e., in the more damaged medium, the stress demonstrates even slower attenuation

In addition, finite element numerical simulations in two dimension were made. The simulation is implemented by ABAQUS (a nonlinear Finite Element Analysis package). We use an 8-node reduced-integration axisymmetric element to solve the problem. The damage constitutive relation is based on the θ-model. In this way we cannot only check the calculations based on the obtained approximate analytic solutions but also examine the transition from elastic to damage models with increasing loading, see Figure 7.

The comparison of all these calculations shows a clear trend of longer range of stress redistribution with increasing index q, although the M-model and the full

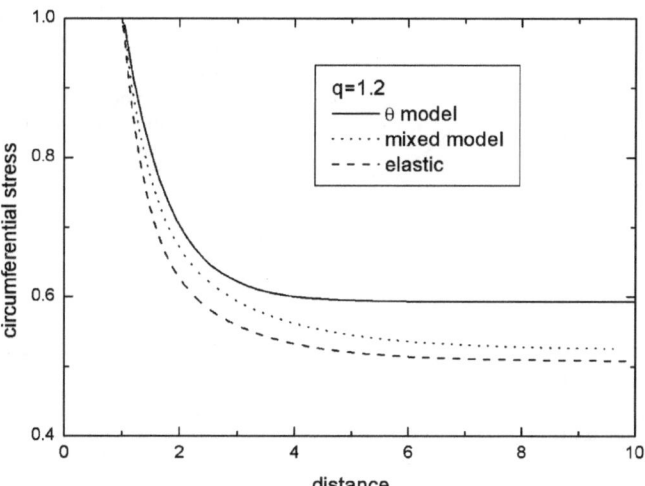

Figure 4
The variations of circumferential stresses with radial distance for $q = 1.2$, in θ-model (solid), mixed model (dotted) and elastic (dashed), respectively. For easy comparison, all stresses are renormalized by the stress value at the inner surface of the hole.

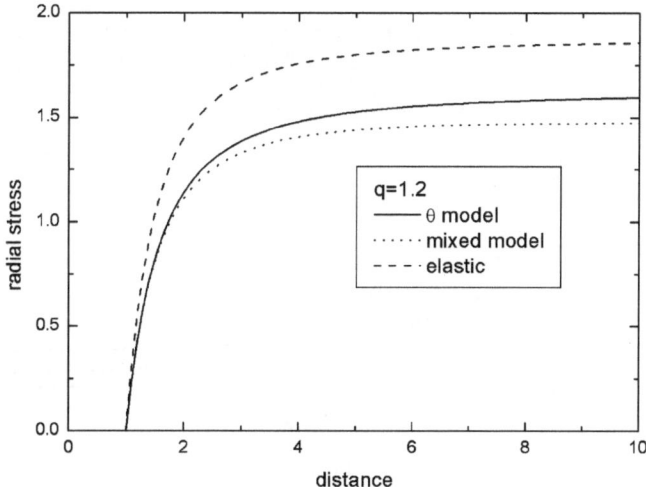

Figure 5
The variations of radial stresses with radial distance for $q = 1.2$, in θ-model (solid), mixed model (dotted) and elastic (dashed), respectively.

numerical simulations show considerably more complicated behavior than the simple power law of stress redistribution in the θ-model. Based on these results, the stress redistribution in heterogeneous media with interaction range longer than in linear

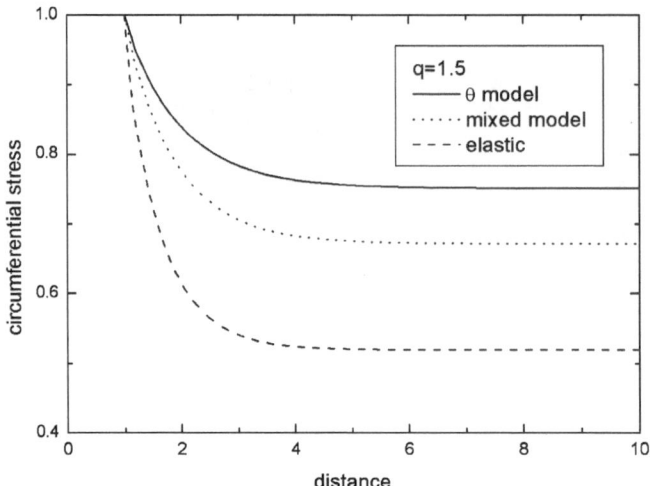

Figure 6
The variations of circumferential stresses with radial distance for $q = 1.5$, in θ-model (solid), mixed model (dotted) and elastic (dashed), respectively. For easy comparison, all stresses are renormalized by the stress value at the inner surface of the hole.

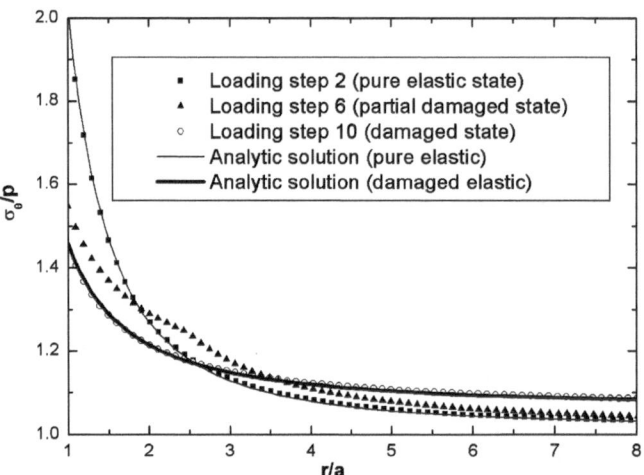

Figure 7
The variations of circumferential stress with radial distance for the elastic-heterogeneous brittle medium with parameters $q = 1.5$ and $v = 0.25$ for various loading steps. The points are the FE results. The two solid lines are the analytical results of elastic and the damaged elastic-brittle models, respectively. The agreements between FE and analytic solutions are very good. The dotted line in between is the FE result for the state of partly elastic (outer part) and partly damaged (inner part).

homogeneous elastic medium (like $p = 3$ in three dimension) might be a reasonable mechanism rather than a virtual assumption. In addition, we could apply these long-range interactions to cellular automata to simulate earthquakes.

Then, what is the physical basis of the obtained results? As mentioned in the introduction, long-range stress redistribution may imply some intrinsic length scales. Physically, intrinsic length scales in heterogeneous media may be cataloged into two groups: small ones δ of mesoscopic heterogeneities or microcracks, and large ones l of macroscopic faults, which follow $\delta \ll \Delta \ll l$, where Δ is the length scale of representative element volume (REV) in the calculation model and tends to become infinitesimally small in the continuum approximation. All intrinsic structures with length scales δ are averaged into the distribution function of heterogeneity and become hidden in the present mean field approximation of the model. Their effect is to bring about the stress redistribution with longer range, owing to less load-supporting ability in the damaged REV. This is what we modeled in the present paper. However, large macroscopic faults with length scales $l \gg \Delta$ should be the free interface in the concerned body. Clearly, these macroscopic free internal boundaries would screen stress field. Therefore, we assume that the long-range stress redistribution resulting from damage in heterogeneous media with intrinsic length scales to quite possibly be a mechanism governing main shocks.

4. Concluding Remarks

In order to understand why a declustered or characteristic large earthquake may occur relevant to some intrinsic length scales observed in earthquake data and with a longer-range stress redistribution observed in CA simulations, we propose a linear-elastic but heterogeneous-brittle model. The stress redistribution owing to damage in the heterogeneous-brittle medium has a power-law exponent in the θ-model, of $p = 2\frac{2-v}{1-v} - \frac{1-v}{1-v}q$ in three dimensions and $p = 3 + v - (1 + v)q$ in two dimensions, respectively instead of 3 and 2 in elastic medium, hence implying longer-range interactions. Other calculations, like finite element simulations, also show a clear trend of longer range of stress redistribution with increasing index q, although more complicated than simple power law. Therefore, it is thought that the long-range stress redistribution resulting from damage in heterogeneous media may quite possibly be a mechanism governing main shocks.

Acknowledgement

This work is granted by the National Natural Science Foundation of China (NSFC-10172084 and NSFC-10232050) and Major State Research Project G200007735.

REFERENCES

BAI, Y. L., XIA, M. F., KE, F. J., and LI, H. L. (2002), *Non-equilibrium evolution of collective microdamage and its coupling with mesoscopic heterogeneities and stress fluctuations*. In *Shock Dynamics and Non-equilibrium Mesoscopic Fluctuations in Solids* (eds. Horie, Y. Thadhani, N. and Davison, L., Springer-Verlag (to appear).

BAI, Y. L., XIA, M. F., KE, F. J., and LI, H. L. (2002), *Closed Trans-scale Statistical Microdamage Mechanics*, Acta Mechanica Sinica *18*, 1–17.

HILL, D. P., REASENBERG, P. A., MICHAEL, A., ARABAZ, W. J., BEROZA, G. and *et al.* (1993), *Seismicity Remotely Triggered by the Magnitude 7.3 Landers, California, Earthquake*, Science *260*, 1617–1623.

KLEIN, W., ANGHEL, M., FERGUSON, C. D., RUNDLE, J. B., and SA MARTINS, J. S. (2000), *Statistical analysis of a model for earthquake faults with long-range stress transfer*. In *Geocomplexity and the Physics of Earthquakes* (eds Rundle J.B., Turcotte, D. and Klein, W.,) AGU 2000.

KNOPOFF, L. (2000), *The Magnitude Distribution of Declustered Earthquakes in Southern California*, PNAS *97*, 11,880–11,884.

RUNDLE, J. B. and KLEIN, W. (1995), *Dynamical segmentation and rupture patterns in a "toy" slider block model for earthquakes*, Nonlinear Proc. In *GeoPhysics, 2*, 61–81.

WEATHERLEY, D., XIA, M. F., and MORA, P. (2000), *Dynamical Complexity in Cellular Automata with Long-range Stress Transfer*, AGU 2000.

(Received September 27, 2002, revised February 28, 2003, accepted March 7, 2003)

 To access this journal online:
http://www.birkhauser.ch

Pure appl. geophys. 161 (2004) 1853–1876
0033–4553/04/101853–24
DOI 10.1007/s00024-004-2536-z

❚ Pure and Applied Geophysics

Review of the Physical Basis of Laboratory-derived Relations for Brittle Failure and their Implications for Earthquake Occurrence and Earthquake Nucleation

N. M. BEELER[1]

Abstract—A laboratory-derived crack growth-based constitutive relation for brittle faulting is developed. The relation consists of two rheologic components, a nonlinear Arrhenius dependence of strain rate on temperature and stress, corresponding to subcritical crack growth, and a linear slip-weakening behavior associated with dilatancy, crack coalescence and supercritical crack growth. The implications of this general behavior for the onset of rapid slip- (earthquake nucleation) are considered. Laboratory observations of static fatigue and time- dependent failure from rock fracture and rock friction experiments are consistent with this simple constitutive description, as are the predictions of rate- and state- dependent equations for the onset of rapid frictional slip between bare rock surfaces. I argue that crack growth is the physical process that controls time-dependent rock fracture and the time-dependent onset of unstable frictional sliding. Some similar and related arguments made in the past 1/2 century in the fields of rock mechanics and earthquake seismology are reviewed. For stressing rates appropriate for the San Andreas fault system, the simple constitutive relation with lab-derived constants predicts a minimum time for nucleation of ~ 1 yr. General predictions are a minimum nucleation patch radius of 0.06 to 0.2 m, and a minimum earthquake moment of 8.5×10^7 Nm.

Key words: Earthquake nucleation, static fatigue, delayed failure, rate- and state-dependent friction.

1. Introduction

It has been nearly 40 years since BRACE and BYERLEE [1966] noted the similarity between repeating earthquake cycles and stick-slip sliding between bare surfaces of rock. In the intervening years earthquake research in rock fracture and rock friction has proceeded assuming that the onset of dynamic stress drop during laboratory rock failure or during frictional stick-slip experiments and the onset of earthquakes are controlled by the same underlying physical processes. Many significant advances have been made in rock friction, e.g., conditions where a fault responds unstably to changes in stress (e.g., DIETERICH, 1978; TULLIS and WEEKS, 1986), rate- and state- dependent constitutive relations to describe the lab observations (DIETERICH, 1979; RUINA, 1983; RICE and RUINA, 1983), and the mechanical conditions for frictional instability (RICE and RUINA, 1983; GU *et al.*, 1984). However, a significant limitation to laboratory-

---

[1] Brown University, Providence, Rhode Island. E-mail: Nicholas-Beeler@brown.edu

based relations, particularly rate and state equations which well describe rock friction under laboratory conditions, is that their exact physical basis is unclear. So long as particular laboratory-observed effects are empirical, i.e., not clearly attributed to thermally activated processes, it is difficult to extrapolate the observations in pressure and temperature to conditions other than those of the particular experiments, establish the dependence of the effects on environmental factors such as the presence of fluids and fluid chemistry, or to determine how the specific effects compare with others which might compete at significant depth in the earth.

This paper attempts to combine the essential experimental observations from brittle rock deformation into simple physically-based constitutive equations for earthquake nucleation. I argue that laboratory observations of rock fracture and rock friction relevant to earthquake occurrence and earthquake nucleation result from subcritical and supercritical crack growth, as has been suggested many times for failure of intact rock, probably first by SCHOLZ (1968b). A qualitative two-component constitutive description for rock fracture which results from considering the physics of crack growth is the same form as proposed by DIETERICH (1992, 1994) as a simplification of rate and state equations appropriate for describing the onset of rapid frictional slip between bare rock surfaces. Thus, I argue that crack growth underlies the rheology of unstable frictional slip between rock surfaces and the rheology of rock fracture. In the context of earthquake nucleation, considering frictional instability and rock failure as both resulting from crack growth is conceptually appealing because fractured rock and highly comminuted wear product are the most obvious of byproducts of brittle rock deformation. The equations resulting from the analysis presented herein, consistent with other recent process-based models of rock friction (LAPUSTA et al., 2000; NAKATANI, 2001, RICE et al., 2001), make specific predictions for the expected size and form of the temperature dependence of rheological parameters associated with earthquake nucleation, and might be used to predict how these effects will change with experimental configuration, the presence of fluids, and fluid chemistry. Using the simple constitutive equations and laboratory-derived model parameters I summarize new and review previously published predictions for earthquake occurrence and earthquake scale, in particular the duration of nucleation, nucleation zone size, and minimum earthquake size.

2. Simple Description of Brittle Deformation

To model the onset of rapid slip at low temperature I assume that all inelastic deformation requires cooperative slip and crack growth within a localized zone. Slip may occur between asperities on opposite sides of a fault, between particles in a fault gouge, or on previously fractured surfaces. Regions which are fractured by crack growth during deformation may be regions of intact rock in front of a propagating rupture (LOCKNER et al., 1991), interlocking asperities on preexisting fault surfaces

(DIETERICH and KILGORE, 1994), or particles which form grain bridges in gouge layers (e.g., SAMMIS and STEACY, 1994). To characterize the net shear resistance at constant deformation rate I use an approach similar to that of MOGI (1962) and SAVAGE et al. (1996) where the shear resistance is

$$\tau_0 = \tau_f \frac{A_f}{A} + \tau_i \frac{A_i}{A}. \tag{1}$$

Here τ_f is the shear resistance to slip, τ_i is the shear stress necessary to break the unfractured material, A is the initial contacting area per unit mass of the deforming zone, and A_f and A_i are the area per unit mass of the slipping (fractured) region and the cross-sectional area per unit mass of the unfractured portions of the fault zone, respectively. Taking $A = A_f + A_i$, assume that the intrinsic resistance to slip is smaller than the resistance to crack growth, $\tau_f < \tau_i$ and that A is constant. During failure, weakening with progressive slip δ results from the increase of A_f at the expense of A_i. For example, for the propagation of a single circular crack, the cracked area increases as $A_f \propto \delta^2$. So in general the fault zones of interest are unable to sustain large shear strain without loss of strength, consistent with general definitions of brittle behavior. Here for simplicity, assume that crack area increases linearly with slip $A_f = C\delta$, where C is a constant with dimensions of length per unit mass. Using the definitions $\Delta\tau = \tau_i - \tau_f$ and $d_* = A/C$, equation (1) becomes

$$\tau_0 = \tau_i - \Delta\tau \frac{\delta}{d_*}. \tag{2}$$

This relationship (Fig. 1a) is consistent with laboratory observations of gradual slip-weakening during failure of intact rock (e.g., WONG, 1986) and stick-slip friction (e.g., OKUBO and DIETERICH, 1981; 1984) (Fig. 1b). Equation (2) is the slip-weakening form previously shown to be a useful general description of quasi-static and dynamic material failure (PALMER and RICE, 1973; IDA, 1972) that has been used extensively in cohesive zone models of shear fault propagation in the geophysical literature.

In previous implementations of slip-weakening the characteristic length d_* describes a gradual loss of material cohesion as in (2). However in many previous applications to earthquake mechanics, slip-weakening is used to model dynamic rupture propagation, that is, only super-critical behavior. In such applications the characteristic length has additional significance, principally, d_* is a model representation, related to the dimension of the region of yielding (the process zone) surrounding the tip of the propagating rupture. The size of the region surrounding the tip in which the stress is high enough to cause yielding increases with crack length or distance of propagation, implying that the process zone increases with the distance of crack propagation (ANDREWS, 1976). Thus, in models of large scale rupture propagation in the earth, the characteristic length should increase with displacement (ANDREWS, 1976). However, in this paper I am interested in the very small

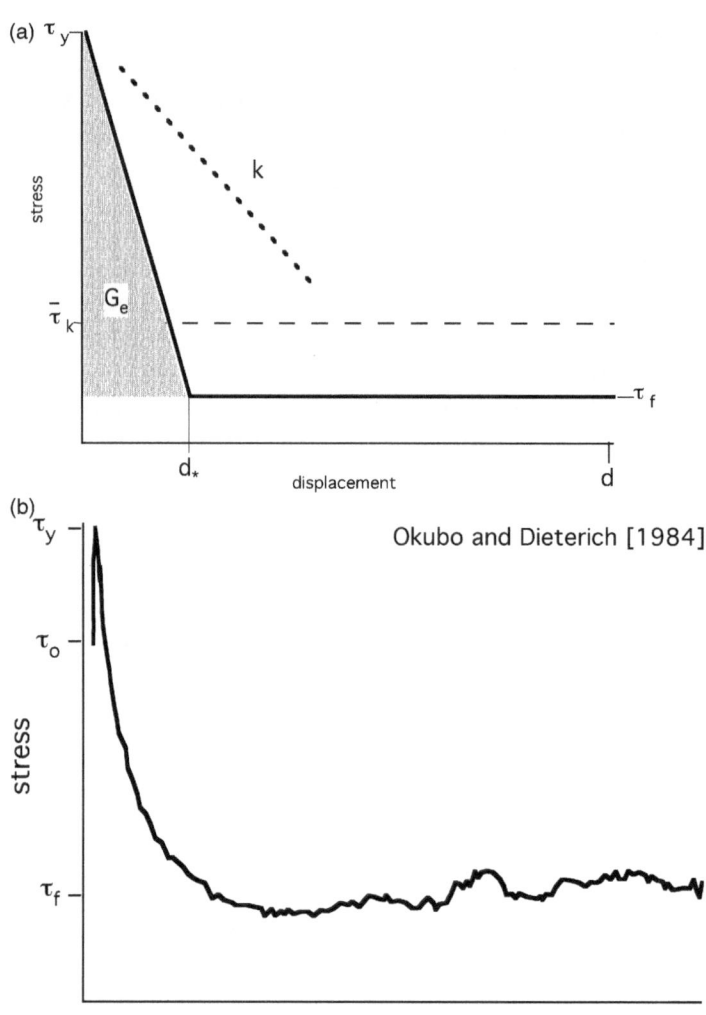

Figure 1

Slip-weakening relationship for fault strength drop. a) Idealized linear relationship after IDA (1972) and PALMER and RICE (1973). Strength drops from a peak (yield) τ_y to a residual level τ_f over a characteristic displacement d_*. The energy per unit area expended in dropping the strength $G_e = (\tau_y - \tau_f)d_*/2$, the effective shear fracture energy (shaded), can be equivalently expressed as $G_e = (\bar{\tau}_k - \tau_f)d$, where d is the total displacement and $\bar{\tau}_k$ is the displacement averaged fault strength. In this example, the initial displacement rate of strength drop (solid line), the slip-weakening rate $(\tau_y - \tau_f)/d_*$, is greater than the rate of stress drop (heavy dashed line) $k = d\tau/d\delta$ defining the conditions necessary for unstable, rapid slip. b) Laboratory observation of slip-weakening of a granite fault surface during dynamic strength drop at 3.45 MPa normal stress (OKUBO and DIETERICH, 1984).

displacements associated with earthquake nucleation; with reference to Figure 1, these displacements $\delta < d_*$ are the pre-peak and immediately post-peak displacements associated with the onset of rapid slip. Initially, I assume that d_* in the qualitative model (2) is constant. Later, in section 4.3 I consider implications of variable d_* for nucleation.

2.1 Slow Crack Growth

In brittle materials the resistance to crack growth depends positively on the growth rate while the growth rate is low. Figure 2 shows examples of this 'subcritical' behavior, rate data from single cracks propagating in selected quartzofeldspathic materials - fused silica glass (LAWN, 1993), synthetic quartz, and Westerly Granite (ATKINSON and MEREDITH, 1987). In Figure 2 the horizontal axis is crack propagation velocity and the vertical axis is the stress intensity K_I; stress intensity is related to the remotely applied driving stress σ_r and to the crack length c as. $K_I \propto \sigma_r \sqrt{c}$. Thus, as shown in Figure 2, subcritical fracture growth has a nonlinear viscous rheology; as the stress on the sample increases crack velocity increases, or alternatively, as the crack propagation velocity increases, the crack strength increases. Subcritical crack growth in minerals is generally similar to that in glass and at comparable temperature crystalline materials (quartz) are often somewhat stronger than their amorphous counterpart (fused silica). Note finally that there is a measurable temperature dependence to crack growth, indicating that a thermally activated mechanism controls growth. Arguably, subcritical crack growth occurs in all instances of natural faulting and in laboratory rock fracture and friction experiments as indicated by damage such as pervasive micro-cracks, macroscopic fracture, and wear material (gouge).

For well-studied minerals and for other brittle materials it is generally accepted that subcritical crack propagation rate is determined by the rate of chemical reaction at the crack tip, i.e., the rate at which bonds are broken at the tip (e.g., CHARLES and HILLIG, 1962; ATKINSON and MEREDITH, 1987). Descriptions of the chemical reaction rate v have the form

$$v = v_0 \, exp\left(\frac{-E + \Omega\sigma_t - \Omega_m\Gamma/\rho}{RT}\right). \tag{3}$$

(CHARLES and HILLIG, 1962) where E is the stress free activation energy, Ω is the activation volume, Ω_m is molar volume, Γ is the interfacial energy, R is the gas constant (8.3144 J/mol K) and ρ is the radius of curvature at the crack tip. Note that as v increases, the crack tip strength σ_t increases. Using the substitutions $\Delta Q = E + \Omega_m\Gamma/\rho$ and $\sigma_0 = \Delta Q/\Omega$ (3) can be written

$$v = v_o \, exp\frac{-\Delta Q(1 - \sigma_t/\sigma_0)}{RT}. \tag{4}$$

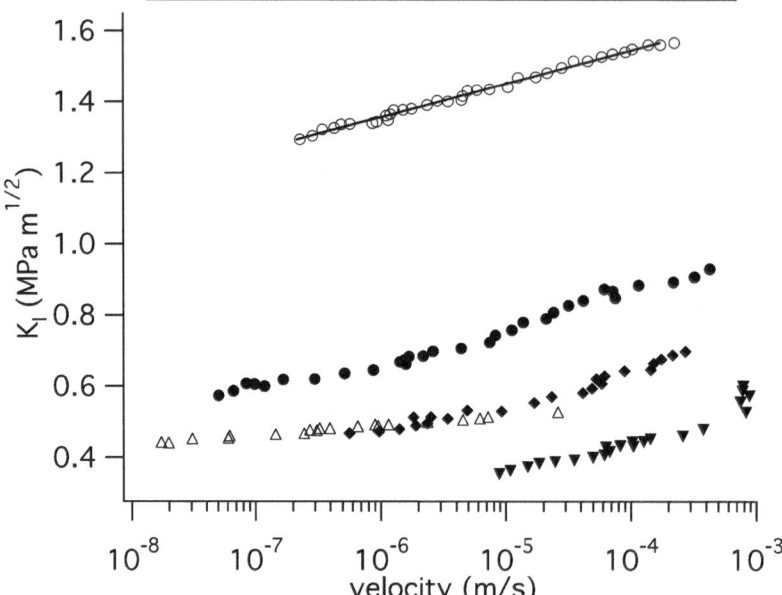

Figure 2

Subcritical crack growth in quartzofeldspathic materials. Crack stress intensity K_I versus velocity for fused silica glass (*Lawn*, 1993), synthetic quartz (ATKINSON and MEREDITH, 1987), and Westerly granite (ATKINSON and MEREDITH, 1987). The stress intensity is proportional to the remotely applied stress; thus the vertical axis reflects the driving stress for subcritical crack extension. Also shown is a fit to the granite data using $K_I = K_0 + (dK_I/d\,lnV)\,\ln V$, with $dK_I/d\,ln\,V = 0.041$ MPa m$^{1/2}$.

Assume the rate of crack propagation is proportional to the reaction rate in (4). I further expect that the propagation velocity of a single crack, or the average propagation velocity of cracks within a localized fault zone to be proportional to the shear deformation rate across the fault zone so that $V = \alpha v$. Finally, assume that the remotely applied shear- stress driving crack propagation τ, equivalently macroscopic shear resistance of regions to fracture, reflects the intrinsic strength of the subcritical cracks within the region, $\sigma_t = \kappa \tau$. Here, κ is like a dimensionless stress intensity factor that depends on the geometry. However, unlike stress intensity, for simplicity I have arbitrarily assumed that κ is constant and does not depend on crack dimension, consistent with the crack growth experiments. After combining all of these assumptions the deformation rate is

$$V = V_0\,exp\frac{-\Delta Q(1 - \tau\kappa/\sigma_0)}{RT} \tag{5a}$$

where $V_* = v_0/\alpha$. Rearranging (5a) with V as the independent variable and using the substitutions $\tau_0 = \sigma_0/\kappa$ and $\zeta = RT\sigma_0/\Delta Q\kappa$ yields

$$\tau = \tau_0 + \zeta \ln \frac{V}{V_*}. \tag{5b}$$

Note that the resulting equation (5b) has the same form of stress dependence as the experimentally measured subcritical crack growth data (Fig. 2); stress depends logarithmically on the deformation rate with a coefficient ζ that is related to temperature and the thermal activation energy (also see, LAPUSTA et al., 2000; NAKATANI, 2001; RICE et al., 2001).

2.2 Simple Failure Equation

Combining the rate-independent relation for brittle deformation (2) with the accounting of the rate dependence of crack growth (5b) leads to a slip and slip rate dependent formulation

$$\tau = \tau_i + \zeta \ln \frac{V}{V_*} - \Delta\tau \frac{\delta}{d_*}. \tag{6a}$$

Observations from studies of rock friction (e.g., DIETERICH, 1981; TULLIS and WEEKS, 1986; BLANPIED et al., 1998) and failure of intact rock (SCHOLZ, 1968a, b; LOCKNER, 1998), over a wide range of deformation rates are generally consistent with (6b); rock shear resistance at low temperature, low strain, and low strain rate depends weakly and logarithmically on the rate of inelastic deformation. Normalizing by normal stress yields a relation for friction, $\mu = \tau/\sigma_n$,

$$\mu = \mu_* + a \ln \frac{V}{V_*} - \frac{\Delta\tau\delta}{\sigma_n d_*}. \tag{6b}$$

Equation (6b) is the same form as a simplification, appropriate for representing the onset of rapid slip, suggested by DIETERICH (1992; 1994), of a particular rate and state variable constitutive equation widely used in studies of rock friction and earthquake mechanics (RUINA, 1983; LINKER and DIETERICH, 1992). Equivalence with (6a) requires $a = \zeta/\sigma_n$ and $\mu_* = \tau_i/\sigma_n$.

2.3 Expected Values of Coefficients

According to equation (6), strength during loading to brittle failure depends only on slip rate and slip. Expected values from laboratory experiments for the two coefficents a and, $\Delta\tau/\sigma_n d_*$ which control the respective dependencies, are listed in Table 1. Both friction and intact failure tests yield values of a ranging 0.003 to 0.009 for granite at low temperature. Agreement between a for friction and intact rock was found by TULLIS and WEEKS (1987) for carbonate rocks. In the case of granite, this similarity lead LOCKNER (1998) to suggest that subcritical crack growth is the

Table 1

Parameters for Granite

	Rate dependence	$\Delta\tau/\alpha_n d_*$	Reference
Friction	$\xi = 0.003\sigma_n$ to $0.009\sigma_n$		BLANPIED *et al.* (1998)
Friction		0.014 to 0.044/μm	OKUBO and DIETERICH (1984)
Fracture		0.096×10^{-3} to $0.21 \times 10^{-3}/\mu$m	WONG (1986)
Fracture	$\xi = 0.007\sigma_n$ to $0.009\sigma_n$ (20°C)		LOCKNER (1998)
Crack growth	$dK_I/d \ln V = 0.041$ MPa m$^{1/2}$ (20°C)	N/A	ATKINSON and MEREDITH (1987)

mechanism underlying this effect in both friction and rock failure of granite. While the rate dependence of shear strength from subcritical crack growth experiments in granite $dK_I/d \ln V = 0.041$ MPa m$^{1/2}$ (Fig. 2, Table 1, data from ATKINSON and MEREDITH, (1987)) is available for comparison with friction tests at low normal stress, there are many complications in interpretation. The conclusion from friction and rock fracture tests is that a is a frictional quantity; that is, it does not depend on normal stress, requiring that ξ in (6) scales with normal stress. While this normal stress dependence of ξ is not included in the present qualitative model, it might be incorporated following the approach of LOCKNER (1993). An additional difficulty is that rate dependence inferred from ATKINSON and MEREDITH (1987) is crack length dependent and crack length is ignored in our treatment. At present no method of comparison between friction experiments and crack growth data has been established. A third difficulty is that $dK_I/d \ln V$ inferred from ATKINSON and MEREDITH (1987) applies to mode I (tensile) crack propagation; at high confining pressure appropriate for crustal shear faulting, shear and mixed mode crack propagation is expected and such subcritical crack growth measurements are not available. A final problem is that there are four regions of subcritical fracture growth, all with different rate dependence. The existing data for mode I in rocks and minerals correspond primarily to region 1 (ATKINSON and MEREDITH, 1987); it is unclear whether these are the appropriate data to compare to rock friction and rock failure tests. Given these unknowns, in the remainder of this study equation (6) is used with an empirically determined value of $a = 0.008$ appropriate for both granite friction and failure of intact granite (Table 1).

To infer values of the slip-weakening coefficient I use the slip-weakening strength drop (Fig. 1) from failure experiments on granite summarized by WONG (1986) and from granite stick-slip experiments of OKUBO and DIETERICH (1981). The slip-weakening component of (6b) can be equated with the dynamic strength drop from the experimental measurements, $\Delta\tau = (\tau_y - \tau_f)$ (Fig. 1). In both cases I estimate the effective dynamic strength drop from the measured shear fracture energy G_e and the characteristic slip-weakening distance d_* according to $(\tau_y - \tau_f) = 2G_e/d_*$ Knowing the strength drop, normal stress and slip-weakening distance specifies the displacement rate of slip-weakening $(\tau_y - \tau_f)/\sigma_n d_*$. If d_* scales with A, the initial contact area per unit mass within the fault zone, then the slip-weakening coefficient $\Delta\tau/\sigma_n d_*$ should be very different for intact failure tests and friction. As friction involves areas of asperity contact which are typically a few percent or less of the total fault area, and intact failure involves fracture of the entire fault, d_* is expected to be two to three orders of magnitude larger for intact rock failure. Thus, excepting differences in stress drop $\Delta\tau$, the slip-weakening component $\Delta\tau/\sigma_n d_*$ should be about two to three orders of magnitude larger for friction than it is for intact failure. This is consistent with the observations (Table 1); the representative displacement rate of slip-weakening for fault slip in terms of friction is $0.029/\mu m$ whereas for rock fracture it is $0.00015/\mu m$.

However, $\Delta\tau$ and d_* are probably coupled. For rock fracture the "fault" initially is as strong as the surroundings, having the strength of intact rock. On the other extreme, simulated laboratory faults are weak with respect to the surrounding rock and are cohesionless and artificially flat except at small wavelengths. Thus the strength drop for rock fracture should exceed that for friction. Similarly, the amount of strain necessary to weaken intact rock exceeds that necessary to break asperities on a laboratory fault surface. Natural seismic faults should have roughness much greater than in laboratory friction experiments (POWER et al., 1987), and have non-zero cohesion less than that of intact rock. Considering the displacement rate of slip-weakening from laboratory friction and laboratory rock fracture experiments as end members, I use a value intermediate between the two different types of laboratory tests, $\Delta\tau/d_*\sigma_n = 0.002$ to $0.003/\mu m$ in applying laboratory results to model earthquake nucleation and small earthquakes.

3. Physical Characteristics, Significance, Interpretation

If one allows that earthquake occurrence involves crack growth at some scale, constitutive models for failure must involve two components, a ductile response at small strains, represented in (6b) by the rate-dependent term $a \ln V/V_*$, and a brittle response at large strain represented by the slip-weakening term $-\Delta\tau\delta/\sigma_n d_*$, in (6b). The implied behavior is easily illustrated; consider a fault loaded at a constant displacement rate V_L (Fig. 3a); for small values of slip, $\delta << d_*\sigma_n/\Delta\tau$, the third term on the right-hand side of (6b) is negligible and the second term dominates. That is, at small inelastic strains the fault strength depends positively on deformation rate, and the fault is ductile in the classic sense that strain can be accommodated without loss of strength. The ductile component of (6b) $\mu = \mu_* + a \ln V/V_*$ characterizes temporal variation of slip rate very well over most of the loading to failure (Fig. 3b). During this time, labeled ductile in Figure 3c, the sliding velocity increases (Fig. 3b), and yet the stress on the fault also increases (Figure 3a and 3c). As more strain accrues, the third

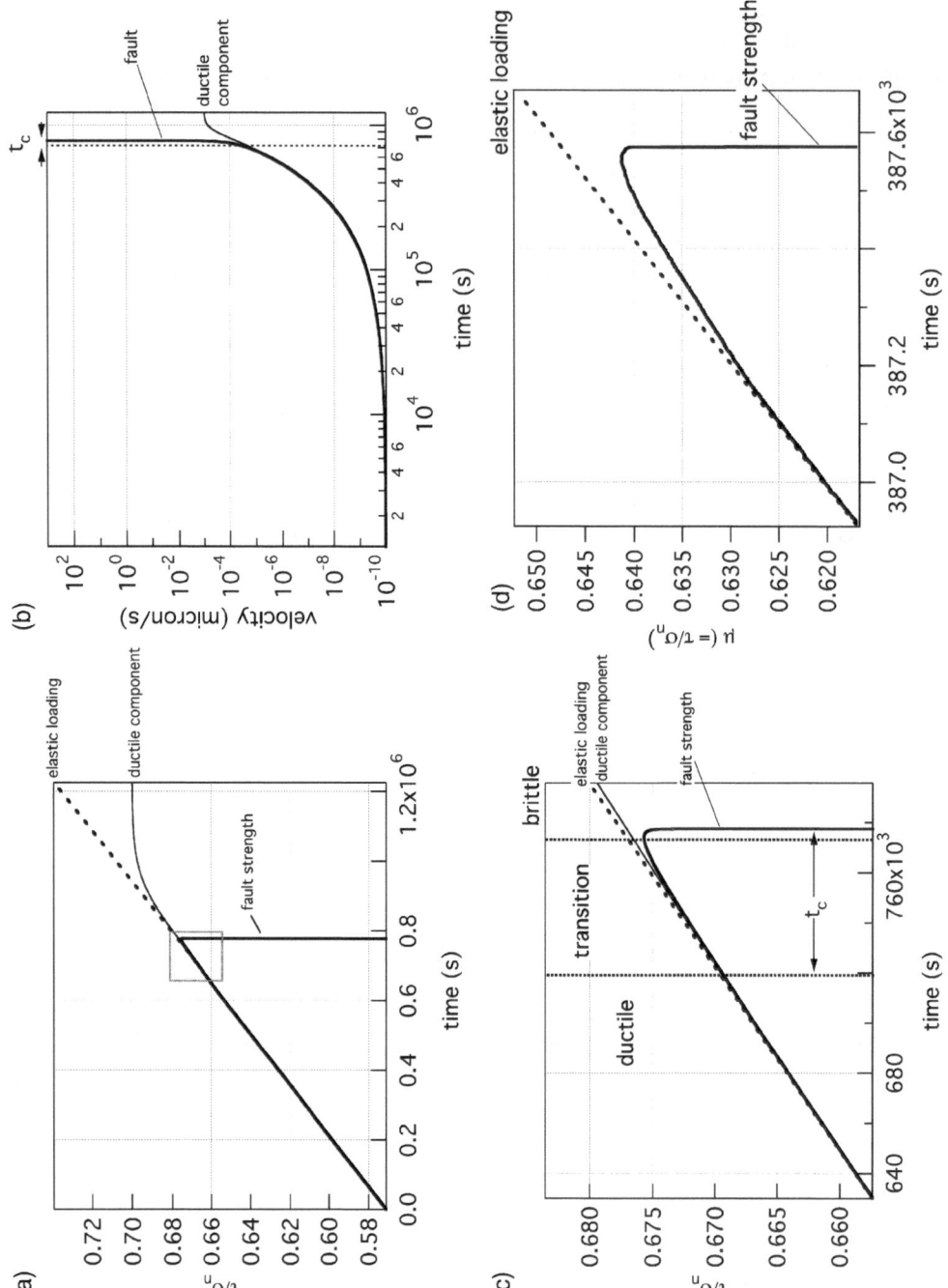

term of (6b) becomes significant and leads to a deviation from a linear (elastic) increase in shear stress τ with time t (Figure 3a and 3c); shear stress on the fault is $\tau = k(V_L t - \delta)$, where k is the stiffness of the surroundings. In this transition stage the slip rate also deviates from the ductile curve (Figure 3b). Eventually the slip weakening term in (6b) dominates, resulting in truly brittle behavior at large strains, and subsequently rapid stress drop occurs. Figure 3d depicts a laboratory example of this gradual onset of failure, loading of a granite fault at 50 MPa confining pressure at $V_L = 0.1$ μm/s (BEELER and LOCKNER, 2003). Thus the crack-based model of faulting, and the laboratory observations on which it is based, exhibit delayed failure— a lag time between the onset of detectable slip, or the time of peak stress, and the time of failure.

The ductile response represents the stable growth of subcritical cracks while the eventual brittle response reflects critical and super-critical crack growth, the coalescence of previously isolated cracks, dilatancy and the associated loss of material cohesion. More sophisticated two-component models for rock fracture, based on the same conceptual approach taken in the previous section, are common in the international rock mechanics and fracture mechanics literature dating back at least to the 1960s (e.g., ZHURKOV, 1965; ZHURKOV et al., 1977).

4. Implications for Time of Earthquake Occurrence

4.1 Static fatigue and delayed failure

As reflected in equation (6b) and in laboratory experiments, neither rock failure nor frictional instability occur at a threshold stress. This is well illustrated by the

◄

Figure 3

Characteristics of laboratory failure during loading at a constant rate. Parts a) – c) show the results of simulations with equation (6b) where the fault is loaded elastically $d\tau/dt = k(V_L - V)$ with $V_L = 0.001\mu$m/s, stiffness $k/\sigma_n = 0.000137/\mu$m, $\mu_* = 0.7$, $a = 0.008$, $\Delta\tau/\sigma_n d_* = 0.0024$ /μm, and initial velocity $V_s = 1.0 \times 10^{-10}$ μm/s at $t = 0$. a) Shear stress in the plane of the fault in the shear direction versus time. Shown are the elastic loading $d\tau/dt = kV_L$ that would result if the fault did not slip (dashed), the ductile component of the failure relation $\tau/\sigma_n = \mu_* + a \ln V/V_*$(black) and the complete solution to (6b) (heavy black). The gray box denotes the region shown in Figure 3c. b) Sliding velocity versus time for the calculation shown in Figure 3a. The ductile component of the failure equation is shown for comparison (black). During most of the loading time the fault behavior is well described by the ductile solution because the total displacement is extremely small and the slip weakening term $\Delta\tau\delta/\sigma_n d_*$ in (6b) is negligible. c) Blow up of gray box in Figure 3a. The elastic loading (dashed) and ductile component of the failure relation are shown again for comparison. Also shown are the approximate boundaries (dotted lines) between the ductile and brittle regions of (6b). The onset of truly brittle behavior occurs at the peak stress when $V = V_L$. An approximation to the upper limit of truly ductile behavior is shown (see discussion in section 4.3). Over the duration of time labeled transition, the fault shows mixed behavior; it strengthens as the sliding rate increases but is weaker than the purely ductile component. d) Actual laboratory observation of the onset of unstable sliding between surfaces of Westerly granite at 50 MPa confining pressure and $V_L = 0.1$ μm/s (BEELER and LOCKNER, 2003).

laboratory observed phenomenon known in mechanical engineering, material science, and rock mechanics as 'static fatigue' or 'stress corrosion'—an inherent time delay between the application of a stress increase and the occurrence of brittle failure induced by the stress change (e.g., SCHOLZ, 1972; KRANZ, 1980). Here and throughout I refer to changes in the fault shear stress τ using the normalized frictional stress $\mu = \tau/\sigma_n$ (6b). To consider static fatigue of (6b), take the temporal derivative and equate it to zero. If stress is raised to a particular level μ_s at time $t=0$ and then held constant, the sliding rate evolves from the starting velocity V_s at $t=0$ according to

$$V = \frac{V_s a \sigma_n d_*}{a \sigma_n d_* - \Delta \tau t V_s} \tag{7a}$$

where $V_s = V_* exp([\mu_s - \mu_*]/a)$ (Fig. 4a). The delay time t_d between the application of the stress that induces failure and actual time of failure is

$$t_d = \frac{a \sigma_n d_*}{\Delta \tau V_*} exp([\mu_* - \mu_s]/a) \tag{7b}$$

The relationship between stress change and the delay time is such that the larger the stress level, the sooner the induced failure event (Fig. 4b). Thus, if earthquake failure is controlled by the same processes as in laboratory rock failure and frictional instability experiments, the failure time of an earthquake triggered by a particular stress change depends on the size of stress change and on how close the fault is to failure prior to the stress change (proximity to failure). That rock failure time is related to stress change in this general way has been known since GRIGGS (1936). Figure 4c shows a laboratory example of the typical static fatigue behavior of rock, rock fracture of granite at 53 MPa confining pressure (KRANZ, 1980).

As the relationship between stress change and failure time is nonlinear, when one considers populations of faults with an initially random distribution of proximity to failure, a stress change produces a seismicity rate due to triggered events that is initially very high and decreases with time following the stress change. This is the behavior seen in induced acoustic emissions, microseismicity, and Omori aftershock sequences, as was noted by MOGI (1962), and subsequently by many researchers (e.g.,

▶

Figure 4

Static fatigue characteristics. Parts a) and b) are simulations using equation (6b) with $\mu_* = 0.7$, $V_* = 0.001$ μm/s, $a = 0.008$, and $\Delta\tau/\sigma_n d_* = 0.0024$ /μm. In a static fatigue test, the fault shear stress is raised from zero to μ_s nearly instantaneously. In this simulation, the level of stress fixes the initial velocity $V_s = V_*$ exp($[\mu_s-\mu_*]/a$). a) Slip velocity with time as given by equation (7a). Shown are three cases corresponding to $\Delta\mu = \mu_s - \mu_* = 0.03, 0.01$ and -0.01. The higher the stress relative to the reference μ_*, the higher the initial slip rate and the sooner failure occurs. b) The time of failure t_d versus the stress level for a series of simulated static fatigue tests as specified by equation (7b). c) Actual experimental static fatigue data from rock fracture of granite at 53 MPa confining pressure (KRANZ, 1980). Failure stress was converted to friction assuming a 30° angle between the greatest principal stress and the incipient failure plane. $\Delta\mu$ was calculated using μ at 10^4 s as the arbitrary reference μ_*.

(a)   fatigue characteristics of simple fault failure relation

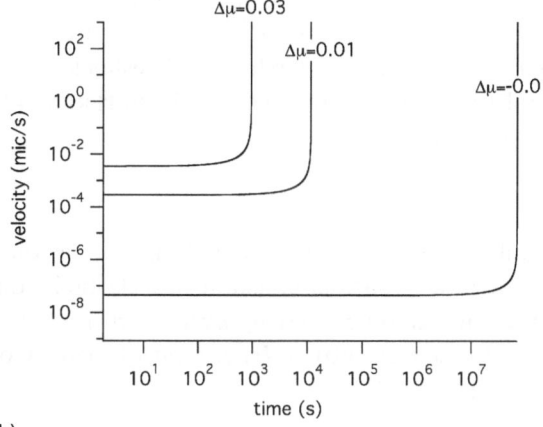

(b)

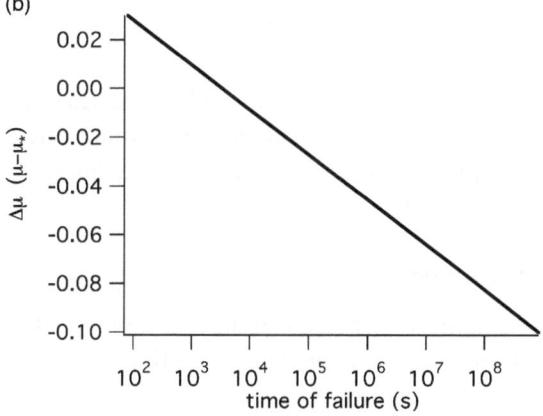

(c)

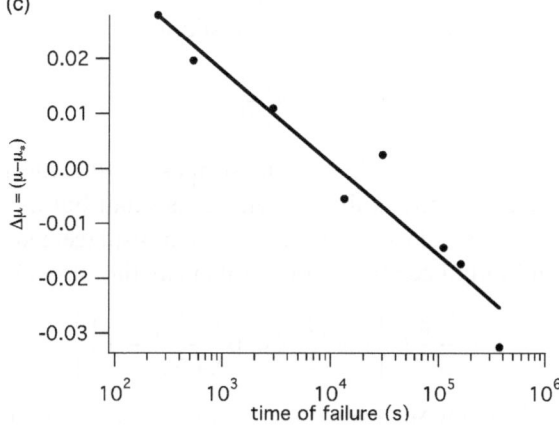

SCHOLZ, 1968b; KNOPOFF, 1972; DAS and SCHOLZ, 1981; YAMASHITA and KNOPOFF, 1987; HIRATA, 1987; REUSCHLE, 1990; and MARCELLINI, 1995, 1997 among others). Furthermore, many of these authors have developed fracture growth based models of aftershocks; successful modeling of aftershocks, foreshocks and general seismicity rate changes with (6b) by DIETERICH (1994) and DIETERICH et al. (2000) being a recent, general and significant advance.

4.2 The Characteristic Time

Because the small strain behavior of equation (6b) is ductile, this equation exhibits a characteristic time. Consider constant rate elastic loading of a fault where the loading rate represents tectonic stressing $d\tau/dt = k(V_L - V)$. At small strain the rate of stress change of equation (6b) is $d\tau/dt \approx a\sigma_n dV/VdV$. Combining these two expressions and integrating yields

$$\frac{1}{V} \approx \left(\frac{1}{V_s} - \frac{1}{V_L}\right) exp\left(-\frac{\dot{\tau}}{a\sigma_n}t\right) + \frac{1}{V_L}.$$

where V_s is the velocity at $t = 0$, and $\dot{\tau} = kV_L$ Thus, slowness (V^{-1}) is exponential with a characteristic time $t_c = a\sigma_n/\dot{\tau}$. This constant loading rate, characteristic time figures prominently in applications of (6b) to earthquake nucleation and earthquake occurrence as I summarize next.

4.3 The Duration of Nucleation

The simple failure equation (6b) has a number of analytical solutions for simple stressing histories that are useful for analysis of earthquake occurrence, earthquake rate, and precursory slip (DIETERICH, 1992, 1994; GOMBERG et al., 1998). For predicting earthquake failure time under a constant rate of stressing $\dot{\tau}$ due to plate motion, time until failure depends only on the sliding velocity

$$t_{tf} = \frac{a\sigma_n}{\dot{\tau}} ln\left[\frac{\dot{\tau}}{\gamma V} + 1\right]. \tag{8a}$$

where $\gamma = (\Delta\tau/d_* - k)$ and k is the elastic stiffness (DIETERICH, 1994) (Fig. 5a). During most of the loading, the fault slip velocity is small but as the stress rises slip rate becomes significant. Similarly, the total accrued displacement prior to failure is exceedingly small and only becomes appreciable near the time of failure

$$\delta = -\frac{a}{\gamma} ln\left[1 - \frac{\gamma V_s}{\dot{\tau}}\left\{exp\left(\frac{\dot{\tau}t}{a\sigma_n}\right) - 1\right\}\right]. \tag{8b}$$

(DIETERICH, 1992, 1994; GOMBERG et al., 1998) where V_s is the velocity at $t=0$ (Fig. 5b). Whether or not the fault is slipping at an appreciable rate directly reflects the expected tradeoffs between the two rheologic components of (6b). With reference to Figure 5a, time between the change from steep to shallow slope and the failure time is marked by the time at which the full solution of (6b) deviates from the ductile component of (6b). This time can be thought of as the duration of nucleation, the time during which the fault is undergoing appreciable precursory slip. The duration of nucleation is effectively independent of the slip-weakening component $-\Delta\tau\delta/\sigma_n d_*$ of equation (6b), as is shown in constant loading rate calculations with $\Delta\tau\delta/\sigma_n d_*$ varying over 4 orders of magnitude (Fig. 6). The nucleation time therefore is controlled entirely by the ductile component of (6b) $a\ln V/V_*$ and depends only on a. Furthermore, the nucleation time is on the order of the characteristic time of $a\sigma_n/\dot\tau$. t_n has been approximated as $t_n = t_c = a\sigma_n/\dot\tau$ by DIETERICH (1994) and as $t_n = 2\pi t_c = 2\pi a\sigma_n/\dot\tau$ by BEELER and LOCKNER (2003). These estimates of t_n seem to bound the actual value well (Figs. 5a and 6).

In addition to being the effective duration of nucleation, t_n is the predicted duration of Omori-like aftershock sequences for fault populations obeying (6b) (DIETERICH, 1994). Why the duration of an aftershock sequence is related to the nucleation time can be understood by noting that for constant loading rate t_n is the maximum possible delay between the onset of appreciable slip and the time of failure (Fig. 5b). Therefore for faults that are initially closer to failure than t_n, a small stress change changes the sliding velocity and shortens the delay time by an amount that depends on the initial time to failure. However for faults initially much farther from failure than t_n, a small stress change changes the sliding velocity but will not shorten their delay time. As a result, the seismicity rate at times greater than t_n after a static stress change, is the same as the seismicity rate prior to the stress change (DIETERICH, 1994). t_n is also the expected period of time over which earthquake occurrence rates should be insensitive to continuous small amplitude stress changes such as the solid earth tides (BEELER and LOCKNER, 2003). That the nucleation time determines whether earthquake occurrence is sensitive to tidal stresses can be understood again by noting that t_n is the maximum possible delay time. For an increase in stress of a period considerably longer than t_n, the nucleation time is negligible in comparison to the period; under these circumstances failure is effectively instantaneously triggered by the stress change, and there is a strong correlation between stress change and failure time. At periods shorter than t_n, failure is delayed due to the damping term $a\ln V/V_*$ and the correlation between small stress changes and seismicity rate is very weak (KNOPOFF, 1964; BEELER and LOCKNER, 2003).

To evaluate expected values of t_n for the San Andreas fault system assume a 10 MPa stress drop and recurrence intervals between 30 and 100 years (appropriate for large earthquake recurrence) ($\dot\tau = 1.0 \times 10^{-8}$ to 3.17×10^{-9} MPa/s), a depth range of 5–15 km with $\sigma_n = 18$ MPa/km (appropriate for hydrostatic fluid pressure), and an experimentally derived value of $a = 0.0045$ (BEELER and LOCKNER, 2003) ($a\sigma_e = 0.4$–1.2 MPa). Using $t_n = t_c$ the predicted duration is 1.2 to 12 years and for $t_n = 2\pi t_c$ the typical duration is 7.6 to 76 years. These values are consistent with some aftershock sequence durations and the absence of obvious tidal triggering by the

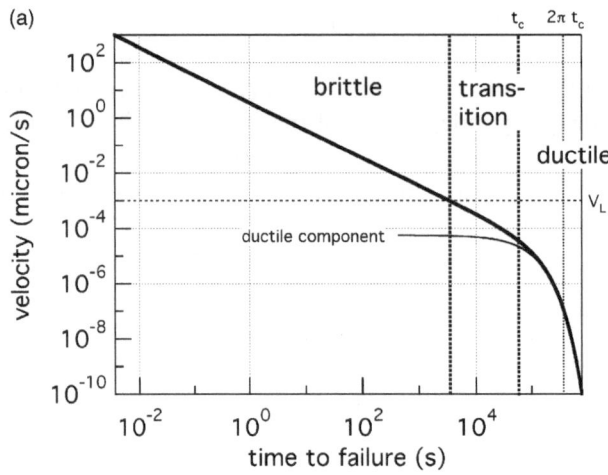

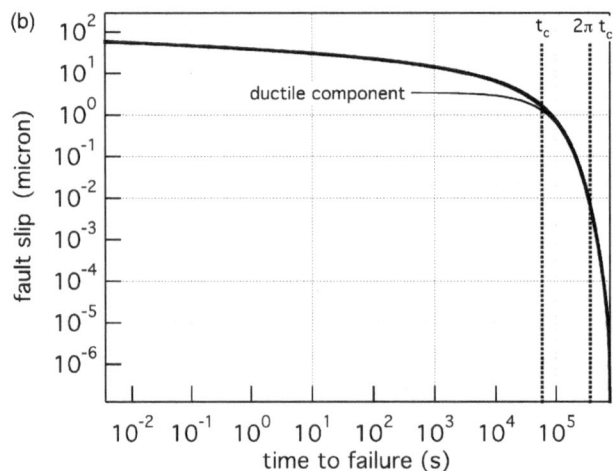

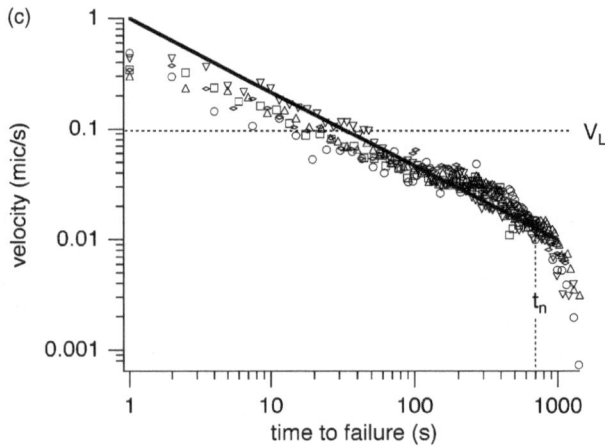

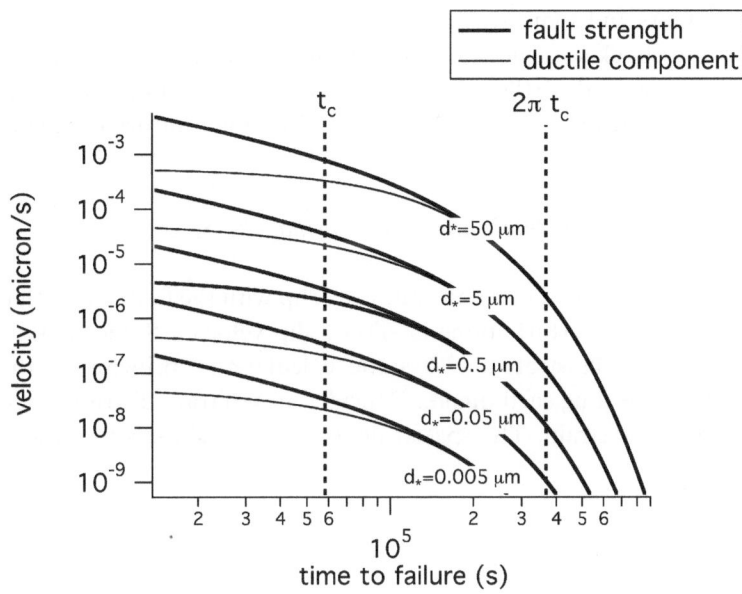

Figure 6

Time to failure characteristics of the simple brittle failure relation equation (6b) during loading at a constant rate for different values of the characteristic length d_*, as labeled. The fault (heavy curves) is loaded elastically $d\tau/dt = k(V_L - V)$ with $V_L = 0.001$ μm/s, stiffness $k/\sigma_n = 0.000137/\mu$m, $\mu_* = 0.7$, $a = 0.008$ and initial velocity $V_s = 1.0 \times 10^{-10}$ μm/s at $t = 0$. Various values of the ratio $\Delta\tau/\sigma_n d_*$ are used. The curve labeled $d_* = 5$ μm corresponds to $\Delta\tau/\sigma_n d_* = 0.0024\mu$m. Also shown is the ductile component of (6b) $\tau/\sigma_n = \mu_* + a \ln V/V_*$. For a particular value of d_*, the intersection of the heavy black and black curves is the duration of nucleation; the 4 sets of curves show that the duration of nucleation is essentially independent of d_*. Also shown are the two estimates of nucleation duration, the characteristic time $t_c = a\sigma_n/\dot{\tau}$ and $2\pi a\sigma_n/\dot{\tau}$.

solid earth tides. However, though shorter aftershock sequences may be evidence of low effective normal stress (super-hydrostatic pore pressure), smaller stress drops, higher stressing rates, or smaller values of a, aftershock durations can be much less

◄

Figure 5

Time to failure characteristics. Parts a) and b) are simulations using equation (6b) during loading at a constant rate. The fault is loaded elastically $d\tau/dt = k(V_L - V)$ with $V_L = 0.001\mu$m/s, stiffness $k/\sigma_n = 0.000137/\mu$m, $\mu_* = 0.7$, $a = 0.008$, $\Delta\tau/\sigma_n d_* = 0.0024$ /μm, and initial velocity $V_s = 1.0 \times 10^{-10}$ μm/s at $t = 0$. a) Slip velocity versus time to failure. At high velocity, time to failure is inversely related to slip velocity; this behavior characterizes the mixed and brittle behavior of (6b). At very low velocity, the behavior is well described by the ductile component of (6b) $\dot{\tau}/\sigma_e = \mu_* + a \ln V/V_*$ (black). The upper boundary of ductile behavior is well approximated by the characteristic time $t_n = a\sigma_n/\dot{\tau}$ (heavy dotted line labeled t_c). Another estimate of the limit of ductile behavior $t_n = 2\pi a\sigma_n/\dot{\tau}$ is shown (dotted line). b) Displacement versus time to failure for the same calculation as shown in a). c) Actual laboratory observation of slip velocity versus time to failure for 5 successive slip events on a granite fault at 50 MPa confining pressure and $V_L = 0.1$ μm/s (BEELER and LOCKNER, 2003], after DIETERICH and KILGORE [1996]. The solid line denotes a slope of -1, and the dashed line t_n is an estimate of the duration of nucleation.

than 12 to 76 years and some are on the order of weeks or months. Further analysis with (6b) of aftershock sequences and of reported correlation between stress changes due to oceanic tides and earthquake occurrence (e.g., WILCOCK, 2001) may provide constraints on t_n that are not available from existing laboratory experiments.

5. Implications for Earthquake Size

On a planar fault loaded at a constant rate, slip with the fault strength (6b) can be inherently unstable to perturbations in slip or slip velocity. Some spatial aspects of the behavior can be illustrated with static calculations, principally, the expected spatial size of a nucleating earthquake. Consider an arbitrary distribution of slip on a planar fault in plane strain. The spatial distribution of shear stress $\tau(x)$ due to slip $\delta(x)$ is

$$\tau(x) = \frac{G}{2\pi(1-v)} \int_{-\infty}^{\infty} \frac{-d\delta/dx(x')}{x-x'} dx' \tag{9}$$

(WEERTMAN, 1979). The temporal derivative of (9)

$$\frac{d\tau(x)}{dt} = \frac{G}{2\pi(1-v)} \frac{d}{dt} \left(\int_{-\infty}^{\infty} \frac{-d\delta/dx(x')}{x-x'} dx' \right) \tag{10a}$$

can be evaluated using Leibniz's Rule to find

$$\frac{d\tau(x)}{dt} = \frac{G}{2\pi(1-v)} \int_{-\infty}^{\infty} \frac{\frac{-d(d\delta/dx(x'))}{dt}}{x-x'} dx' \tag{10b}$$

Noting that, $d(d\delta/dx)/dt = d(d\delta/dt)/dx$, (10b) can be expressed as

$$\frac{d\tau(x)}{dt} = \frac{G}{2\pi(1-v)} \int_{-\infty}^{\infty} \frac{-dV/dx(x')}{x-x'} dx' \tag{10c}$$

Now consider a single wavelength perturbation above a uniform sliding speed V_o, $V(x) = V_0(1 - A \cos \omega x)$ where the wavelength is $2 \pi/\omega$ The spatial derivative of velocity is $dV/dx(x) = V_0 \omega A \sin \omega x$ and so long as there is only a single perturbation, the upper and lower integration limits of (10c) can be replaced by 2π and 0, respectively,

$$\frac{d\tau(x)}{dt} = \frac{G}{2\pi(1-v)} \int_0^{2\pi} \frac{-V_0 \omega A \sin \omega x'}{x-x'} dx' \tag{11a}$$

The temporal derivative of (6b) can be rearranged to give

$$\frac{dV}{dt}(x) = \frac{V(x)}{a} \left(\frac{1}{\sigma_n} \frac{d\tau}{dt}(x) + \frac{\Delta\tau V(x)}{\sigma_n d_*} \right) \tag{11b}$$

When (11a) is substituted into (11b), equation (11b) gives the spatial distribution of instantaneous slip acceleration of the perturbed fault patch, a measure of the growth rate. To determine the characteristic wavelength the procedure is to consider the growth rate of patches of different wavelength. Figure 7 shows that the spatial distribution of growth rate depends on the wavelength with the most uniform growth rate associated with a half wavelength $(l_c = \lambda/2) l_c = Gd_*/\Delta\tau$ (heavy solid line, Fig. 7). For this choice of parameters (see figure caption) $l_c = 0.5$ m. This particular wavelength should grow and maintain approximately constant shape whereas larger and smaller patches will either spread to larger or shrink to smaller wavelengths. Thus, earthquake nucleation on faults with (6b) should show a dominant nucleation patch size l_c. Time dependent calculations by DIETERICH (1992) using the same geometry as employed here show similar behavior. DIETERICH finds that during nucleation, patches with half-length greater than l_c shrink to l_c as slip accelerates, and slip perturbations of smaller wavelength tend to become smooth and coalesce to longer wavelength patches. However, in time-dependent calculations l_c is somewhat larger than inferred from static analysis; DIETERICH (1992) reports $l_c = 1.7Gd_*/\Delta\tau$ as the patch approaches dynamic slip speeds.

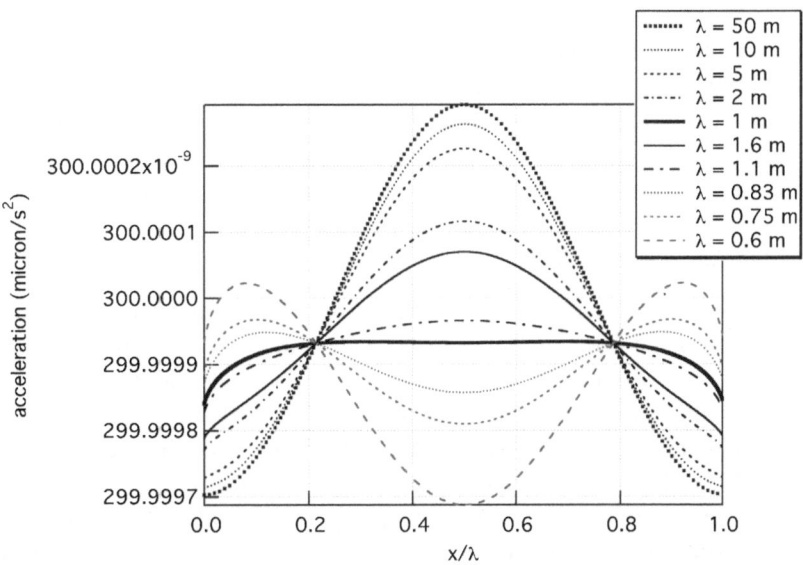

Figure 7
The spatial distribution of instantaneous slip acceleration on patches of a planar fault. Patches with a slip velocity perturbation of the form $V(x) = V_0 (1-A \cos \omega x)$ were calculated with equations (11a) and (11b) using different wavelengths $\lambda = 2\pi/\omega$ for $\mu_* = 0.7$, $a = 0.008$, $\Delta\tau/\sigma_n d_* = 0.0024$ /μm, $\sigma_n = 21$ MPa, $G = 2.5 \times 10^4$ MPa $v = 0.25$ $A = 0.5 \times 10^{-6}$ m/s, and $V_o = 1 \times 10^{-6}$ m/s. The calculated distribution of growth rates depends on the wavelength. The most uniform growth rate being associated with a half wavelength $(l_c = \lambda/2)$ $l_c = Gd_*/\Delta\tau = 0.5$ m.

5.2 Minimum Earthquake Nucleation Size

The minimum earthquake nucleation size can be estimated with a simpler geometry than used in the previous section; for example a fixed length static slipped patch (DIETERICH, 1986). Unstable sliding is only possible if fault strength drops so that stored elastic energy from the surrounding rock can be used to drive further slip. If an increment of slip on the fault surface reduces the fault strength faster than the driving stress is reduced, an unstable condition develops in which fault slip can accelerate into an earthquake and some of the stored elastic energy is radiated as seismic waves. If the elastic stiffness of the loading system is expressed as $k = d\tau/d\delta$, then the requirement for instability at constant normal stress is a critical stiffness (DIETERICH, 1979) $k_c = \sigma_n d\mu/d\delta$. For (6b) slip instability occurs for $k < k_c \approx \Delta\tau/d_*$ (DIETERICH, 1992) (Fig. 1). Using a circular crack of constant stress drop, the stiffness is $k = 7\pi G/16r$. For this circular patch, earthquake nucleation with (6b) is associated with patches of minimum radius $r_c = 7\pi G d_*/16\Delta\tau$. For $G = 2.5 \times 10^4$ MPa, an intermediate value of the ratio $\Delta\tau/d_*\sigma_n = 0.002/$ μm (see section 2.3) and effective normal stress between 90 and 270 MPa (effective normal stress gradient of 18 MPa/km over 5 to 15 km depth), the expected critical patch radii is between 0.06 and 0.2 meters. Nucleating slip on patches of this size would not be detectable at 7 km from the source even with the most sensitive instruments currently available (e.g., borehole strain meters) (JOHNSTON and LINDE, 2002). A more rigorous discussion and analysis of precursory strain and moment is given by DIETERICH (1992). Given that surface and space-based strain measures are much less sensitive to slip at depth than borehole instruments, the laboratory observations imply that precursory strain of earthquakes can only be measured in deep boreholes adjacent to seismic faults (also see, LORENZETTI and TULLIS, 1989).

5.3 Minimum Earthquake Size

Laboratory observations of frictional slip also imply a minimum earthquake size. Comparisons between lab, mining-induced events and other small earthquakes by MCGARR (1994; 1999) indicate that all are energetically similar, implying that fracture energy and radiated energy can be treated as scale-independent percentages of the available energy, at least for small events. Lab stick-slip events (OKUBO and DIETERICH, 1981; LOCKNER and OKUBO, 1983) have ratios $d_*/\Delta\delta_s \approx 0.1$ and the laboratory observed dynamic strength drops are frictional ($(\tau_y - \tau_f)/\sigma_n$ constant) with $(\tau_y - \tau_f)/\sigma_n = 0.09$ (WONG, 1986); using these values to determine the slip weakening coefficient $\Delta\tau/d_*\sigma_n$ leads to $\Delta\delta_s \approx 0.9\sigma_n d_*/\Delta\tau$. A lab-derived intermediate value for the slip-weakening coefficient, $\Delta\tau/d_*\sigma_n = 0.002/\mu m$ predicts a minimum-sized earthquake to have seismic displacements $\Delta\delta_s \approx 450 \mu m$.

For a circular crack, the relationship between radius, static stress drop $\Delta\tau_s$ and seismic slip $\Delta\delta_s$ is $r = 7\pi G\Delta\delta_s/16\Delta\tau_s$. Taking the estimate $\Delta\delta_s = 0.45$ mm and a range of static stress drops 0.1 to 10 MPa yields $r = 155$ to 1.55 m. The seismic moment M_o of a circular rupture of radius r is $M_0 = G\pi r^2 \Delta\delta_s$ and the lab-estimated values of r and $\Delta\delta_s$ yield $M_0 = 8.5 \times 10^{11}$ to 8.5×10^7 Nm. The minimum seismic moment from the smallest friction dominated mining-induced events reported by RICHARDSON and JORDAN (2002), $M_0^{min} = 4.7 \times 10^9$ Nm, are somewhat higher than the laboratory-estimated minimum seismic moment, $M_0^{min} = 8.5 \times 10^7$ Nm, but within the range of the prediction.

6. Conclusions

If crack growth is involved in earthquake nucleation at some scale, then constitutive equations that describe nucleation should have two components resulting in stable deformation at small inelastic strains (subcritical crack growth) and unstable slip- weakening behavior at large strains (critical crack growth). Laboratory observations from rock fracture and rock friction relevant to earthquake occurrence and earthquake nucleation are consistent with such constitutive equations, as are the predictions of the rate- and state- dependent constitutive formulation specific to rock friction. Considering the detailed rheology of friction and rock fracture as both resulting from crack growth is reasonable and generally expected because fractured rock and highly comminuted wear product are the byproducts of brittle rock deformation. A crack growth-based constitutive equation for earthquake nucleation used with lab-derived constants predicts a minimum earthquake nucleation time on the San Andreas fault of ~1 yr. The predicted minimum nucleation patch radius for an earthquake is on the order of 0.2 to 0.06 m and strains associated with nucleation are much too small to be detected by surface or space-borne instruments. The minimum earthquake moment implied by lab observations is ~8.5×10^7 Nm.

Acknowledgements

Discussions and correspondence with T. Tullis, J. Weeks, J. Dieterich, M. Nakatani, and particularly D. Lockner are gratefully acknowledged. This paper was significantly improved owing to the comments of two anonymous reviewers; many thanks to them and to Peter Mora.

REFERENCES

ANDREWS, D. J. (1976), *Rupture Propagation with Finite Stress in Antiplane Strain*, J. Geophys. Res. *81,* 3575–3582.

ATKINSON, B. K. and MEREDITH, P. G. (1987), Experimental fracture mechanics data for rocks and minerals. In *Fracture Mechanics of Rock* (B. K. Atkinson ed.), (Academic Press Geology Series, Academic Press, New York, New York), pp. 477–525.

BEELER, N. M. and LOCKNER, D. L. (2003), *Why Earthquakes Correlate Weakly with the Solid Earth Tides*, J. Geophys. Res., *108*, 2391.

BLANPIED, M. L., MARONE, C. J., LOCKNER, D. A., BYERLEE, J. D., and KING, D. P. (1998). *Quantitative Measure of the Variation in Fault Rheology due to Fluid-rock Interactions*, J. Geophys. Res. *103*, 9691–9712.

BRACE, W. F. and BYERLEE, J. D. (1966), *Stick-slip as a Mechanism for Earthquakes*, Science *153*, 990–992.

CHARLES, R. J. and HILLIG, W. B. (1962), *The kinetics of glass failure by stress corrosion*. In Symposium sur la resistance due verre et les moyens de l'ameliorer, *Union Scientific Continentale du Verr*, Charleroi, Belgium, pp. 511–527.

DAS, S. and SCHOLZ, C. H. (1981), *Theory of Time-dependent Rupture in the Earth*, J. Geophys. Res. *86*, 6039–6051.

DIETERICH, J. H. (1978), *Time-dependent Friction and the Mechanics of Stick Slip*, Pure Appl. Geophys. *116*, 790–806.

DIETERICH, J. H. (1979), *Modeling of Rock Friction. 1, Experimental Results and Constitutive Equations*, J. Geophys. Res. *84*, 2161–2168.

DIETERICH, J. H., *Constitutive properties of faults with simulated gouge*. In *Mechanical Behavior of Crustal Rocks: The Handin Volume*, Geophys. Monogr. Ser., vol. 24 (eds N. L. Carter *et al.*)AGU, Washington 1981, pp. 103–120.

DIETERICH, J. H. *A model for the nucleation of earthquake slip*. In *Earthquake Source Mechanics, Geophys. Monogr. Ser.*, vol. 37 (eds S. Das *et al.*, (AGU, Washington, D.C. 1986), pp. 37–49.

DIETERICH, J. H. *Earthquake nucleation on faults with rate- and state-dependent strength*. In *Earthquake Source Physics and Earthquake Precursors* (eds. T. Mikumo *et al.*, Elsevier, New York 1992), pp. 115–134.

DIETERICH, J. (1994), A *Constitutive Law for Rate of Earthquake Production and its Application to Earthquake Clustering*, J. Geophys. Res. *99*, 2601–2618.

DIETERICH, J. H., and KILGORE, B. D. (1994), *Direct Observation of Frictional Contacts: New Insights for State-dependent Properties*, Pure Appl. Geophys. *143*, 283–302.

DIETERICH, J. H. and KILGORE, B. D. (1996), *Implications of Fault Constitutive Properties for Earthquake Prediction*, Proc. Natl. Acad. Sci., *93*, 3787–3794.

DIETERICH, J., CAYOL, V., and OKUBO, P. (2000), *The Use of Earthquake Rate Changes as a Stress Meter at Kilauea Volcano*, Nature, *408*, 457–460.

GOMBERG, J., BEELER, N. M., BLANPIED, M. L., and BODIN, P. (1998), *Earthquake Triggering by Transient and Static Deformations*, J. Geophys. Res. *103*, 24411–24426.

GRIGGS, D. (1936), *Deformation of Rocks under High Confining Pressure*, J. Geol. *44*, 541.

GU, J., RICE, J. R., RUINA, A. L., and TSE, S. (1984), *Stability of Frictional Slip for a Single Degree of Freedom Elastic System with Nonlinear Rate and State-dependent Friction*, J. Mech. Phys. Solids, *32*, 167–196.

HIRATA, T. (1987), *Omori's Power Law Aftershock Sequences of Microfracturing in Rock Fracture Experiment*, J. Geophys. Res. *92*, 6215–6221.

IDA, Y. (1972), *Cohesive Force Across the Tip of a Longitudinal Shear Crack and Griffith's Specific Surface Energy*, J. Geophys. Res. *77*, 3796–3805.

JOHNSTON, M. J. S. and LINDE, A. T. (2002), *Applications of Crustal Strain during Conventional, Slow, and Silent Earthquakes*, International Handbook of Earthquake seismology, Part A, Elsevier W. Lee editor.

KNOPOFF, L. (1964), *Earth Tides as a Triggering Mechanism for Earthquakes*, Bull. Seismol. Soc. Am. *54*, 1865–1870.

KNOPOFF, L. (1972), *Model of aftershock occurrence*. In *Flow and Fracture of Rocks, Geophys. Monograph 16*, 259–263.

KRANZ, R. L. (1980), *The Effects of Confining Pressure and Stress Difference on Static Fatigue of Granite*, J. Geophys. Res. *85*, 1854–1866.

LAWN, B., *Fracture of Brittle Solids* (Cambridge University Press, New York 1993, 378 *pp*.

LAPUSTA, N., RICE, J. R., BEN-ZION, Y., and ZHENG, G. (2000), *Elastodynamic Analysis for Slow Tectonic Loading with Spontaneous Rupture Episodes on Faults with Rate- and State-dependent Friction*, J. Geophys. Res. *105*, 23765–23789.

LINKER, M. F. and DIETERICH, J. H. (1992), *Effects of variable normal stress on rock friction: observations and constitutive equations,* J. Geophys. Res. *97,* 4923–4940.

LOCKNER, D. A. (1993), *Room Temperature Creep in Saturated Granite,* J. Geophys. Res. *98,* 475–487.

LOCKNER, D. A. (1998), *A Generalized Law for Brittle Deformation of Westerly Granite,* J. Geophys. Res. *103,* 5107–5123.

LOCKNER, D. A., and OKUBO, P. G. (1983), *Measurements of Frictional Heating in Granite,* J. Geophys. Res. *88,* 4313–4320.

LOCKNER, D. A., BYERLEE, J. D., KUKSENKO, V., PONOMAREV, A., and SIDORIN, A. (1991), *Quasi-static Fault Growth and Shear Fracture,* Nature *350,* 39–42.

LORENZETTI, E. and TULLIS, T. E. (1989), *Geodetic Predictions of a Strike-slip Fault Model: Implications for Intermediate- and Short-term Earthquake Prediction,* J. Geophys. Res. *94,* 12,343–12,361.

MARCELLINI, A. (1995), *Arrhenius Behavior of Aftershock Sequences,* J. Geophys. Res. *100,* 6463–6468.

MARCELLINI, A. (1997), *Physical Model of Aftershock Temporal Behavior,* Tectonophysics, *277,* 137–146.

McGARR, A. (1994), *Some Comparisons between Mining-induced and Laboratory Earthquakes,* Pure Appl. Geophys. *142,* 467–489.

McGARR, A. (1999), *On Relating Apparent Stress to the Stress-causing Earthquake Fault Slip,* J. Geophys. Res. *104,* 3003–3011.

MOGI, K. (1962), *Study of Elastic Shocks Caused by the Fracture of Heterogeneous Materials and their Relation to Earthquake Phenomenon,* Bull. Earthquake Res. Inst. Univ. Tokyo *40,* 1438.

MOGI, K. (1974), *On the Pressure Dependence of Strength of Rocks and the Coulomb Fracture Criterion,* Tectonophysics 21, 273–285.

NAKATANI, M. (2001), *Conceptual and Physical Clarification of Rate and State Friction: Frictional Sliding as a Thermally Activated Rheology,* J. Geophys. Res. *106,* 13,347–13,380.

OKUBO, P. G. and DIETERICH, J. H. (1981), *Fracture Energy of Stick-slip Events in a Large Scale Biaxial Experiment,* Geophys. Res. Lett. *8,* 887–890.

OKUBO, P. G. and DIETERICH, J. H. (1984), *Effects of Physical Fault Properties on Frictional Instabilities Produced on Simulated Faults,* J. Geophys. Res. *89,* 5817–5827.

PALMER, A. C. and RICE, J. R. (1972), *The Growth of Slip Surfaces in the Progressive Failure of Over-consolidated Clay,* Proc. Roy. Soc. Lond. *A332,* 527–548.

POWER, W. L., TULLIS, T. E., BROWN, S. R., BOITNOTT, G. N., and SCHOLZ, C. H. (1987), *Roughness of Natural Fault Surfaces,* Geophys. Res. Lett. *14,* 29–32.

REUSCHLE, T. (1990), *Slow Crack Growth and Aftershock Sequences,* Geophys. Res. Lett. *17,* 1525–1528.

RICE, J. R. and RUINA, A. L. (1983), *Stability of Steady Frictional Slipping,* J. Appl. Mech. *50,* 343–349.

RICE, J. R., LAPUSTA, N., and RANJITH, K. (2001), *Rate-and State-dependent Friction an the Stability of Sliding between Elastically Deformable Solids,* J. Mech. Phys. Sol *49,* 1865–1898

RICHARDSON, E. and JORDAN, T. H. (2002), *Seismicity in Deep Gold Mines of S. Africa: Implications for tectonic Earthquakes,* Bull. Seismol. Soc. Am. *92,* 1766.

RUINA, A. L. (1983), *Slip Instability and State Variable Friction Laws,* J. Geophys. Res. *88,* 10,359–10,370.

SAMMIS, C. G. and STEACY, S. J.(1994), *The Micromechanics of Friction in a Granular Layer,* Pure Appl. Geophys. *142,* 777–794.

SAVAGE. J. C., BYERLEE, J. D., and LOCKNER, D. A.(1996), *Is Internal Friction, Friction?* Geophys. Res. Lett. *23,* 487–490.

Scholz, C., (1968a). The frequency magnitude relation of microfracturing in rock and its relation to earthquakes, Bull. Seismol. Soc. Am. *58,* 399.

SCHOLZ, C. (1968b), *Microfractures, Aftershocks and Seismicity,* Bull. Seismol. Soc. Am. *58,* 117–130.

SCHOLZ, C. H. (1968c), *Mechanism of Creep in Brittle Rock,* J. Geophys. Res. *73,* 3295–3302.

SCHOLZ, C. H. (1972), *Static Fatigue in Quartz,* J. Geophys. Res. *77,* 2104–2114.

TULLIS, T. E. and Weeks, J. D. (1987), *Micromechanics of Frictional Resistance of Calcite,* EOS Trans. Am. Geophys. Un. *68,* 405.

TULLIS, T. E. and WEEKS, J. D. (1986), *Constitutive Behavior and Stability of Frictional Sliding of Granite,* Pure Appl. Geophys. *124,* 384–414.

WEERTMAN, J. (1979), *Inherent Instability of Quasi-static Creep Slippage on a Fault,* J. Geophys. Res. *84,* 2146–2152.

WILCOCK, W. S. D. (2001), *Tidal Triggering of Microearthquakes on the Juan de Fuca Ridge*, Geophys. Res. Lett. *28*, 3999–4002.

WONG, T.-f. *On the normal stress dependence of the shear fracture energy.* In *Earthquake Source Mechanics*, Geophys. Monogr. Ser., vol. 37 (eds. S. Das *et al.* AGU, Washington, D.C.1986) pp. 1–11.

YAMASHITA, Y. and KNOPOFF, L. (1987), *Models of Aftershock Occurrence*, Geophys. J. R. astr. Soc. *91*, 13–26.

ZHURKOV, S. N. (1965), *Kinetic Concept of the Strength of Solids*, Int. J. Fracture Mech. *1*, 311–323.

ZHRUKOV, S. N., KUKSENKO, V. S., PETROV, V. A., SAVLYEV, V. N., and SULTANOV, U. (1977), *On the Problem of Prediction of Rock Fracture*, Izv. Acad.Sci. USSR, Phs Solid Earth (in English) *13*, 374–379.

(Received September 27, 2002, revised February 28, 2003, accepted March 7, 2003)

To access this journal online:
http://www.birkhauser.ch

Pure appl. geophys. 161 (2004) 1877–1891
0033–4553/04/101877–15
DOI 10.1007/s00024-004-2537-y

❘ Pure and Applied Geophysics

Particle Dynamics Simulations of Rate- and State-dependent Frictional Sliding of Granular Fault Gouge

JULIA K. MORGAN[1]

Abstract—Recent simulations using the particle dynamics method (PDM) have successfully captured many features of natural faults zones as illuminated in laboratory studies. However, 2-D simulations conducted on idealized assemblages of particles using simple elastic-frictional contact laws, yield friction values considerably lower than natural materials, and lack time- and velocity-dependent changes in strength that influence dynamic fault slip. Here, preliminary results of new PDM simulations are described, in which particle motions are restricted as a proxy for particle interlocking and out of plane contacts, and time-dependent contact healing is introduced to capture temporal strengthening of granular assemblages. Frictional strength is increased, and in the absence of interparticle rolling, can attain values observed in the laboratory. The resulting mechanical behavior is qualitatively similar to that described by empirically-based rate-state friction laws, providing new physical insight into the discrete mechanics of natural faults.

Key words: Fault friction, particle dynamics, discrete element method, granular shear zones, contact laws.

Introduction

Particle dynamics method (PDM) simulations represent a unique tool for studying the mechanics of gouge-bearing shear zones, providing views into active and evolving faults in ways never imagined in the field or in the laboratory. PDM simulations have been shown to successfully capture many aspects of shear deformation documented in laboratory experiments, including stick-slip like friction variations with shear (e.g., MORA and PLACE, 1994; MORGAN, 1999; SPARKS and AHARONOV, 2001), frictional dependence on normal stress (AHARONOV and SPARKS, 1999; MORGAN, 1999) and on dilation rate (MORGAN, 1998), and strain localization (ANTONELLINI and POLLARD, 1995; MORA and PLACE, 1998; MORGAN and BOETTCHER, 1999; SPARKS and AHARONOV, 2001). However, several fundamental experimental results thought to be important in the earthquake generation process, have not been well reproduced by PDM simulations. In particular, friction coefficients determined for numerical systems are commonly much lower than ∼0.6

[1] Department of Earth Science, MS-126, Rice University, Houston, TX 77005, U.S.A.
E-mail: morganj@rice.edu

predicted by Byerlee's law (e.g., MORGAN, 1999), a result that can be partially attributed to enhanced particle rolling in 2-D assemblages (e.g., SCOTT, 1996; MORGAN and BOETTCHER, 1999; MAIR et al., 2002). In addition, due to the simplified contact laws employed, PDM simulations do not show 2nd order changes in friction with strain, thought to influence stability of sliding and seismogenesis (e.g., DIETERICH, 1979; MARONE et al., 1990; MARONE, 1998a, 1998b; BEELER et al., 1994, 1996). These differences between experimental and numerical friction need to be reconciled before we can apply PDM simulations to the study of more complicated, natural fault systems.

New studies have been carried out to better implement these physical properties in shearing fault models. Preliminary results of recent studies are presented here, focused on improving and updating PDM simulations using the discrete element method (DEM; CUNDALL and STRACK, 1979) to better reproduce laboratory observations of shear strength and variations in friction with strain. By restricting particle mobility as a proxy for grain roughness, interlocking, and out-of-plane contacts, and by implementing time-dependent healing at interparticle contacts, it is possible to qualitatively reproduce the scale and phenomenology of rate and state constitutive laws for friction. These DEM results bring us one step closer to building a consistent, physics-based description of fault friction phenomena that can be extrapolated to natural faults and an understanding of their seismogenic behavior.

Particle Dynamics Methods

Particle dynamics methods (PDM) are based on molecular dynamics (MD), a numerical technique well established in physics and fluid dynamics (e.g., ALLEN and TILDESLEY, 1987). Several different PDM codes have been developed and applied to study fault zones and granular friction, including the discrete element method (DEM, CUNDALL and STRACK, 1979; CUNDALL, 1988) and lattice solid model (LSM, MORA and PLACE, 1993; PLACE et al., 2002), but all are similar in form and output. A granular system is represented as an assemblage of discrete particles that interact with each other according to fundamental contact physics, thereby preserving its discontinuous nature. The discrete processes of granular shear, such as formation, propagation, and dissipation of microstructures, can be observed in "real-time" (e.g., MORGAN and BOETTCHER, 1999), and the micromechanics of particle interactions explored and correlated with global behavior (MORGAN, 1999; MORA and PLACE, 1998; AHARONOV and SPARKS, 1999).

The DEM approach (CUNDALL and STRACK, 1979), used for this study, is a soft-sphere particle dynamics method that includes the effects of elastic particle deformations during contact (Fig. 1). It employs an explicit, central difference, time-stepping, finite-difference approach to solve Newton's equations of motion simultaneously for every particle:

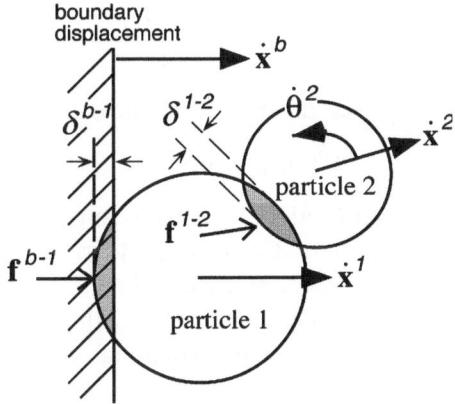

Figure 1

Schematic diagram of particle interactions in DEM. Boundary displacement, $\dot{\mathbf{x}}^b$ causes apparent overlap with particle 1, δ^{b-1}, introducing a contact force, $\mathbf{f}^{b-1}$ Particle 1, in turn, is displaced by $\dot{\mathbf{x}}^1$, and impinges upon particle 2 by δ^{1-2}, imparting an oblique contact force, f^{1-2}. Particle 2 responds through displacement, $\dot{\mathbf{x}}^2$ and rotation, $\dot{\boldsymbol{\theta}}^2$ If the contact shear force is high enough, sliding will also occur. (Modified after MORGAN and BOETTCHER, 1999.)

$$\mathbf{F}^a = m^a \ddot{\mathbf{x}}^a, \tag{1a}$$

$$\mathbf{M}^a = I^a \ddot{\boldsymbol{\theta}}^a, \tag{1b}$$

where $\mathbf{F}^a$ and $\mathbf{M}^a$ are the net force and moments acting on particle a, m^a and I^a are the mass and moment of inertia, and $\ddot{x}^a$ and $\ddot{\boldsymbol{\theta}}^a$ are linear and angular accelerations, respectively. Particle motions are induced by gravitational forces, external forces prescribed by stress or strain rate boundary conditions, and by forces resolved at interparticle contacts. The net force and moment on a particle are substituted into the equations of motion to obtain velocities and displacements. The disequilibrium of forces drives particle displacements. Particle displacements are unrestricted, so the system can accumulate large strains.

Interparticle contact laws play an important role in the resulting assemblage behavior. The DEM implementation here employs cohesionless particles that obey Hertz-Mindlin (elastic-frictional) contact theory (e.g., JOHNSON, 1985). Particles respond elastically to normal forces at their boundaries, leading to a nonlinear repulsion scaled to the contact area. The components of the normal force are given by

$$f^a(n)_i = \frac{4\sqrt{RG}}{3(1 - \vee)} \delta_i^{3/2}, \tag{2}$$

where R is the effective radius of the contacting particles, G is the bulk modulus, v is Poisson's ratio, and δ_i gives the i-th component of relative particle approach ($i = 1,2,3$). Similarly, the components of shear force resolved at the contact are

$$f^a(S)_i = (f^a(n))^{1/3} \frac{(4G)^{2/3}(3R(1 - \vee))^{1/3}}{(2 - \vee)} \delta_i. \qquad (3)$$

If the magnitude of $f^a(s)$ exceeds the critical shear force for frictional sliding,

$$f^a(s)_{\max} - \mu_p f^a(n), \qquad (4)$$

where μ_p is interparticle contact friction, then slip occurs. Energy is dissipated at the contact by viscous or force damping (CUNDALL and STRACK, 1979; CUNDALL, 1987; WALTON, 1995). One of the key objectives of this study is to examine the consequences of more complicated contact laws that preserve time-dependent changes in contact strength inferred for real frictional materials.

PDM Simulations of Granular Shear Zones

Previous PDM studies of shear zones have yielded intriguing insights into the mechanics and deformation of granular shear zones, and in particular, controls on fault friction. Stick-slip behavior during shear experiments reflects variations in shear strength due to formation and collapse of load-bearing particle bridges (e.g., MORGAN, 1999; MORGAN and BOETTCHER, 1999; PLACE and MORA, 2000; SPARKS and AHARONOV, 2001), as described for natural fault zones (e.g., SAMMIS et al., 1987; BIEGEL et al., 1989). The rapid strength drop upon collapse of the stress bridge is accompanied by localization of strain, yielding characteristic fracture arrays with forms consistent with those observed in the natural and experimental fault zone (e.g., MORGAN and BOETTCHER, 1999). These shear fractures are very transient features, however, due to the lack of fracture and contact healing which would impart a memory to the system. The cumulative strain field tends to be relatively homogeneous across the shear zone (MORGAN and BOETTCHER, 1999).

Changes in particle size distribution (PSD), generated by grain fracture and cataclasis during shear, can also influence the strength and frictional behavior of real shear zones (e.g., SAMMIS et al., 1987; BIEGEL et al., 1989; MARONE and SCHOLZ, 1989; SAMMIS and STEACY, 1994). DEM investigations of the effects of PSD variations on shear zone strength demonstrate that they impart a second-order influence on granular friction for 2-D simulations. Shear strength and stress drop decrease with increasing abundance of small particles (MORGAN and BOETTCHER, 1999; MORGAN, 1999). Therefore, progressive grain fracture leading to PSD evolution defines a slip-weakening mechanism that may partially explain enduring strain localizations. This phenomenon, however, cannot explain strength recovery that accompanies stick-slip behavior, which must depend on time-dependent healing of fault surfaces or particle contacts (e.g., SCHOLZ, 1990).

Many of the PDM simulations of granular shear zones yield anomalously low shear strengths, between ~0.3–0.4 (e.g., MORGAN, 1999), falling significantly below the typical values of ~0.6–0.7 documented in laboratory experiments, consistent with Byerlee's Law (e.g., SCHOLZ, 1990). The numerical studies typically employ 2-D assemblages of rounded particles, which lack out of plane contacts and particle interlocking. The low shear strength in the PDM simulations of granular shear results from high levels of particle rolling in idealized 2-D assemblages (MORGAN and BOETTCHER, 1999), as laboratory investigations on the role of particle geometry and dimensionality of idealized glass beads and rods have recently confirmed (FRYE and MARONE, 2002a). Particle rolling is not as common in natural 3-D systems, due the presence of irregularly shaped, interlocking particles over a wide range of sizes (e.g., MAIR et al., 2002).

The results of previous PDM simulations demonstrate that many key aspects of granular shear and deformation can be reproduced, but also expose several weaknesses that should be addressed before the technique can be broadly applied to the study of natural fault zones. Below, we introduce restrictions to particle dynamics and modified contact laws that can be easily implemented in PDM simulations that begin to overcome these limitations.

Particle Rolling

As argued by MAIR et al. (2002), the prevalence of particle rolling in PDM simulations results from use of idealized rounded particles in the models, and is not representative of natural materials. Simulations using non-spherical particles (e.g., elliptical disks, ellipsoids) show higher strength (e.g., ROTHENBURG and BATHURST, 1992; TING et al., 1993; LIN and NG, 1997; WILLIAMS and PENTLAND, 1991). However, irregular particle shapes greatly complicate contact detection and force calculation, significantly increasing simulation times. Irregular and "rough" particle shapes can be constructed by fusing round grains together (e.g., WALTON and BRAUN, 1993; MORA and PLACE, 1998), leading to increased fault strength in the presence of particle rolling (MORA and PLACE, 1998), particularly in the case of wide shear zones (e.g., MORA et al., 2000). Geologic materials composed of 3-D assemblages of interlocking particles, and characterized by a full distribution of particle sizes (e.g., AN and SAMMIS, 1994), however, clearly tend to prevent coordinated particle rolling, enhancing interparticle sliding and dilation during shear deformation (e.g., MAIR et al., 2002). To approach this behavior, we develop alternative proxies to restrict particle rolling in simplified 2-D assemblages.

One way to restrict particle mobility is to damp angular rotations calculated from equation (1b). In 2-D systems, this can be accomplished by introducing a linear damping term, α that scales the angular rotation in the plane of the section, $\dot{\theta}_3^a$ at each time step as follows:

$$\dot{\theta}_3^a = \left(\dot{\theta}_3^a + M_3^a \frac{I^a}{m^a} \Delta t\right)^* \alpha, \tag{5}$$

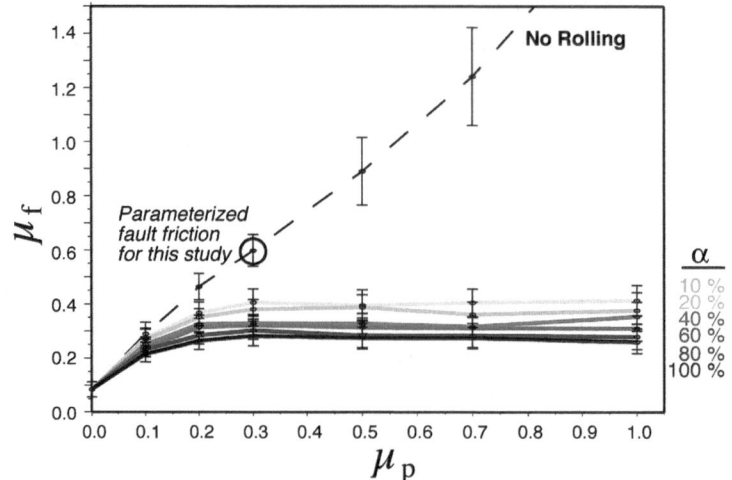

Figure 2

Average value of sliding friction, μ_f is plotted as a function of interparticle friction, μ_p for 2-D simulations with damped angular rotations. Damping parameter, α ranges from 0% (no rolling) to 100% (no damping).

where Δt is the numerical time step. The damping term, α ranges from 0% (no rotation) to 100% (no damping). A suite of 2-D numerical simulations shows the effect of rotational damping for a wide range of interparticle friction, μ_p values. Fault friction, μ_f is seen to increase with decreasing α, but still yields unreasonably low values for μ_f with as α low as 10% (90% damping) for all μ_p (Fig. 2). Clearly, if rolling is possible, this becomes the preferred deformation mechanism. Full damping of particle rotations (no rolling), however, leads to a nearly linear increase in μ_f with μ_p (Fig. 2). To develop a numerical proxy for realistic particle roughness, angularity, and interlocking, interparticle friction of idealized particles can be parameterized to gain realistic values for μ_f. In the experiments presented below, we selected μ_p of 0.3 to yield μ_f of 0.6, consistent with Byerlee's law.

Time-dependent Contact Healing

Due to the simplified elastic-frictional contact laws used in most of the existing PDM simulations, second-order time- and velocity-dependent frictional effects documented in the laboratory (e.g., DIETERICH, 1979; BEELER et al., 1994; BIEGEL et al., 1989; MARONE et al., 1990) have not been observed. In order to reproduce such effects, contact laws must be modified. The origin of time-dependent strengthening and related rate and state friction effects is thought to derive from growth of the contact area (DIETERICH and KILGORE, 1994) via a chemically-assisted mechanism (FRYE and MARONE, 2002b). Under contact loading, particle surfaces may deform inelastically, increasing particle interpenetration or contact surface area, and physico-chemical reactions may increase cohesion between the contacting surfaces. In the laboratory, contact healing has been shown to depend on the logarithm of contact duration (e.g., DIETERICH, 1972; BEELER et al., 1994; MARONE, 1998a). Therefore, one of the most important features to include in PDM simulations is contact memory, such that interparticle friction and cohesion vary in an appropriate way with the duration of static contact.

Resulting changes in friction at interparticle contacts can be implemented in several ways: (1) Through an instantaneous transition from a constant static friction to a lower dynamic value upon slip, followed by an instantaneous reset when slip stops. This results in rapid strength loss during dynamic slip, and instantaneous strength recovery. However, the static to dynamic friction transition is arbitrary and precludes strength evolution during shear. (2) The other extreme is to impose rate- and state-friction laws directly at particle contacts. ABE et al. (2002) show that this produces rate- and state-friction like behavior at the assemblage scale, but yields different parameters at the two scales. This approach is very CPU intensive (ABE, personal communication, 2002), and not ideal for large systems representative of geologic fault zones. (3) An intermediate solution is to introduce contact healing alone, e.g., by tracking contact duration and calculating friction over time, t as:

$$\mu_p^a = \mu_p^0, \qquad\qquad t = 0, \tag{6a}$$

$$\mu_p^a = \mu_p^0 + b\ln(t), \quad t > 0, \tag{6b}$$

where μ_p^a is the time-dependent interparticle friction coefficient, and b is a scalar coefficient. This approach is used below, and is shown to reproduce many aspects of rate- and state-friction phenomenology.

Experimental Method and DEM Results

All experiments described here were initialized similarly, according to the method described by MORGAN and BOETTCHER (1999). The initial particle assemblage, consisting of 1750 particles with identical elastic parameters, was created by randomly generating a specified number of particles of different radii: 250 particles of 200 μm, 500 particles of 150 μm, and 1000 particles of 100 μm; shear modulus was set to 2.9 GPa, and Poisson's ratio, $\vee$ to 0.2. The 2-D domain was then consolidated isotropically to a mean stress of 70 MPa by moving the boundaries inward. Particles can pass freely across periodic boundaries during consolidation.

Within the consolidated domain, two parallel bands of particles were fixed relative to each other 900 μm apart, to serve as shear zone walls; the lateral boundaries remain periodic (Fig. 3a). To impose shear conditions, the upper wall was moved at a constant velocity, under constant vertical stress of 70 MPa. For the experiments carried out here, the upper wall was displaced at a constant velocity

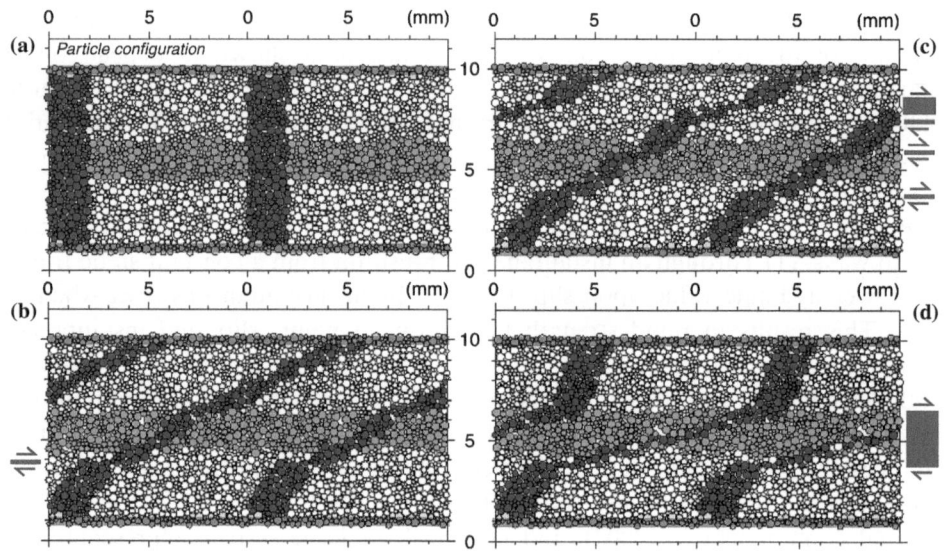

Figure 3

Particle configurations for three simulations. Each image consists of two identical assemblages, merged along the periodic boundary. Sidebars show degree of localization. (a) Initial configuration for all simulations. (b) Distributed bulk strain field at 150% shear strain for control test. (c) Distributed strain field at 150%, with persistent slip planes, sheared at 0.001 μm/s with contact healing. (d) Localized strain field at 150% shear strain, sheared at 0.01 μm/s with contact healing.

ranging from 0.1 to 0.001 μm/s. Upon shearing, equations (6) were employed to calculate μ_p^a, with μ_p^0 set to 0.3, and b set to 0.0075. For comparison, simulations were also conducted without time-dependent contact healing. Without healing, deformation tended to be distributed across the shear zone (Fig. 3b; e.g., MORGAN and BOETTCHER, 1999). With contact healing, shear strain tended to localize and endure within narrow bands (Figs. 3c and d).

The mechanical behavior of the shear zone depends on sliding velocity. Low wall velocities produced characteristic stick-slip behavior, with repeated sequences of elastic loading, pre-failure plastic yield, and finally, failure with a sudden strength drop (Figs. 4a and b). Each event was accompanied by gradual dilation followed by sudden contraction. Peak μ_f was ~0.6–0.62 in the control system and ~0.7 with contact strengthening. Higher velocities resulted in contrasting behavior, reminiscent of the oscillating mode described by NASUNO et al. (1997). Strength and volume strain still fluctuated throughout the experiment, but stick-slip cycles were poorly developed and dilation-contraction sequences were subdued (Figs. 4c and d).

Plots of mean sliding friction and dilation for all experiments, show significant variability, but indicate a direct velocity dependence (Figs. 5a and b). This velocity strengthening behavior is consistent with laboratory studies of granular gouge during initial shearing, and implies stable sliding (e.g., MARONE et al., 1990; BEELER et al.,

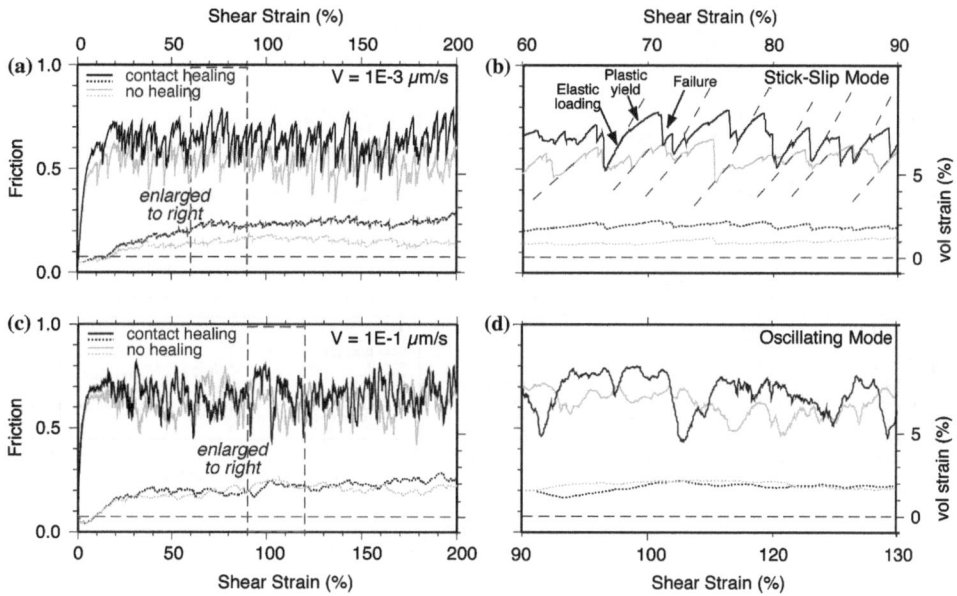

Figure 4

Friction, μ_f (solid) and volume strain (dotted) data for four numerical experiments. (a) Wall velocity of 0.001 μm/s results in repeated stick-slip events with consistent peak strengths. (b) Enlarged view of boxed region in (a) shows form of stick-slip cycles. Elastic modulus is nearly constant for a test, but appears to be higher with contact healing. (c) Wall velocity of 0.1 μm/s produces similar fluctuations in friction and volume strain, however, in detail, (d) stick-slip cycles are absent and friction and volume-strain exhibit nearly symmetric oscillations.

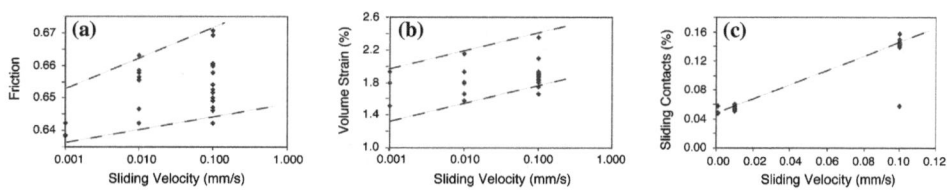

Figure 5

Mean values of steady-state experimental quantities for all experiments. (a) Friction and (b) volume strain show logarithmic increases with sliding velocity, indicative of velocity strengthening behavior. (c) Percentage of sliding contacts is also observed to increase with velocity, but the functional form of the relationship is not clear.

1996). An associated increase in percentage of sliding contacts at higher velocities (Fig. 5c) suggests the mechanism responsible for this behavior.

To examine shear zone restrengthening following stress drops, static hold tests were carried out for different durations, using identical starting configurations. Both frictional relaxation during holds and peak strength upon reloading increased logarithmically with hold time (Figs. 6a and d; Fig. 7). Changes in the number of

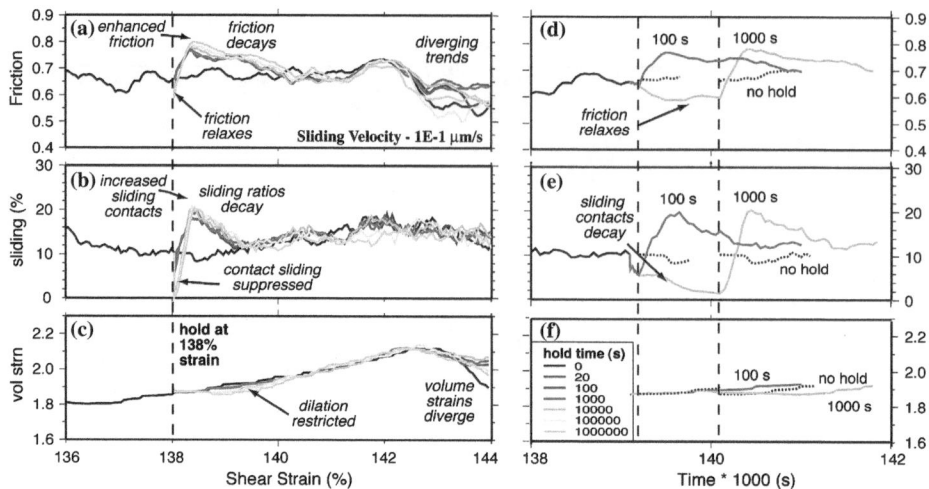

Figure 6

Static loading tests carried out at 138% strain for wall velocity of 0.1 μm/s. (a) Magnitude of friction, μ_f decays logarithmically with time during static holds, and reaches high static values upon reloading. (b) Percentage of sliding contacts mirrors friction trends. (c) Dilation is suppressed and delayed upon reloading as a function of hold time. Plots (d), (e), and (f) show friction, sliding contacts percentage, and volume strain against time for two static holds: 100 s and 1000 s.

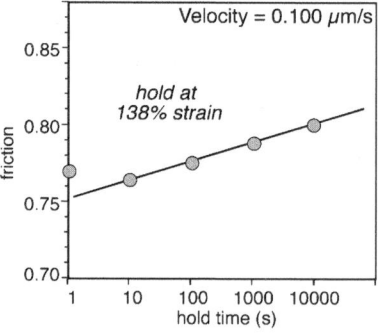

Figure 7

Plot of friction, μ_f against log(hold time) for static loading tests carried out at 138% strain for wall velocity of 0.1 μm/s (e.g., Fig. 6).

sliding contacts mirrored the frictional response (Figs. 6b and e). Shear zone dilation was suppressed during and after holds (Figs. 6c and f). The results of the static hold tests are qualitatively similar to laboratory experiments (e.g., MARONE, 1998a; KARNER and MARONE, 1998). Immediately following static holds, friction and volume strain paths were very similar, but these trends diverged and became effectively uncorrelated within ~5% strain, despite the identical initial conditions (Fig. 6).

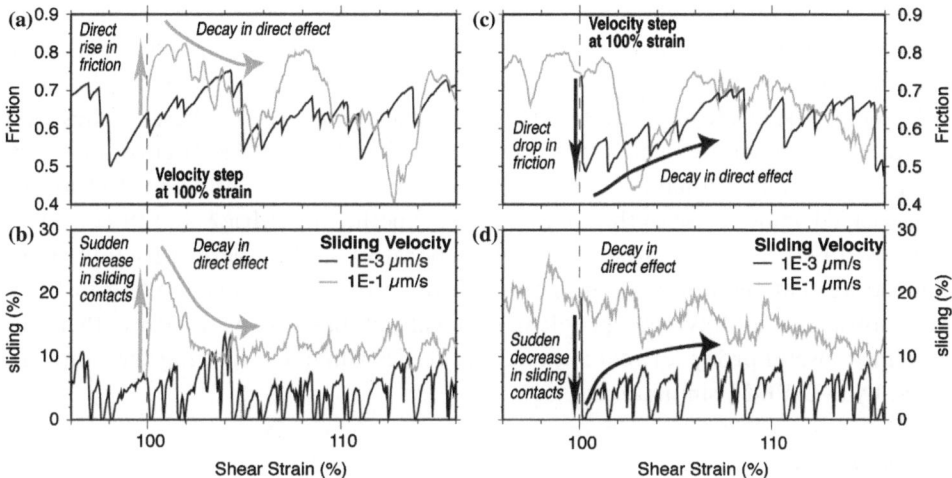

Figure 8

Friction, μ_f and volume strain data for instantaneous changes in sliding velocity, between 0.1 μm/s (black) and 0.001 μm/s (gray). Constant velocity tests shown for comparison. (a) Friction and (b) percentage of sliding contacts show direct increase in response to velocity jump, then decay to higher mean value over 6% strain. (c) Friction and (b) percentage sliding contacts declines in response to velocity drop, then gradually climb to lower mean value over 7–8% shear strain.

Another manifestation of rate- and state-friction is the variation in friction during velocity stepping experiments (e.g., MARONE et al., 1990; MAIR et al., 2002). In our simulations, fluctuations in strength produced by irregular stick-slip events tended to mask transient changes in strength when comparing experiments, but the use of identical starting configurations revealed both direct and evolving changes in friction during velocity steps (Figs. 8a and c). The percentage of sliding contacts mirrored changes in friction (Figs. 8b and d), strongly arguing that velocity changes produced an immediate readjustment of contact loads, forcing a reconfiguration of the system through contact sliding and assemblage dilation. The characteristic slip length for these examples is on the order of 6–8% shear strain, but it is not yet known what governs this scaling.

Discussion

Effects of Particle Rolling

As shown by our study, restricting particle rolling during PDM simulations of granular shear can yield more realistic values of assemblage friction than are gained if free rolling is allowed. This approach is designed to recapture some of the limitations on particle motion in real materials, in particular, due to out-of-plane contacts that frustrate rolling in 3-D systems, and to particle interlocking during shear. The validation for this approach comes from results of laboratory experiments designed to match simplified numerical conditions (FRYE and MARONE, 2002a; MAIR *et al.*, 2002). Using idealized materials (glass spheres and rods), these studies were able to reproduce and extend the numerical results. Assemblage friction increased with particle "dimensionality," and with particle angularity, and the 2-D assemblages yielded values close to that of numerical experiments (MORA and PLACE, 1998; MORGAN, 1999). The comparisons between laboratory and numerical friction values demonstrate that the approximate approach of restricting particle rolling in PDM simulations, can address the problem of idealized particle geometries, therefore providing a useful proxy for real material properties while maintaining the simplicity offered by 2-D PDM simulations.

Dynamic Friction

The initial results of PDM simulations employing time-dependent contact healing yield very rich results that reproduce many aspects of rate- and state-friction phenomenology (e.g., recent review by MARONE, 1998a,b). This includes steady-state velocity strengthening, log(time) healing of the system during static holds, direct change in friction with velocity steps, and strength evolution with slip as a function of static contact duration. Many of these results can be understood intuitively as a result of increasing resistance to slip as contacts build strength during holds. This is nicely demonstrated by plots of velocity stepping experiments (Fig. 8b), where a sudden increase in velocity initially activates a very large number of contacts, finally decaying to a value still above that for the lower sliding velocity. The higher the velocity, the less strength a contact can recover before renewed slip, thereby maintaining a high ratio of sliding contacts. This enhances the dilatancy of the assemblage (Fig. 8c), increasing assemblage friction (Fig. 8a). Similarly, static hold tests (Fig. 6) show a direct linkage between sliding friction and percentage of sliding contacts. Longer holds see a greater reduction in sliding contacts during the hold as the system relaxes, followed by higher ratios of sliding contacts during renewed shear (Fig. 6b). Apparently, deactivation of most contacts during holds erases the memory of sliding contacts, so that upon renewed shear, new contacts distributed across the assemblage are activated. This leads to assemblage strengthening, and divergence of stress-strain plots with continuing shear (Fig. 6a).

Strain localization is also related to the distribution of sliding contacts within the assemblage. Systems lacking contact healing tend to distribute deformation across the system throughout the length of the experiments (Fig. 3b; e.g., MORGAN and BOETTCHER, 1999); with contact healing, strain localization is observed, but the degree depends on sliding velocity (Figs. 3c and d). This can be understood as strengthening contacts peripheral to zones of concentrated shear increasingly resist slip, focusing deformation into actively deforming zones. This behavior is considerably more consistent with observations of strain localization in the laboratory and in the field (e.g., LOGAN et al., 1992; MARONE and SCHOLZ, 1989; BEELER et al., 1996).

The simple experiments carried out here demonstrate the potential offered by PDM simulations using more realistic contact laws, for greater understanding of dynamic friction in granular shear zones. Here, we have used very small systems, and applied only one possible implementation of contact strengthening, but have qualitatively reproduced many aspects of the empirically-based rate- and state-friction laws. Future work will be focused on better constraining the scaling relationships between PDM contact behavior and assemblage properties, such as critical slip distance, and the parameters that define rate- and state-friction laws.

Conclusions

In summary, recent PDM studies show that simple modifications to particle contact laws and restrictions on particle mobility through rotational damping, can approach laboratory estimates for sliding friction, and successfully reproduce many time- and state-dependent second-order changes in friction, lending insight into the micromechanisms responsible for these phenomena. Even though interparticle contact strength depends only on the time of static contact in these models, the bulk assemblage response reveals velocity- and slip-dependent behavior associated with changes in deformation mechanism, particle configuration and packing, and contact orientation. These results demonstrate the complexity of, and range of factors that control, granular deformation under dynamic conditions. Further work using this promising approach is necessary to more fully characterize the physics of this system.

Acknowledgments

This work benefited from constructive discussions with C. Marone, K. Mair, T. Tullis, and many others. Careful reviews by Peter Mora, David Place, and Hans Mühlhaus greatly improved the final manuscript. The research was supported by National Science Foundation Grant # EAR-0096005, issued to J. Morgan at Rice University.

REFERENCES

ABE, S., DIETERICH, J., MORA, P., and PLACE, D. (2002), *Simulation of the Influence of Rate-and State-Dependent Friction on the Macroscopic Behaviour of Complex Fault Zones with the Lattice Solid Model*, Pure Appl. Geophys. *159*, 1967–1983.

AHARONOV, E. and SPARKS, D. (1999), *Rigidity Phase Transition in Granular Packings*, Phys. Rev. E *60*, 6890–6896.

ALLEN, M. P. and TILDESLEY, D. J., *Computer Simulations of Liquids* (Clarendon Press, Oxford 1987).

AN, L-J. and SAMMIS, C. G. (1994), *Particle Size Distribution of Cataclastic Fault Materials from Southern California: A 3-D Study*, Pure Appl. Geophys. *143*, 203–227.

BEELER, N. M., TULLIS, T. E., and WEEKS, J. D. (1994), *The Roles of Time and Displacement in Evolution Effect in Rock Friction*, Geophys. Res. Lett. *21*, 1987–1990.

BEELER, N. M., TULLIS, T. E., and WEEKS, J. D. (1996), *Frictional Behavior of Large Displacement Experimental Faults*, J. Geophys. Res. *101*, 8697–8715.

BIEGEL, R. L., SAMMIS, C. S., and DIETERICH, J.H. (1989), *The Frictional Properties of a Simulated Gouge Having a Fractal Particle Distribution*, J. Struct. Geol. *11*, 827–846.

CUNDALL, P. A. and STRACK, O. D. L. (1979), *A Discrete Numerical Model for Granular Assemblies*, Géotechnique *29*, 47–65.

CUNDALL, P. A., *Distinct Element Models of Rock and Soil Structure*. In *Analytical and Computationl Methods in Engineering Rock Mechanics* (ed. Brown, E.T.) (Allen & Unwin, London, 1987) pp. 129–163.

CUNDALL, P. A., *Computer Simulations of Dense Sphere Assemblies*. In *Micromechanics of Granular Materials* (eds. Satake, M. and Jenkins, J.T.) (Elsevier Science Publishers, New York, 1988) pp. 343–352.

DIETERICH, J. H. (1972), *Time-dependent Friction in Rocks*, J. Geophys. Res. *77*, 3690–3697.

DIETERICH, J. H. (1979), *Modeling of Rock Friction: 1. Experimental Results and Constitutive Equations*, J. Geophys. Res. *84*, 2161–2168.

DIETERICH, J. H. and KILGORE, B. D. (1994), *Direct Observation of Frictional Contacts; New Insights for State-dependent Properties*, Pure Appl. Geophys. *143*, 283–302.

FRYE, F. M. and MARONE, C. (2002a), *The Effect of Particle Dimensionality on Granular Friction in Laboratory Shear Zones*, Geophys. Res. Lett. *29*, doi:10.01029/2002GL015709.

FRYE, K. M. and MARONE, C. (2002b) *The Effect of Humidity on Granular Friction at Room Temperature*, J. Geophys. Res. *107*, doi:10.1029/2001JP000654.

JOHNSON, K. L., *Contact Mechanics* (Cambridge University Press, Cambridge, 1985).

KARNER, S. L. and MARONE, C. (1998), *The Effect of Shear Load on Frictional Healing in Simulated Fault Gouge*, Geophys. Res. Lett. *25*, 4561–4564.

LIN, X. and NG, T.-T. (1997), *A Three-dimensional Discrete Element Model Using Arrays of Ellipsoids*, Geotechnique *47*, 319–329.

LOGAN, J. M., DENGO, C. A., HIGGS, N. G., and WANG, Z. Z., *Fabrics of Experimental Fault Zones: Their Development and Relationship to Mechanical Behavior*. In *Fault Mechanics and Transport Properties in Rocks* (eds. Evans, B., and Wong, T.-F.) (Academic Press, London 1992) pp. 33–67.

MAIR, K., FRYE, K. M., and MARONE, C. (2002), *Influence of Grain Characteristics on the Friction of Granular Shear Zones*, J. Geophys. Res. *107*, 2219,doi:10.1029/2001JB000516.

MARONE, C. (1998a), *The Effect of Loading Rate on Static Friction and the Rate of Fault Healing during the Earthquake Cycle*, Nature *391*, 69–72.

MARONE, C. (1998b), *Laboratory-derived Friction Constitutive Laws and their Application to Seismic Faulting*, Ann. Rev. Earth Planet. Sci. *26*, 643–696.

MARONE, C., RALEIGH, C. B., and SCHOLZ, C. H. (1990), *Frictional Behavior and Constitutive Modeling of Simulated Fault Gouge*, J. Geophys. Res. *95*, 7007–7025.

MARONE, C. and SCHOLZ, C. H. (1989), *Particle Size Distribution and Microstructure within Simulated Fault Gouge*, J. Struct. Geol. *11*, 799–814.

MORA, P., PLACE, D., ABE, S. and JAUMÉ, S. (2000), *Lattice Solid Simulation of the Physics of Earthquakes:The Model, Results and Directions*. In *GeoComplexity and the Physics of Earthquakes* (Geophysical Monograph series; no. 120), (eds. Rundle, J.B., Turcotte, D.L. and Klein, W.) (Am. Geophys. Union, Washington, DC), pp. 105–125.

MORA, P. and PLACE, D. (1993), *A Lattice Solid Model for the Nonlinear Dynamics of Earthquakes*, Intl. J. of Modern Physics C *4*, 1059–1074.

MORA, P. and PLACE D. (1994), *Simulation of the Frictional Stick-slip Instability*, J. Pure Appl. Geophys. *143*, 61–87.

MORA, P. and PLACE, D. (1998), *Numerical Simulation of Earthquake Faults with Gouge: Toward a Comprehensive Explanation for the Heat Flow Paradox*, J. Geophys. Res. *103*, 21,067–21,089.

MORGAN, J. K. (1998), *The Micromechanics of Localization and Dilation in Granular Shear Zones Revealed by Distinct Element Simulations,* EOS Trans. AGU, Spring Meeting Suppl. *79,* 222.

MORGAN, J. K. (1999), *Numerical Simulations of Granular Shear Zones Using the Distinct Element Method: I Shear Zone Kinematics and the Micromechanics of Localization,* J. Geophys. Res. *104,* 2703–2719.

MORGAN, J. K. and BOETTCHER, M. S. (1999), *Numerical Simulations of Granular Shear Zones Using the Distinct Element Method: II Effects of Particle Size Distribution and Interparticle Friction on Mechanical Behavior,* J. Geophys. Res. *104,* 2721–2732.

NASUNO, S., KUDROLLI, A., and GOLLUB, J. P. (1997), *Time-resolved Studies of Stick-slip Friction in Sheared Granular Layers,* Phys. Rev. Lett. *85,* 1428–1431.

PLACE, D., LOMBARD, F., MORA, P., and ABE, S. (2002), *Simulation of the Microphysics of Rocks Using LSMearth,* Pure Appl. Geophys. *159,* 1911–1932.

PLACE, D. and MORA, P. (2000), *Numerical Simulation of Localisation Phenomena in a Fault Zone,* Pure Appl. Geophys. *157,* 1821–1845.

ROTHENBURG, L. and BATHURST, R. J. (1992), *Micromechanical Features of Granular Assemblies with Planar Elliptical Particles,* Geotechnique *39,* 601–614.

SAMMIS, C. G., KING, G., and BIEGEL, R. (1987), *The Kinematics of Gouge Deformation,* Pure Appl. Geophys. *125,* 777–812.

SAMMIS, C. G. and STEACY, S. J. (1994), *The Micromechanics of Friction in a Granular Layer,* Pure Appl. Geophys. *142,* 777–794.

SCHOLZ, C. H., *The Mechanics of Earthquakes and Faulting* (Cambridge Univ. Press, New York, 1990).

SCOTT, D. R. (1996), *Seismicity and Stress Rotation in a Granular Model of the Brittle Crust,* Nature *381,* 592–595.

SPARKS, D. and AHARONOV, E., *Anatomy of a Slip event in an Idealized Fault Gouge.* In *2nd ACES Workshop Proc.* (eds. Matsu'ura, M., Nakajima, K., and Mora, P.) (APEC Cooperation for Earthquake Simulation, Brisbane, Australia, 2001) pp. 77–82.

TING, J. M., KHWAJA, M., MEACHUM, L., and ROWELL, J. (1993), *An Ellipse-based Discrete Element Model for Granular Materials,* Int. J. Numer. and Anal. Meth. in Geomech. *17,* 603–623.

WALTON, O. R. *Force Models for Particle-dynamics Simulations of Granular Materials.* In *Mobil Particulate System* (eds. Guazzelli, E. and Oger, L.) (Kluwer Academic Publishers, Dordrecht, 1995) pp. 366–378.

WALTON, O. R. and BRAUN, R. L. (1986), *Viscosity, Granular-temperature, and Stress Calculations for Shearing Assemblies of Inelastic, Frictional Disks,* J. Rheol. *30,* 949–980.

WILLIAMS, J. R. and PENTLAND, A. P. (1991), *Superquadrics and Model Dynamics for Discrete Elements in Concurrent Design,* Technical Report Order No. IESL91-12, Intelligent Engineering Systems Laboratory, Massachusetts Institute of Technology.

(Received September 27, 2002, revised February 28, 2003, accepted March 7, 2003)

To access this journal online:
http://www.birkhauser.ch

B. Scaling Physics

Pure appl. geophys. 161 (2004) 1895–1913
0033–4553/04/101895–19
DOI 10.1007/s00024-004-2538-x

▌Pure and Applied Geophysics

The Dependence of Constitutive Properties on Temperature and Effective Normal Stress in Seismogenic Environments

Aitaro Kato[1,2], Shingo Yoshida[2], Mitiyasu Ohnaka[2,3],
and Hiromine Mochizuki[2]

Abstract—We have evaluated how the parameters prescribing the slip-dependent constitutive law are affected by temperature and effective normal stress, by conducting the triaxial fracture experiments on Tsukuba-granite samples in seismogenic environments, which correspond to a depth range to 15 km. The normalized critical slip displacement D_c almost remains constant below 300°C (insensitive to both temperature and effective normal stress σ_n^{eff}); D_c increases with increasing temperature above 300 °C, and the rate of D_c increase with temperature tends to be largest at higher σ_n^{eff}. The breakdown stress drop $\Delta\tau_b$ for the granite at constant σ_n^{eff} is roughly 80 MPa below 300 °C, and does not depend on σ_n^{eff}. Above 300 °C, $\Delta\tau_b$ decreases gradually with increasing temperature, and the rate of $\Delta\tau_b$ reduction with temperature increases at higher σ_n^{eff}. The peak shear strength τ_p increases nearly linearly with increasing σ_n^{eff} below 300 °C. However, τ_p becomes lower above 300 °C, deviating from the linear relation extrapolated from below 300 °C. This is consistent with the onset of crystal plastic deformation mechanisms of Tsukuba granite.

Key words: Constitutive law, shear fracture experiments, seismogenic environments, Tsukuba-granite.

1. Introduction

Earthquake ruptures at shallow crustal depths result from shear failure instabilities occurring on pre-existing faults. There is convincing evidence, however, that the earthquake rupture process involves concurrent frictional slip failure on a pre-existing fault and fracturing of intact rock. Since an earthquake rupture includes not only frictional slip failure but also the shear failure of intact rock, it is critical to understand the constitutive properties of shear failure in intact rock. Further, we especially must know how ambient conditions affect the constitutive properties of the fault zone (e.g., OHNAKA, 1992; OHNAKA *et al.*, 1997; BLANPIED *et al.*, 1998; LOCKNER, 1998). KATO

[1] Institute for Frontier Research on Earth Evolution, Japan Marine Science and Technology Center, 3173-25 Showa-machi, Kanazawa-ku, Yokohama, Kanagawa 236-0001, Japan.
E-mail: aitaro@jamstec.go.jp
[2] Earthquake Prediction Research Center, Earthquake Research Institute, The University of Tokyo, 1-1-1 Yayoi, Bunkyo-ku, Tokyo 113-0032, Japan. E-mails: akato@eri.u-tokyo.ac.jp; shingo@eri.u-tokyo.ac.jp; h-mochi@eri.u-tokyo.ac.jp.
[3] Department of Earth Sciences, University College London, London, UK.
E-mail: ohnaka@fd.catv.ne.jp.

et al. (2003) demonstrates that the constitutive properties of shear failure in intact granite vary with depth in the seismogenic layer under a limited range of environmental conditions, based on the slipdependent constitutive formulation. To apply the results observed in the laboratory to a wider range of environmental conditions, we must investigate the general relation between the constitutive properties and the environmental conditions (including temperature and effective normal stress), rather than depth. To date, a quantitative expression has not been obtained.

In view of the problem mentioned above, we need to know how the constitutive property of intact rock varies with temperature and effective normal stress in the seismogenic layer. We, therefore, have conducted an additional series of triaxial fracture experiments using intact granite under the temperature-pressure conditions that correspond to a crustal depth range down to 15 km. The dependence of constitutive law parameters on temperature and effective normal stress was evaluated by combining the present data set and the data in KATO et al. (2003).

2. Procedure

2.1 Experimental Method

Tsukuba granite from Ibaraki Prefecture, Japan, was chosen for this study. It is a fine-grained (0.5–2.0 mm), homogeneous rock with a fairly isotropic fabric, and has been used in previous studies on stick-slip and fracture experiments (e.g., OHNAKA and KUWAHARA, 1990; OHNAKA et al., 1997; ODEDRA et al., 2001). The details regarding the sample and the preparation are described by KATO et al (2003). Cylindrical test specimens (length = 40 mm, diameter = 16 mm) were cored from a block of Tsukuba granite, to an accuracy of within 0.02 mm. Because we conducted the experiments under wet conditions, it was necessary to saturate those samples with water before the experiments.

A triaxial testing apparatus recently constructed at ERI was used for the present experiments. A detailed description of this apparatus and experimental methods are found in OHNAKA et al. (1997) and KATO et al. (2003). A cylindrical rock specimen is jacketed by a silver sleeve, and loaded in the apparatus. The confining pressure P_c, pore water pressure P_p and temperature T range up to 500 MPa, 400 MPa and 500 °C, respectively. Temperature was raised at a constant rate of 3°C/min to the test run value, after which P_c, P_p and temperature were held constant by servo - control for about two hours. In the present experiments, the strain rate was fixed at 10^{-5}/s, which corresponds to an axial displacement rate of 0.4 μm/s. To investigate the dependence of the constitutive property on temperature and effective normal stress σ_n^{eff}, a series of fracture experiments was conducted under simulated crustal conditions, where temperature ranges up to 480 °C, lithostatic pressure up to 480 MPa and pore water pressure up to 390 MPa.

2.2 Quantification of Constitutive Properties

Since the fracture of intact rock occurs by shear mode under compressive stress conditions, it is appropriate to represent failure strength, in terms of the resolved shear strength along the macroscopic fracture plane, as a function of slip displacement. The corresponding slip displacement is the relative displacement between both walls of the fault zone thickness, which may include the combined effects of slip associated with the deformation of asperity fractures, micro-cracking and local displacement between contacting gouge fragments. Note therefore that both shear strength and slip displacement used here for the constitutive formulation are defined in a macroscopic sense. The resolved shear strength τ, the slip displacement D and the resolved normal stress σ_n across the failure surface are calculated from the following equations

$$\tau = \frac{1}{2}(\sigma_1 - \sigma_3)\sin 2\theta , \tag{1}$$

$$D = D_{app} - D_{el} = \frac{\Delta l}{\cos \theta} - \frac{\tau - a}{b} , \tag{2}$$

$$\sigma_n = \frac{1}{2}(\sigma_1 - \sigma_3)(1 - \cos 2\theta) + \sigma_3 . \tag{3}$$

Here, θ is the failure angle between the macroscopic failure plane and the sample axis, σ_1 is the maximum principal stress, σ_3 is the minimum principal stress (which is equal to confining pressure in the present experiments) and Δl is the axial displacement of the sample. The first term ($D_{app} = \Delta l/\cos \theta$) on the right-hand side in equation (2) represents apparent slip along the failure plane, and the second term ($D_{el} = (\tau - a)/b$) indicates the correction for elastic deformation of the rock sample, when the linear line in Figure 13 in OHNAKA *et al.* (1997) is represented by $\tau = a + bD_{el}$, where a and b are constants.

The failure angle of a fractured sample was measured with a protractor after the silver jacket was removed from the sample. Figure 1 shows how the failure angle θ varies with temperature. . It is found from this Figure that θ varies within the range of 26° to 34°, and shows a weak tendency to increase with increasing temperature. The existing criterion for bulk plastic deformation predicts that the shear zone makes an angle of 45° against the maximum principal stress axis (e.g., JAEGER and COOK, 1976). Thus the increase in failure angle is consistent with the observations of KATO *et al.* (2003), who showed that some mineral grains on the failure surfaces experienced plastic deformation.

Figure 2 displays a typical example of the observed constitutive relation (thick curve) between τ and D under conditions simulating crustal depths of 8 km (P_c, P_p, T) = (240 MPa, 192 MPa, 240 °C). In Figure 2, τ_p is the peak shear strength, τ_r is the residual frictional strength, τ_{io} is the critical stress beyond which τ deviates from the linear-elastic line, D_a is the critical displacement required for τ to reach its peak

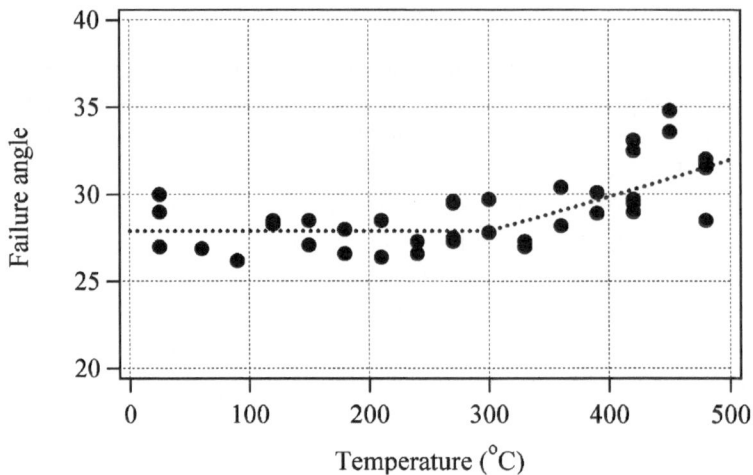

Figure 1
Failure angle θ observed in the present experiments versus temperatures. The dotted line shows the variation of the optimum angle θ_{opt}, which is defined by $\theta_{\text{opt}} = \frac{\pi}{4} - \frac{1}{2}\tan^{-1}(\mu_p(T))$ with temperature, where $\mu_p(T)$ is given by equation (10).

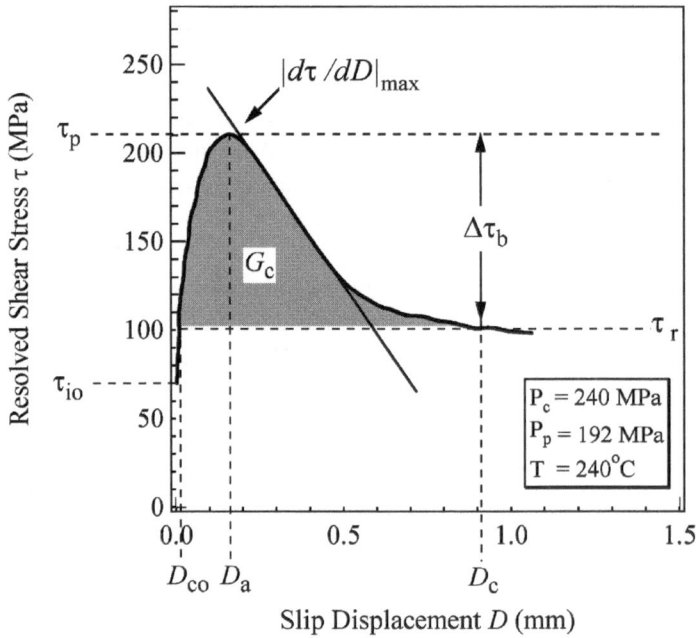

Figure 2
A typical example of the observed constitutive relation (thick curve) between τ and D under the conditions $(P_c, P_p, T) = (240 \text{ MPa}, 192 \text{ MPa}, 240 \,°C)$, corresponding to a crustal depth of 8 km (Reproduced from KATO *et al.*, 2003). For an explanation of the symbols for constitutive law parameters, see the text.

value. D_c is the critical slip displacement required for τ to decrease to τ_r, $\Delta\tau_b$ is the breakdown stress drop defined as the difference between τ_p and τ_r, $|d\tau/dD|_{\max}$ is the maximum slip-weakening rate during the breakdown process. Since the post-failure process is stabilized in the present experiments, note that the value of $|d\tau/dD|_{\max}$ does not depend on the stiffness of the loading system. The apparent fracture energy G_c is defined as (PALMER and RICE, 1973; OHNAKA and YAMASHITA, 1989)

$$G_c = \int_{D_{co}}^{D_c} [\tau(D) - \tau_r]dD = \frac{\Gamma}{2}\Delta\tau_b D_c \quad [J/m^2] \ , \tag{4}$$

where D_{co} represents the relative displacement at which τ intersects with an extrapolation of the residual stress level τ_r, $\tau(D)$ is a function of representing the constitutive relation between τ and D, and Γ is a dimensionless parameter dependent on a specific form of the function $\tau(D)$. Γ has been estimated to be around 1.0 for the shear fracture of intact granite (OHNAKA et al., 1997). Note that G_c defined by equation (4) equals the area of the shaded portion in Figure 2.

In order to quantify the constitutive properties we evaluate the constitutive parameters $(D_c, \Delta\tau_b$ and $\tau_p)$, from the observed constitutive relations. The method of evaluating constitutive parameters is the same as that described in KATO et al. (2003). The constitutive parameters $(D_c, \Delta\tau_b, |d\tau/dD|_{\max}$ and $G_c)$, determined for the entire data are listed in Table 1. The uncertainties of these obtained values were estimated to be $\pm (3 \sim 8)\%$ for D_c and $\pm (3 \sim 10)\%$ for $\Delta\tau_b$

3. Results

We discuss here how the constitutive law parameters vary with temperature and effective normal stress σ_n^{eff}. The constitutive law parameters at each condition of temperature and σ_n^{eff} are evaluated from the constitutive curves. The purpose of this section is to quantify the variation of constitutive law parameters $(\tau_p, \Delta\tau_b,$ and $D_c)$ as a function of temperature and σ_n^{eff} corresponding to seismogenic layer.

3.1 Peak Shear Strength τ_p

To allow prediction of the peak shear strength τ_p of the intact portion on a fault in seismogenic environments, we obtain the quantitative formulation for τ_p as a function of temperature and effective normal stress $(\sigma_n^{\mathrm{eff}} = \sigma_n - P_p)$. In Figure 3, a portion of data on peak shear strength τ_p is plotted against σ_n^{eff}. In this figure, black circles, white squares and black triangles denote data for τ_p tested at temperatures below 300 °C, at 420 °C and at 480 °C, respectively. We notice from Figure 3 that below 300 °C, τ_p increases monotonically, but at a very weak decreasing rate of τ_p as a function of σ_n^{eff}. It is also found that τ_p above 300 °C is

Table 1

List of constitutive law parameters evaluated under the seismogenic environmental conditions at a constant strain rate of 10^{-5}/s. For details on the method of evaluating the constitutive law parameters, see Kato et al (2003)

Pc (MPa)	Pp (MPa)	Temp ()	θ (deg)	τ_p (MPa)	τ_1 (MPa)	D_a (mm)	D_c (mm)	$\Delta\tau_b$ (Mpa)	$\sigma_n^{eff\,(peak)}$ (MPa)	τ_r (Mpa)	$\sigma_n^{eff\,(residual)}$ (MPa)	
380	100	25	30.0	492.5	175	0.484	*	*	564.3	*	*	
270	216	25	27.0	268.0	120.0	0.250	1.00	125.1	190.6	142.9	126.8	
378	216	25	29.0	384.5	180.0	0.310	1.29	148.3	374.9	236.2	292.9	
60	20	60	26.9	215.0	65.0	0.181	0.80	128.0	149.1	87.0	84.1	Kato et al. (2003)
90	30	90	26.2	232.0	102.0	0.176	1.00	140.0	174.2	92.0	105.3	"
120	40	120	28.3	268.2	126.0	0.222	1.23	121.2	224.4	147.0	159.2	"
120	40	120	28.5	259.7	129.0	0.170	1.32	132.6	221.0	127.1	149	"
150	50	150	28.5	313.7	118.5	0.200	0.82	123.7	270.3	190.0	203.2	"
150	50	150	27.1	290.4	124.0	0.215	0.85	114.0	248.6	176.4	190.3	"
180	60	180	28.0	320.0	131.0	0.236	0.83	106.4	290.1	213.6	233.6	"
180	144	180	26.6	187.9	76.0	0.167	0.80	106.6	130.1	81.3	76.7	"
210	70	210	28.5	350.0	164.5	0.223	1.05	130.0	330.0	220.0	259.5	"
210	168	210	26.4	195.5	85.0	0.181	0.83	102.2	139.0	93.3	88.3	"
240	80	240	27.3	367.4	171.0	0.210	0.95	128.4	349.6	239.0	286.5	"
240	192	240	26.6	210.7	70.0	0.162	0.73	103.0	153.5	107.7	101.9	"
270	90	270	27.3	367.1	182.5	0.192	0.85	112.4	369.5	254.7	311.4	"
270	216	270	29.6	232.4	108.0	0.209	0.97	125.6	186.0	106.8	114.6	"
270	216	270	27.5	237.1	110.0	0.185	1.00	116.9	177.4	120.3	116.6	"
378	216	270	27.5	341.5	200.0	0.250	1.19	102.3	339.8	239.2	286.5	
460	216	270	29.5	413.0	210.0	0.340	1.80	82.0	477.7	331.0	431.3	
300	100	300	27.8	377.9	182.0	0.234	1.02	111.0	399.2	266.9	340.7	Kato et al. (2003)

300	240	29.7	258.5	140.0	0.165	0.99	105.8	207.4	152.7	147.1	"
330	110	27.3	398.7	183.0	0.240	1.08	104.7	425.8	294.0	371.7	"
330	264	27.0	241.6	91.0	0.213	0.95	115.6	189.1	126.0	130.2	"
360	120	28.2	398.9	220.0	0.280	1.60	87.4	453.9	311.5	407	"
360	288	30.4	240.8	135.0	0.180	1.33	108.7	213.3	132.1	149.5	"
390	130	30.1	423.4	210.0	0.345	1.84	102.5	505.4	320.9	446	"
390	312	28.9	246.0	130.0	0.210	1.77	111.0	213.8	135.0	152.5	"
300	240	29.0	231.7	110.0	0.220	0.98	108.0	188.4	123.7	128.6	"
420	140	33.1	445.9	220.0	0.385	2.36	92.4	570.7	353.5	510.4	Kato et al. (2003)
420	210	32.5	351.7	140.0	0.370	2.50	56.2	433.8	295.5	398	
420	252	29.5	320.5	145.0	0.310	2.20	81.2	349.4	239.3	303.4	Kato et al. (2003)
420	336	29.7	252.8	116.0	0.265	2.15	133.4	228.2	119.4	152.1	
450	150	33.6	426.5	200.0	0.463	3.00	83.5	583.9	343.0	528.3	"
450	360	34.8	284.0	155.0	0.260	2.29	109.2	287.4	174.8	211.5	"
270	216	28.5	205.0	105.0	0.250	2.68	100.5	165.3	104.5	110.7	
378	216	31.5	306.5	150.0	0.290	2.60	120.3	349.8	186.3	276.1	
420	210	31.8	307.2	135.0	0.390	3.04	55.8	401.9	251.4	365.6	
480	160	32.0	397.0	168.0	0.428	3.53	61.3	568.1	335.7	529.8	Kato et al. (2003)

* It was impossible to estimate these parameters because the post-failure process was unstable.

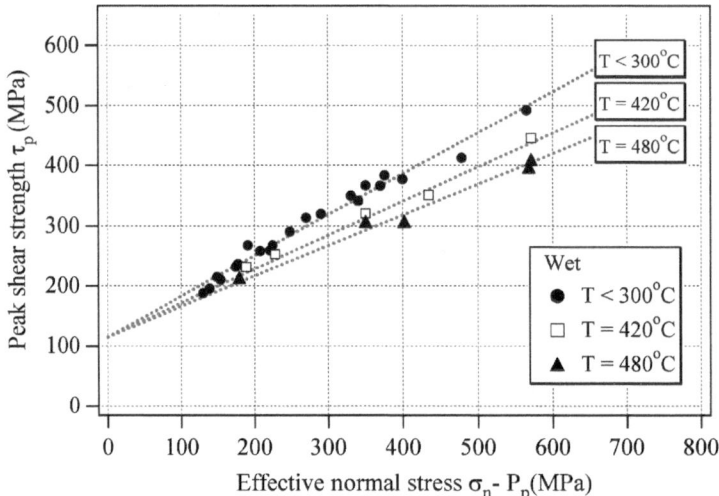

Figure 3
Plot of the relationship between the peak shear strength τ_p and effective normal stress σ_n^{eff} (KATO *et al.*, 2003). Black circles, white squares and black triangles denote data for τ_p tested at temperatures below 300 °C, at 420 °C, and at 480 °C, respectively. Dependence of τ_p on σ_n^{eff} is fitted by equation (5) at each temperature condition (dotted lines).

lower than that expected compared to its value at a comparable effective normal stress below 300 °C. In order to quantify the dependence of τ_p on temperature and σ_n^{eff} shown above, we assume here that the shear strength τ_p can be a function of σ_n^{eff} and temperature T. Although we could have expressed τ_p as a polynomial of σ_n^{eff}, a straight line also provides a good fit to the experimental data when σ_n^{eff} is less than 600 MPa. We thus assume that τ_p is expressed approximately as a linear function of σ_n^{eff} given by

$$\tau_p(T, \sigma_n^{eff}) = C_{po} + \mu_p(T) \quad \sigma_n^{eff} \quad , \tag{5}$$

where $\mu_{pn}(T)$ is written as follows;

$$\mu_p(T) = \begin{cases} \mu_{po} & T < T_o (= 300 \text{ °C}) \\ \mu_{po}(1 - \alpha[T - T_o]) & T > T_o. \end{cases}$$

Here C_{po} is the cohesive strength and $\mu_p(T)$ is the internal frictional coefficient, which is a function of temperature. $\mu_p(T)$ has the constant value μ_{po} below 300 °C. Above 300 °C, $\mu_p(T)$ decreases linearly from μ_{po} with increasing temperature. Unknown parameters are cohesive strength C_{po}, internal frictional coefficient μ_{po} below 300 °C and coefficient of temperature-dependence α. For all of the τ_p data, an iterative least-squares method was used to determine values for C_{po}, μ_{po} and α. The values determined for the best-fitted model are $C_{po} = 115$ MPa, $\mu_{po} = 0.7$, $\alpha = 1.5 \times 10^{-3}$.

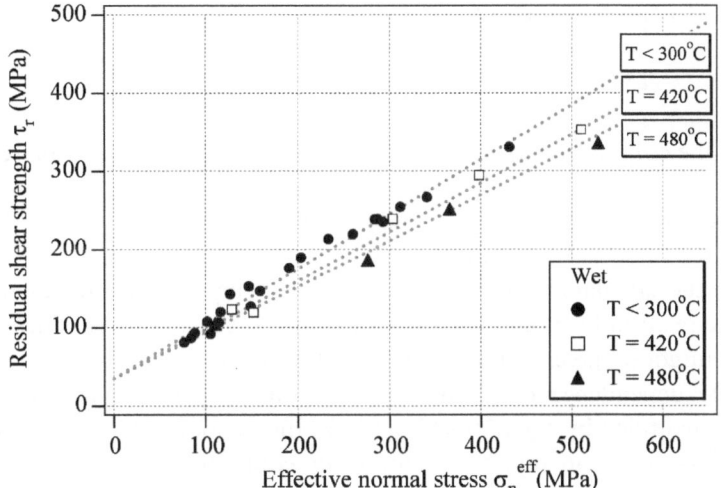

Figure 4

Plot of the relation between the residual stress level τ_r and σ_n^{eff}. Black circles, white squares and black triangles denote data for τ_r tested at temperatures below 300 °C, at 420 °C, and at 480 °C, respectively. Dependence of τ_r on σ_n^{eff} (dotted line) is fitted by equation (6) at each temperature condition.

By inserting the calculated value for $\mu_p(T)$ at each temperature into equation (5), we obtained the relationship between τ_p and σ_n^{eff} with temperature shown in Figure 3 (dotted lines). Note that each line fits the τ_p data well within the range of σ_n^{eff} < 600 MPa.

3.2 Breakdown Stress Drop $\Delta\tau_b$

The breakdown stress drop $\Delta\tau_b$ is defined as the difference between τ_p and τ_r. To evaluate how $\Delta\tau_b$ at constant σ_n^{eff} varies with temperature and σ_n^{eff}, it is necessary to investigate the dependence of residual frictional stress level τ_r on temperature and σ_n^{eff}. In Figure 4, data for τ_r is plotted as a function of σ_n^{eff}. In this figure, black circles, white squares and black triangles denote τ_r tested at below 300 °C, at 420 °C and at 480 °C, respectively. We notice from Figure 4 that τ_r tested below 300 °C monotonically increases at a very weak decreasing rate of τ_r as a function of σ_n^{eff}. It is also found that τ_r above 300 °C is lower than that expected for comparable effective normal stress below 300 °C, and decreases even further at higher temperatures. This indicates that τ_r also becomes sensitive to both temperature and σ_n^{eff} above 300 °C. These behaviors are fundamentally similar to the dependence of τ_p on temperature and σ_n^{eff} as described in the previous section. We thus assume that τ_r is approximately expressed as a function of temperature and σ_n^{eff} by

$$\tau_r(T, \sigma_n^{eff}) = C_{ro} + \mu_r(T)\sigma_n^{eff}, \qquad (6)$$

where $\mu_r(T)$ is written as follows;

$$\mu_r(T) = \begin{cases} \mu_{ro} & T < T_o (= 300\,°C) \\ \mu_{ro}(1 - \beta[T - T_o]) & T > T_o \end{cases}.$$

Here C_{ro} is the cohesive strength. Also, $\mu_r(T)$ has the same temperature-dependence as $\mu_p(T)$. Unknown model parameters are C_{ro}, μ_{ro} and β. For all data for τ_r, an iterative least-squares method was used to determine these values. The values determined for the best-fitted model are $C_{ro} = 35$ MPa, $\mu_{ro} = 0.7$, $\beta = 1.0 \times 10^{-3}$. By inserting the calculated values for $\mu_r(T)$ at each temperature into equation (6), we obtained the relation between τ_r and σ_n^{eff} shown in Figure 4 (dotted lines). Note that each line is in good agreement with the data for τ_r.

Since the dependence of τ_p and τ_r on both temperature and σ_n^{eff} has been empirically quantified by equations (5) and (6), the breakdown stress drop $\Delta\tau_b$ at the constant σ_n^{eff} is thus evaluated by the following equation;

$$\begin{aligned} \Delta\tau_b(T, \sigma_n^{eff}) &= \tau_p(T, \sigma_n^{eff}) - \tau_r(T, \sigma_n^{eff}) \\ &= \Delta\tau_{bo} - \gamma(T)\sigma_n^{eff}, \end{aligned} \tag{7}$$

where the value of $\Delta\tau_{bo}$ is 80 MPa, and $\gamma(T)$ is given as follows;

$$\gamma(T) = \begin{cases} 0 & T < T_o (= 300\,°C) \\ 5 \times 10^{-4}[T - T_o] & T > T_o \end{cases}.$$

Figure 5 shows the dependence of $\Delta\tau_b$ on temperature at σ_n^{eff} of 100 MPa and 300 MPa. Note that $\Delta\tau_b$ has constant value of 80 MPa below 300 °C, which is derived from the result of $\mu_{po} = \mu_{ro} = 0.7$. This result indicates that $\Delta\tau_b$ below 300 °C is not affected by either temperature or σ_n^{eff}. Above 300 °C, $\Delta\tau_b$ decreases linearly with increasing temperature. Above 300 °C, $\Delta\tau_b$ at higher σ_n^{eff} decreases more rapidly with increasing temperature than that at lower σ_n^{eff} (Figure 5).

Note that $\Delta\tau_b$ has roughly a constant value below 300 °C. It was also previously observed that $\Delta\tau_b$ has a roughly constant value across a range of confining pressure from 100 MPa to 300 MPa, by experiments conducted using dry intact Westerly and Fichtelbirge granite at room temperatures (WONG, 1982, 1986). In Figure 6, we plot the values of τ_p (black square) and τ_r (white square) as a function of σ_n^{eff} tested using Fichtelbirge granite (WONG, 1986). This Figure shows that the slopes of τ_p and τ_r plotted against σ_n^{eff} are almost identical and that $\Delta\tau_b$ at constant σ_n is roughly 60 MPa for Fichtelbirge granite. It is thus apparent that $\Delta\tau_b$ of intact granite is almost constant under conditions simulating the brittle regime at temperatures below 300 °C, despite the differences between wet and dry conditions. In Figure 6, note that the cohesive strength of Tsukuba granite (our study) is about 20 MPa larger than that of Fichtelbirge granite (Wong's study). The reason for this is mainly due to the dissimilarity of τ_p between Tsukuba and Fichtelbirge granite, although differences in τ_r are not significant. If the residual stress level τ_r is not

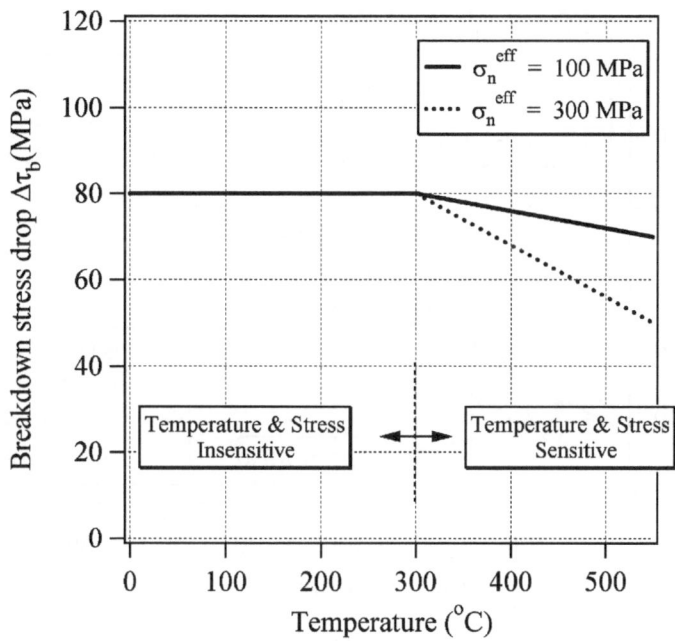

Figure 5
Variations in $\Delta\tau_b$ against temperature for two different conditions of σ_n^{eff} (100 MPa and 300 MPa), which are obtained using equation (7).

affected by the difference of rock type, we suggest that the cohesive strength C_{po} determines the amount of $\Delta\tau_b$ below 300 °C.

3.3 Critical Slip Displacement D_c

To evaluate the dependence of D_c on temperature and σ_n^{eff}, we plot data for D_c against the average effective normal stress in Figure 7. During one shear fracture experiment, σ_n^{eff} varies with increasing slip displacement as a result of the change in differential axial stress obeying the equation (3). Here, the average effective normal stress is defined as the average value of σ_n^{eff} at peak shear strength ($D = D_a$) and σ_n^{eff} at residual stress level ($D = D_c$). It is found from Figure 7 that D_c below 300 °C (open circle) has an almost constant value at σ_n^{eff} below 400 MPa. Note that D_c increases gradually with increasing temperature above 300 °C, except for the data tested at the conditions $(T, \sigma_n^{\text{eff}}) = (420 \text{ °C}, 158 \text{ MPa})$. We can also see that D_c tested at high σ_n^{eff} tends to be greater than that tested at low σ_n^{eff}, although the data are scattered.

To quantify the dependence of D_c on temperature and σ_n^{eff}, we assume here that D_c can be approximately expressed as a function of temperature and σ_n^{eff} by

$$D_c(T, \sigma_n^{eff}) = D_{co}\,\rho(T, \sigma_n^{eff}), \tag{8}$$

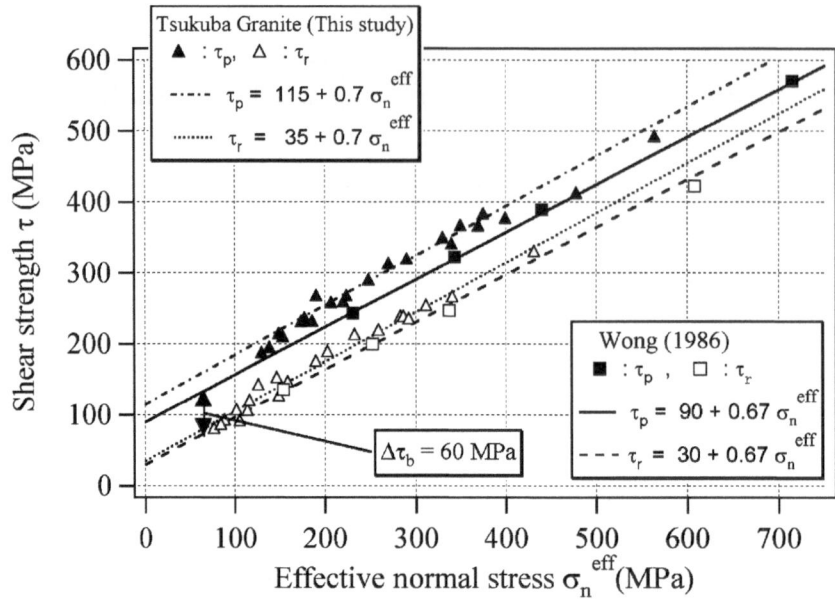

Figure 6

Plot of τ_p (black square) and τ_r (white square) as a function of σ_n^{eff} estimated from the experiments conducted using dry Fichtelbirge granite at room temperature (WONG 1986). Solid line and dashed line are obtained by the least-squares fit to τ_p and τ_r, respectively. The difference between τ_p and τ_r corresponds to the value of $\Delta\tau_b$ (= 60 MPa) for Fichtelbirge granite at constant σ_n^{eff}. For the purpose of reference, the data on τ_p (black triangle) and τ_r (white triangle) for Tsukuba-granite are also plotted (below 300 °C). A dashed-and-dotted line and dotted line correspond to the equations (5) and (6), respectively.

where

$$\rho(T, \sigma_n^{eff}) = \begin{cases} 1 & T < T_o(= 300 \text{ °C}) \\ 1 + (a + b\,\sigma_n^{eff})[T - T_o] & T > T_o \end{cases}.$$

Above 300 °C, the increasing rate of $\rho(T, \sigma_n^{\text{eff}})$ against temperature becomes large at high σ_n^{eff}. Unknown model parameters here are D_{co}, a and b. For all data for D_c, an iterative least-squares method was used to determine values for D_{co}, a and b. The values determined for the best-fitted model are $D_{co} = 1.0$ mm, $a = 0.005$ and $b = 1.5 \times 10^{-5}$. By inserting the calculated values for D_{co}, a and b into equation (8), the dependence of D_c on temperature and σ_n^{eff} is obtained as shown in Figure 7. Each broken line shows the variation of D_c against σ_n^{eff} at each temperature. Note that the observational data for D_c are well fitted by equation (8).

WONG (1982) investigated the effect of temperature and normal stress on D_c using dry Westerly granite. Although the number of his data points for D_c was very limited, there was a rough tendency for D_c below 300 °C to be insensitive to temperature and for D_c above 300 °C to be sensitive to temperature. Even at temperature of 670 °C, D_c measured by WONG (1982) reached values only about 2~3 times larger than those

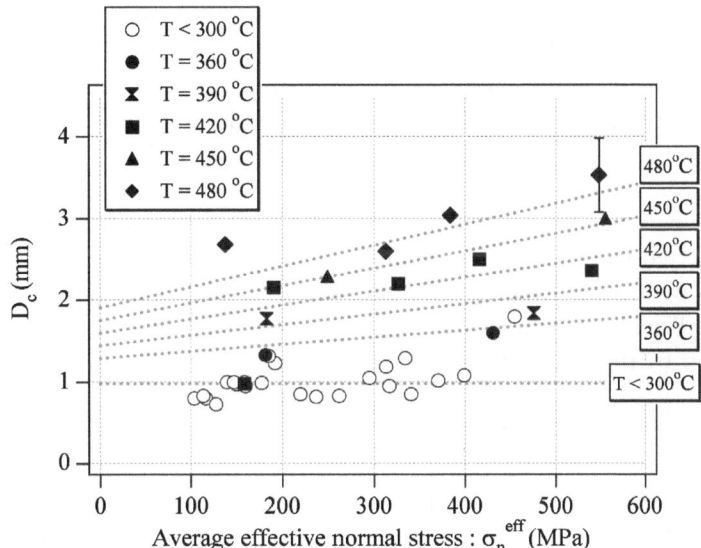

Figure 7

Data on D_c at various temperatures is plotted as a function of σ_n^{eff}. The error bar indicates the error accompanied by the fitting procedure. The dotted lines denote the dependence of D_c on temperature and σ_n^{eff} predicted by equation (8).

below 300 °C. The predicted value of D_c under wet conditions at 670 °C by equation (8) becomes 5 times larger than that below 300 °C. This suggests that D_c under wet conditions is more sensitive to temperature than that under dry conditions above 300 °C.

Each constitutive law parameter clearly shows a transition from a weak temperature-dependent regime to a strong temperature-dependent regime at around 300 °C. This transition at 300 °C corresponds to the change in deformation mode at the brittle-plastic transition regime, where brittle cracking and plastic flow coexist as deformation mechanism. The mechanism of crystal plasticity involving the dislocation glide is presumed to be thermally activated above 300 °C as shown in KATO *et al.* (2003).

4. Discussion

4.1 Mechanism for Variation of Constitutive Properties above 300 °C

Information pertaining to the operative deformation mechanisms during the shear failure of intact Tsukuba can be gained by microscopic examination of thin sections from tested samples. KATO *et al.* (2003) showed that both full plastic deformation of biotite and a slight plastic deformation of quartz occurred above 300 °C in the presence of pore water.

Within the range of temperature and effective normal stress tested in this study, the plastic deformation of biotite and quartz grains is considered to be mainly caused by the dislocation glide (KRONENBERG *et al.*, 1990; PASSCHIER and TROUW, 1996). As the mechanism of plastic flow deformation, the Peierls stress model was suggested to be applicable to the case of rock crystals (WEERTMAN, 1957). As shown in the previous section, it was observed that both τ_p and τ_r decrease almost linearly as temperature increases above 300 °C at constant σ_n^{eff} (equations (5) and (6)). It is therefore inferred from this linear dependence of shear strength on temperature that one of the operative mechanisms for plastic deformation is the dislocation glide derived from Peierls stress model. At high Peierls stresses, the shear strength τ can be approximated by (POIRIER, 1985)

$$\tau = \frac{2\tau^*}{\pi} - \frac{2\tau^*}{\pi}\frac{kT'}{Q}\log\left(\frac{C}{V}\right), \tag{9}$$

where τ^* is Peierls stress, k is Boltzman's constant, Q is the activation energy, V is the slip velocity, T' is absolute temperature, and C is a constant. In the brittle-plastic transitional regime, portions of biotite and quartz are assumed to deform in the manner of plastic flow obeying equation (9). According to this equation, the shear strength decreases linearly with increasing temperature at constant slip velocity. This property is consistent with the dependence of τ_p, or τ_r on temperature above 300 °C. It is, thus, suggested that the dislocation glide derived from Peierls stress model occurs concurrently with the brittle cracking in the brittle-plastic transition regime.

It is difficult to explain the operative mechanism for the increase of D_c above 300 °C. The increase of D_c implies that the necessary slip displacement for τ to decrease to τ_r becomes large at high temperatures. It is thought that the resistance force against the slip displacement increases with temperature, which may be caused by such nucleation of kink-bands or pile-up of dislocations as the deformation mechanism concerning plastic flow. This resistance force is considered to bring about the increase of D_c, although the quantitative discussion can't be given here.

4.2 Depth Variations of Constitutive law Parameters

In order to simulate the process by which large earthquakes are generated in the crust, it is crucial to know the depth variation of constitutive law parameters. We discuss here how constitutive law parameters vary with depth in the seismogenic layer by using the equations ((5), (7), (8)). We assume the stress condition for a thrust fault at the optimum angle

$$\theta_{opt} = \frac{\pi}{4} - \frac{1}{2}\tan^{-1}(\mu_p(T)). \tag{10}$$

It is found that equation (10) explains well the variation of failure angle with temperature as shown in Figure 1 (solid line). The stress condition for a thrust fault is evaluated according to Andersonian fault theory (ANDERSON, 1942). In the compressional field, the peak shear strength τ_p is taken to obey equation (5). Equation (5) is rewritten in terms of principal stresses ($\sigma_1 > \sigma_2 > \sigma_3$), denoting the effective stress (σ - P_p) as σ', for faulting at an optimum angle (SIBSON, 1977):

$$\sigma_1' = 2C_{po}\sqrt{K} + K\,\sigma_3',$$

where

$$K = \left(\sqrt{1 + \mu_p(T)^2} + \mu_p(T)\right)^2.$$

(The model is applicable to normal, strike-slip, and thrust faults.) The effective overburden pressure (σ_v') is given by

$$\sigma_v' = \rho_r g z - P_p = \rho_r g z - \rho_w g z = (\rho_r - \rho_w)g z,$$

where ρ_r is the rock density, ρ_w is the density of water, g is the gravitational acceleration, and z is crustal depth. For a thrust fault, if a commonly made assumption (e.g., SIBSON, 1974) is adopted

$$\sigma_v' = \sigma_3',$$

then we have

$$\sigma_1' = 2C_{po}\sqrt{K} + K\sigma_v', \quad \sigma_3' = \sigma_v'.$$

Thus, the critical shear stress and effective normal stress at which failure is expected to occur is

$$\tau_p = \frac{\sigma_1' - \sigma_3'}{2}\sin(90 - \tan^1 \mu_p(T))$$

$$\sigma_n^{eff} = \sigma_3' + \frac{\sigma_1' - \sigma_3'}{2}\{1 - \cos(90\tan^{-1}\mu_p(T))\}. \tag{11}$$

We assume that ρ_r is 2700 kg/m^3, ρ_w is 1000 kg/m^3, and the geothermal gradient is 30 °C /km. These values are close to those measured in the borehole drilled down to 9.1 km achieved by the German Continental Deep Drilling Program (KTB) (e.g., CLAUSER et al., 1997; HUENGES et al., 1997; ZOBACK and HARJES, 1997).

The peak shear strength τ_p is plotted against the simulated crustal depth in Figure 8(a). It is found that τ_p increases linearly with depth at depths shallower than 10 km. At depths greater than 10 km however, τ_p is lower than that extrapolated from the linear relation observed at shallower depths and slightly decreases with increasing depth. This is because wet granite is weakened by the effect of temperatures and σ_n^{eff} above 300 °C, which corresponds to the brittle-plastic transitional regime. Figure 8(a) represents the depth profile of τ_p for the intact rock portion, of which cohesive strength is 115 MPa. If the cohesive strength is weak

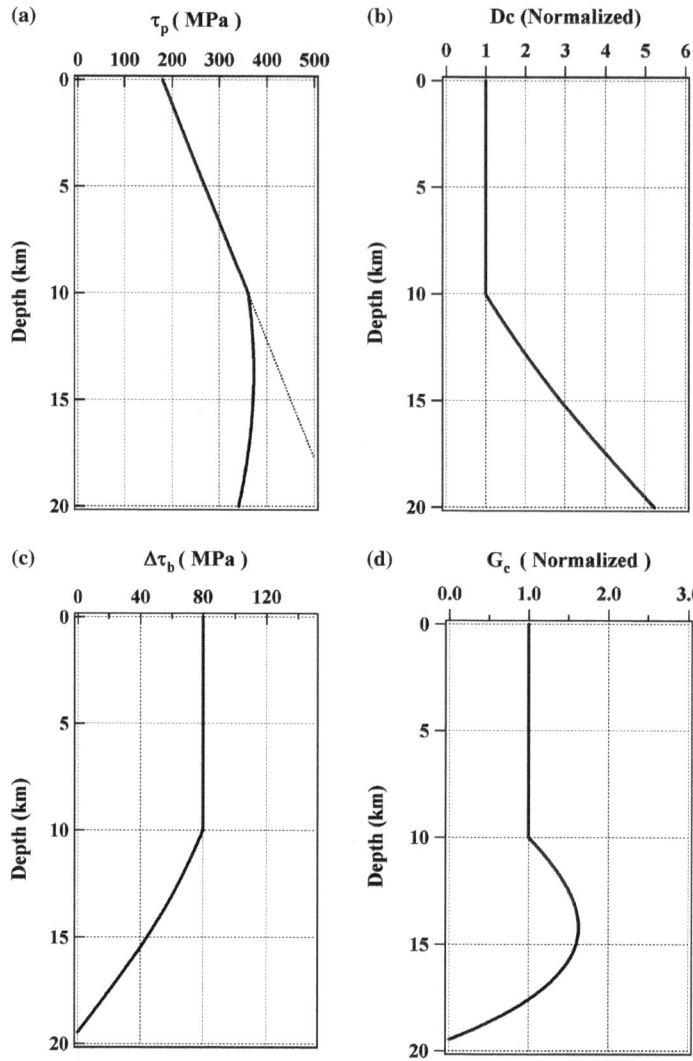

Figure 8
Variation of constitutive law parameters against the simulated crustal depth under the stress conditions of a thrust fault. The gradient of lithostatic pressure, pore water pressure and temperature is assumed to be 27 MPa/km, 10 MPa/km, 30 °C /km, respectively. (a) τ_p (b) Normalized D_c (c) $\Delta\tau_b$ (d) Normalized G_c.

(healing of the fault has not proceeded far), τ_p as a whole is reduced by the amount of decrease in cohesive force.

Figure 8(b) shows the variation in D_c against crustal depth. D_c has been normalized to the value of $D_{co} = 1.0$ mm. The reason for the normalization is that the value of D_c is inherently scale-dependent, and that the value of D_c in small-scale systems, for example, in the laboratory, is considerably smaller than that for large-scale earthquake rupture in the field (OHNAKA, 1996; OHNAKA and SHEN, 1999; OHNAKA, 2003). It is thus improper to plot the absolute value of D_c in the laboratory against a simulated crustal depth without considering its scale-dependence. It is found from Figure 8(b) that D_c is almost constant under the conditions corresponding to crustal depths shallower than 10 km. On the other hand, it is also found from this Figure that D_c increases with increasing depth at depths greater than 10 km.

In Figure 8(c), $\Delta\tau_b$ is plotted against crustal depth. It is found that $\Delta\tau_b$ is roughly 80 MPa throughout the crustal depth to 10 km. It is also seen from this Figure that $\Delta\tau_b$ at depths greater than 10 km shows a tendency to decrease gradually as depth increases.

Figure 8(d) shows a plot of apparent fracture energy G_c against simulated crustal depth. Since G_c is calculated by equation (4), the value of G_c is necessarily scale-dependent. For this reason, G_c is normalized to the value of 40 kJ/m². The value of G_c represents the resistance to rupture growth, which influences the progress of the rupture front. It is found from Figure 8(d) that G_c is constant under the conditions of crustal depths shallower than 10 km, whereas it tends to increase at greater depths, becoming nearly two times larger around the depth of 15 km. The increase in G_c at depths greater than 10 km indicates that the resistance to rupture growth increases at these depths, which may contribute to the arrest of the rupture front at the bottom of the seismogenic zone (the brittle-plastic transition regime).

5. Conclusion

In this paper we successfully quantify how constitutive law parameters for the shear failure of intact Tsukuba granite are described as a function of temperature and effective normal stress σ_n^{eff} in seismogenic environments. The granite samples were deformed at a strain rate of 10^{-5}/s.

The critical slip displacement D_c almost remains constant below 300 °C (insensitive to both temperature and σ_n^{eff}); D_c increases with increasing temperature above 300 °C, and the increasing rate of D_c against temperature tends to be substantial at higher effective normal stress. The breakdown stress drop $\Delta\tau_b$ at constant σ_n^{eff} is roughly 80 MPa below 300 °C, and does not depend on σ_n^{eff}. Above 300 °C, $\Delta\tau_b$ decreases gradually with increasing temperature, and the rate of $\Delta\tau_b$ reduction against temperature increases at higher σ_n^{eff}. Although the peak shear strength τ_p increases linearly with increasing σ_n^{eff} below 300 °C, τ_p becomes lower above 300 °C, deviating from the linear relation extrapolated at temperatures below 300 °C. This is consistent with the microscopic observation that the mechanical behavior of Tsukuba granite is semi-brittle (mainly brittle and slightly plastic).

Based on the present results, it is possible to evaluate the distributions of constitutive law parameters of intact granite in seismogenic environments. This study is, thus, crucial for providing the constraints on the distribution of constitutive law parameters on a fault zone in a seismogenic environment.

Acknowledgements

We are grateful to Julia K. Morgan for her constructive comments and corrections in our text, which led to substantial improvement in the original manuscript.

REFERENCES

ANDERSON, E. M., *The Dynamics of Faulting, first edition* (Oliver and Boyd, Edinburg, 1942).

BLANPIED, M. L., MARONE, J. C., LOCKNER, D. A., BYERLEE, J. D., and KING, D. P. (1998), *Quantitative Measure of the Variation in Fault Rheology due to Fluid-rock Interactions*, J. Geophys. Res. *103*, 9691–9712.

CLAUSER, C., GIESE, P., HUENGES, E., KOHL, T., LEHMANN, H., RYBACH, L., SAFANDA, J., WILHELM, H., WINDLOFF, K., and ZOTH, G. (1997), *The Thermal Regime of the Crystalline Continental Crust: Implications from KTB*, J. Geophys. Res. *102*, 18, 417–18,441.

HUENGES, E., ERZINGER, J., KUCK, J., ENGESER, B. and KESSELS, W. (1997), *The Prmeable Crust: Geohydraulic Properties down to 910 1m Depth*, J. Geophys. Res. *102*, 18255–18265.

JAEGER, J. C. and COOK, N. G. W., in *Fundamentals of Rock Mechanics*, 2nd ed. London (Chapman and Hall, 1976).

KATO, A., OHNAKA, M., and MOCHIZUKI, H. (2003), *Constitutive Properties for the Shear Failure of Intact Granite in Seismogenic Environments*, J. Geophys. Res. *108* B1, 2060, doi:10.1029/2001JB000791.

KRONENBERG, A. K., KIRBY, S. H., and PINKSTON, J. (1990), *Basal Slip and Mechanical Anisotropy of Biotite*, J. Geophys. Res. *95*, 19,257–19,278.

LOCKNER, A. D. (1998), *A Generalized Law for Brittle Deformation of Westerly Granite*, J. Geophys. Res. *103*, 5107–5123.

ODEDRA, A., OHNAKA, M., MOCHIDUKI, H., and SAMMONDS, P. (2001), *Temperature and Pore Water Pressure Effects on the Shear Strength of Granite in the Brittle-Plastic Transition Regime*, Geophys. Res. Lett. *28*, 3011–3014.

OHNAKA, M. and YAMASHITA, T. (1989), *A Cohesive Zone Model for Dynamic Shear Faulting based on Experimentally Inferred Constitutive Relation and Strong Motion Source Parameters*, J. Geophys. Res. *94*, 4089–4104.

OHNAKA, M. and KUWAHARA, Y. (1990), *Characteristic Features of Local Breakdown near Crack-tip in the Transition Zone from Slick-slip Experiments to Natural Earthquakes*, Tectonophysics *175*, 197–220.

OHNAKA,, M. (1992), *Earthquake Source Nucleation: A Physical Model for Short-term Precursors*, Tectonophysics *211*, 149–178.

OHNAKA, M. (1996), *Nonuniformity of the Constitutive Law Parameters for Shear Rupture and Quasistatic Nucleation to Dynamic Rupture: A Physical Model of Earthquake Generation Processes*, Proc. Natl. Acad. Sci. U. S. A. *93*, 3795–3802.

OHNAKA, M., AKATSU, M., MOCHIZUKI, ODERA, H., TAGASHIRA, F., and YAMAMOTO, Y. (1997), *A Constitutive Law for the Shear failure of Rock under Lithospheric Conditions*, Tectonophysics *277*, 1–27.

OHNAKA, M. and SHEN, L.-F. (1999), *Scaling of the Shear Rupture Process from Nucleation to Dynamic Propagation: Implications of Geometric Irregularity of the Rupturing Surfaces*, J. Geophys. Res. *104*, 817–844.

OHNAKA, M. (2003), *A Constitutive Scaling Law and a Unified Comprehension for Frictional Slip Failure, Shear Fracture of Intact Rock, and Earthquake Rupture*, J. Geophys. Res. *108* B2, 2080, doi: 10.1029/2000JB000123.

PALMER, A. C. and RICE, J. R. (1973), *The Growth of Slip Surfaces in the Progressive Failure of Overconsolidated Clay*, Proc, R. Soc. London, Ser. A *332*, 527–548.

PASSCHIER, C. W. and TROUW, R. A. J., *Microtectonics* (Springer, Berlin, 1996).

POIRIER, J.-P., *Creep of Crystals* (Cambridge: Cambridge University Press, 1985).

SIBSON, R. H. (1974), Frictional Constraints on Thrust, Wrench and Normal faults, Nature *249*, 542–544.

SIBSON, R. H. (1977), Fault Rocks and Fault Mechanisms, J. Geol. Soc. London *133*, 191–213.

WEERTMAN, J. (1957), *Steady-state Creep of Crystals*, J. Appl. Phys. *28*, 1185–1189.

WONG, T.-F. (1982), *Shear Fracture Energy of Westerly Granite from Post-failure Behavior*, J. Geophys. Res. *87*, 990–1000.

WONG, T.-F., *On the normal stress dependence of the shear fracture energy*. In *Earthquake Source Mechanics* (eds. S. Das, J. Boatwright, and C. H. Scholz) (Geophys. Monograph 37, Am. Geophys. Union 1986) pp.1–11.

ZOBACK, M. D. and HARJES, H. P. (1997), *Injection-induced Earthquakes and Crustal Stress at 9 km Depth at the KTB Deep Drilling Site, Germany*, J. Geophys. Res. *102*, 18,477–18,491.

(Received September 27, 2003, revised January 27, 2003, accepted February 10, 2003)

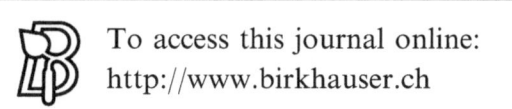

To access this journal online:
http://www.birkhauser.ch

Pure appl. geophys. 161 (2004) 1915–1929
0033–4553/04/101915–15
DOI 10.1007/s00024-004-2539-9

▌Pure and Applied Geophysics

A Constitutive Scaling Law for Shear Rupture that is Inherently Scale-dependent, and Physical Scaling of Nucleation Time to Critical Point

MITIYASU OHNAKA[1]

Abstract—It is shown that the rupture nucleation length increases up to the critical length with time according to a power law, and that the accelerating phase of nucleation leading up to the critical point is scaled in the framework of fracture mechanics based on slip-dependent constitutive formulation. Geometric irregularity of the rupturing surfaces plays a fundamental role in scaling the accelerating phase of nucleation up to the critical point. A power-law scaling relation between the rupture growth length and the nucleation time to the critical point is derived from theoretical consideration based on laboratory data. This power-law scaling relation has no singularity, and hence it may be useful for the predictive purpose of an imminent, large earthquake.

Key words: Inhomogeneous fault, characteristic length, accelerating phase of nucleation, power law, universal scaling relation.

Introduction

Unstable, dynamic high-speed rupture on an inhomogeneous fault in the brittle regime is in general preceded by an initial, quasistatic rupture growth, and the subsequent, accelerating rupture growth. The transition process consisting of such an initial, quasistatic phase and the subsequent, accelerating phase is referred to as the nucleation process, and this has been revealed and conclusively demonstrated with recent high-resolution laboratory experiments on propagating shear failure (OHNAKA and SHEN, 1999). The present paper deals with physical scaling of the shear rupture nucleation in the accelerating phase.

In the framework of fracture mechanics based on slip-dependent constitutive formulation, it has been derived theoretically (OHNAKA, 2000) that the critical size of earthquake nucleation scales with the ensuing earthquake size. In other words, the main shock seismic moment is proportional to the third power of the critical size of the nucleation zone, which in turn scales with the breakdown displacement. This

[1] The University of Tokyo, Tokyo, Japan, and University College London, London, UK.
Present address: Utsukushigaoka-Nishi 3-40-19, Aoba-Ku, Yokohama 225-0001, Japan.
E-mail: ohnaka@gos.itscom.net

theoretical scaling relation explains observed seismological data in quantitative terms (see OHNAKA, 2000). We wish to show in this paper that the nucleation time to the critical point is also consistently scaled in quantitative terms in the same framework. The critical point is defined here as the critical time t_c at which the critical length L_c of the nucleation zone is attained.

The purpose of this paper is to demonstrate, on the basis of laboratory data and theoretical consideration, that the rupture growth length L in the accelerating phase of nucleation increases up to the critical length L_c with time t according to a power law, and to derive a universal scaling relation from the power law. The universal scaling relation will be capable of unifying not only laboratory data on slip-failure nucleation, but also field data on earthquake rupture nucleation.

Geometric Irregularity of Rupture Surfaces and Characteristic Length Scales

Real rupture surfaces of heterogeneous materials such as rock cannot be flat planes, but inherently exhibit geometric irregularity (or roughness). The shear rupture process on such irregular rupturing surfaces is necessarily affected by the geometric irregularity, because the rupturing surfaces are in mutual contact and interact during slip. Indeed, it has been demonstrated that the breakdown displacement D_c required for the shear strength to degrade to a residual frictional stress level is severely affected by the geometric irregularity (OHNAKA, 2003). D_c is a scale-dependent constitutive law parameter, so that the constitutive law for shear rupture is also necessarily affected by geometric irregularity of the rupturing surfaces. Therefore, the effect of the geometric irregularity (or roughness) must be incorporated into the governing constitutive law, and to do so, the geometric irregularity needs to be quantified properly (see OHNAKA, 2003).

Irregular shear rupture surfaces of heterogeneous materials such as rock in general exhibit self-similarity; however, the self-similarity is not present at all scales (OHNAKA, 2003). The self-similarity is necessarily within a finite scale range, because the slipping process during the breakdown period is the process that smoothes away geometric irregularity of the rupturing surfaces. This means that self-similar rupture surfaces are scale-invariant within a finite scale range, and that there is at least one characteristic length scale which represents geometric irregularity of the rupture surfaces. The characteristic length λ_c can be defined as the critical wavelength beyond which geometric irregularity of the rupture surfaces no longer exhibits the self-similarity (OHNAKA and SHEN, 1999; OHNAKA, 2003). The characteristic length defined as such represents a predominant wavelength component contained in geometric irregularity (roughness) of the rupturing surfaces. When rupture surfaces have band-limited self-similarities, a different fractal dimension can be calculated for each band, and a characteristic length λ_c can be defined as the corner wavelength that separates the neighboring two bands with different fractal dimensions. The characteristic length defined as such also represents a predominant wavelength component of geometric irregularity of the rupture surfaces (OHNAKA, 2003).

Geometric irregularity of the rupturing surfaces may thus in general be quantified and characterized in terms of the following two parameters: the fractal dimension of each band, and the characteristic length defined as the corner wavelength that separates the neighboring two bands. Of these two parameters, it is the characteristic length λ_c that plays a fundamental role in scaling the scale-dependent constitutive parameter D_c. This has been demonstrated with laboratory experiments (OHNAKA and SHEN, 1999; OHNAKA, 2003).

A Constitutive Scaling Law

The earthquake rupture that occurs along a pre-existing fault in the Earth's crust characterized by inhomogeneities is a mixed process between what is called frictional slip failure and the fracture of initially intact rock. The fracture of initially intact rock during the earthquake rupture will occur at interlocking asperities on the fault surfaces with topographic irregularity, and/or at portions of cohesion healed between the mating fault surfaces during the inter-seismic period. Fracture of initially intact rock may also occur at portions of fault stepover. Indeed, seismological analyses indicate that the breakdown stress drop at local patches on a fault is as high as 50 to 100 MPa (e.g., PAPAGEORGIOU and AKI, 1983; ELLSWORTH and BEROZA, 1995; BOUCHON, 1997). Such a high breakdown stress drop roughly equals the breakdown stress drop of intact rock (KATO et al., 2003, 2004; OHNAKA, 2003). Hence, the governing law for the earthquake rupture must be a unifying constitutive law that governs both frictional slip failure and the shear fracture of intact rock. In addition, rupture phenomena including earthquakes are inherently scale-dependent, so that the governing law must also be formulated so as to account for scale-dependent physical quantities inherent in the rupture over a broad scale range. These two requirements can be met, if the governing law is formulated as a slip-dependent constitutive law, and if the effect of geometric irregularity of the rupturing surfaces is properly incorporated into the law.

The slip-dependent constitutive law is a unifying law that governs both frictional slip failure and the shear fracture of intact rock (OHNAKA, 2003). Both slip-dependent constitutive formulation and incorporation of the effect of the geometric irregularity into the law are the key to physical scaling of scale-dependent physical quantities inherent in the earthquake rupture. The slip-dependent constitutive formulation assumes the slip displacement to be an independent and fundamental variable, and the rate- or time-dependence to be of secondary significance. This assumption is justified by the fact that the dynamic rupture regime of high slip velocities for stick-slip failure and earthquakes does not exhibit any rate- or time-dependence (OKUBO and DIETERICH, 1986; BEROZA and MIKUMO, 1996; DAY et al.,

1998). In the slip-dependent formulation, the shear traction is expressed as a function of the slip displacement (Fig. 1), with parameters prescribing the law that implicitly depend on the rate or time (OHNAKA et al., 1997; OHNAKA, 2003). The slip-dependent constitutive law is commonly specified by the following five parameters: τ_i, τ_p, $\Delta\tau_b$, D_a, and D_c. Here, τ_i is the initial strength on the verge of slip at the rupture front, τ_p is the peak shear strength, $\Delta\tau_b$ is the breakdown stress drop defined as the stress difference between τ_p and the residual frictional stress τ_r, D_a is the critical displacement at which the peak shear strength is attained, and D_c is the breakdown displacement defined as the critical slip required for the shear strength to degrade to the residual frictional stress (see Fig. 1).

Recent laboratory experiments (OHNAKA, 2003) on the shear fracture of intact rock and frictional slip failure on a pre-cut fault have demonstrated that data on both shear fracture and frictional slip failure are unified consistently in terms of a single constitutive law, and that the constitutive law parameters τ_p, $\Delta\tau_b$, and D_c (or equivalently $D_{wc} = D_c - D_a$) are constrained by the following relation (see Fig. 2):

$$\Delta\tau_b/\tau_p = \beta(D_c/\lambda_c)^M \tag{1}$$

or equivalently

$$\Delta\tau_b/\tau_p = \beta'(D_{wc}/\lambda_c)^M \tag{2}$$

where β, β' and M are numerical constants. The double-error regression analysis of these data on both shear fracture and frictional slip failure gives the following values for β, β' and M with their standard deviations: $\beta = 1.64 \pm 0.29$, $\beta' = 2.26 \pm 0.38$,

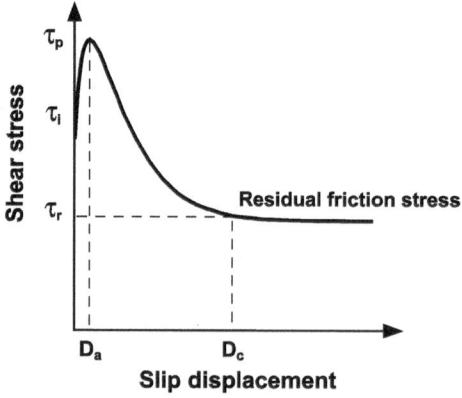

Figure 1
A slip-dependent constitutive relation for the shear rupture. τ_i is the initial strength on the verge of slip, τ_p is the peak shear strength, $\Delta\tau_b$ is the breakdown stress drop defined as the stress difference between τ_p and the residual frictional stress τ_r, D_a is the critical displacement at which the peak shear strength is attained, and D_c is the breakdown displacement defined as the critical slip required for the shear strength to degrade to the residual frictional stress.

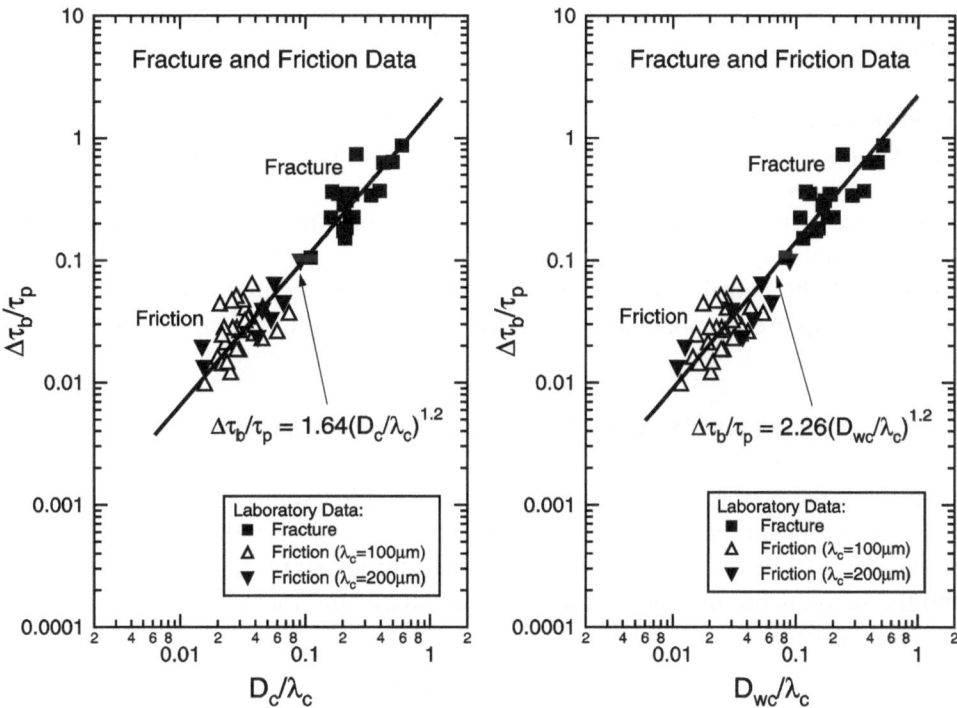

Figure 2

A plot of the logarithm of $\Delta\tau_b/\tau_p$ against the logarithm of D_c/λ_c or D_{wc}/λ_c for laboratory data on shear fracture and frictional slip failure. Reproduced from OHNAKA (2003).

and $M = 1.20 \pm 0.06$ for Tsukuba granite rock in the brittle regime (OHNAKA, 2003). Figure 2 clearly shows that laboratory data on both shear fracture and frictional slip failure are unified by a single relation (1) or (2).

Rewriting relation (1) or (2) leads to (OHNAKA, 2003)

$$D_c = m(\Delta\tau_b/\tau_p)\lambda_c \tag{3}$$

or

$$D_{wc} = (\beta/\beta')^{1/M} m(\Delta\tau_b/\tau_p)\lambda_c = 0.766 \times m(\Delta\tau_b/\tau_p)\lambda_c \tag{4}$$

where $m(\Delta\tau_b/\tau_p)$ is a dimensionless parameter expressed as a function of $\Delta\tau_b/\tau_p$ as follows:

$$m(\Delta\tau_b/\tau_p) = (1/\beta)^{1/M}(\Delta\tau_b/\tau_p)^{1/M} = 0.662(\Delta\tau_b/\tau_p)^{0.833}. \tag{5}$$

Similarly we have

$$D_a = \left[1 - (\beta/\beta')^{1/M}\right] m(\Delta\tau_b/\tau_p)\lambda_c = 0.234 \times m(\Delta\tau_b/\tau_p)\lambda_c. \tag{6}$$

We find from relations (3), (4) and (6) that all the displacement constitutive parameters D_c, D_{wc}, and D_a scale with λ_c, since $\Delta\tau_b/\tau_p$ is scale-independent. The scale-dependency of these constitutive parameters is particularly important to note. The slip-dependent constitutive law includes these scale-dependent displacement parameters D_c, D_{wc}, and D_a. Hence, the constitutive law itself has an inherent property of scale-dependence; otherwise, scale-dependent physical quantities inherent in the rupture over a broad scale range from 10^{-3} to 10^5 m cannot be treated consistently and quantitatively in terms of a single constitutive law. This scale-dependent property of the constitutive law is fundamentally important for describing scale-dependent rupture phenomena in quantitative terms (see OHNAKA, 2003).

Physical Scaling of Nucleation Time to Critical Point

It has been conclusively demonstrated from high-resolution laboratory experiments on propagating shear failure (OHNAKA and SHEN, 1999) that the nucleation process consists of the following two phases: an initial, stable and quasistatic phase, and the subsequent, unstable and accelerating phase, and that the accelerating phase obeys a power law of the form:

$$V/V_s = \alpha(L/\lambda_c)^n, \tag{7}$$

where V is the rupture growth velocity, V_S is the shear wave velocity, L is the rupture growth length in the accelerating phase, λ_c is the characteristic length representing geometric irregularity (or roughness) of the rupturing surfaces in the slip direction, and α and n are numerical constants ($\alpha = 8.87 \times 10^{-29}$, and $n = 7.31$; see OHNAKA and SHEN, 1999). Noting that $dt = dL/V$, we have from equation (7):

$$t = \int_{L_{SC}}^{L} \frac{dL}{V(L)} = \frac{\lambda_c^n}{\alpha V_S} \int_{L_{SC}}^{L} L^{-n}dL = \frac{\lambda_c^n}{\alpha V_s(1-n)}(L^{-n+1} - L_{sc}^{-n+1}) \quad (n \neq 1) \tag{8}$$

and

$$t_c = \int_{L_{SC}}^{L_C} \frac{dL}{V(L)} = \frac{\lambda_c^n}{\alpha V_S} \int_{L_{SC}}^{L_C} L^{-n}dL = \frac{\lambda_c^n}{\alpha V_s(1-n)}(L_C^{-n+1} - L_{sc}^{-n+1}) \quad (n \neq 1) \tag{9}$$

In the above equations, L_{sc} is the critical length beyond which the rupture nucleation begins to proceed spontaneously, at accelerating speeds L_c is another critical length beyond which the rupture propagates dynamically at a steady, high speed close to sonic velocities, and t_c is the critical time at which $L = L_c$ is attained. Note that relation (7) holds for L in the interval $[L_{sc}, L_c]$. Subtracting equation (8) from equation (9) leads to

$$L(t) = L_c \left(\frac{t_a - t_c}{t_a - t} \right)^{1/(n-1)} \quad (t \le t_c) \tag{10}$$

where

$$t_a = \frac{1}{\alpha(n-1)} \frac{\lambda_c}{V_S} \left(\frac{\lambda_c}{L_c} \right)^{n-1} + t_c. \tag{11}$$

The nucleation process in the interval from L_{sc} to L_c is a spontaneous rupture growth driven by the release of the elastic strain energy stored in the surrounding medium, whereas the nucleation process in the interval below L_{sc} is a quasistatic rupture growth controlled by the rate of an applied load. Thus, the behavior of rupture growth changes at the critical length L_{sc} from a quasistatic phase controlled by the loading rate to a self-driven, accelerating phase controlled by the inertia. The behavior of the rupture also changes at the critical length L_c from the accelerating phase to the phase of a steady propagation at a high-speed close to elastic wave velocities. In this paper, focus is placed on L_c (not placed on L_{sc}). Note, therefore, that the critical length L_c defined here is physically not identical with the critical length defined by ANDREWS (1976a, b).

The critical length L_c can be derived from a specific rupture nucleation model (OHNAKA, 2000), and L_c is expressed in terms of D_c as follows (OHNAKA and SHEN, 1999; OHNAKA, 2000):

$$L_c = \frac{1}{k} \frac{\mu}{\Delta \tau_b} D_c, \tag{12}$$

where k is a well-defined dimensionless quantity (OHNAKA and YAMASHITA, 1989; OHNAKA, 2000), and μ is the rigidity. From (3), (5) and (12), we have

$$L_c = \frac{1}{k} \left(\frac{1}{\beta} \right)^{1/M} \frac{\mu}{\Delta \tau_b} \left(\frac{\Delta \tau_b}{\tau_p} \right)^{1/M} \lambda_c. \tag{13}$$

If appropriate values are substituted for β, M, μ and k; that is, $\beta = 1.64$, $M = 1.20$, $\mu = 30000$ MPa and $k = 3$ (see OHNAKA, 2000), equation (13) is reduced to:

$$L_c = 6.62 \times 10^3 \frac{1}{\Delta \tau_b} \left(\frac{\Delta \tau_b}{\tau_p} \right)^{0.833} \lambda_c. \tag{14}$$

Relation (13) or (14) shows that L_c scales with λ_c, since $\Delta \tau_b$ and $\Delta \tau_b/\tau_p$ are scale-independent. In the laboratory experiments on propagating frictional slip failure (OHNAKA and SHEN, 1999), $\Delta \tau_b$ had a value of the order of 0.1 MPa, and $\Delta \tau_b/\tau_p$ had a value ranging from 0.04 to 0.07. If these values are substituted for $\Delta \tau_b$ and $\Delta \tau_b/\tau_p$ in equation (14), equation (14) is further reduced to:

$$L_c = (4.5 - 7.2) \times 10^3 \lambda_c \tag{15}$$

which agrees with previous estimates (OHNAKA and SHEN, 1999).

Equation (10) predicts that the nucleation zone length $L(t)$ in the accelerating phase increases with time t according to a power law. It can be confirmed directly from laboratory data that this power-law relation is valid for the accelerating phase of frictional slip failure nucleation. Figure 3 presents an example of laboratory data (black circles) representing the relation between the rupture growth length L and time t, for the accelerating phase observed during the nucleation process of frictional slip failure on a smooth fault with $\lambda_c = 46$ μm. Figure 4 is another example of laboratory data (black rhombuses) representing the same relation for the accelerating phase observed during the nucleation process of frictional slip failure on an extremely smooth fault with $\lambda_c = 10$ μm. These laboratory data were taken from OHNAKA and SHEN (1999), and for their experimental procedures, see OHNAKA and SHEN (1999). It has been discussed specifically in the paper how the fault surface roughness

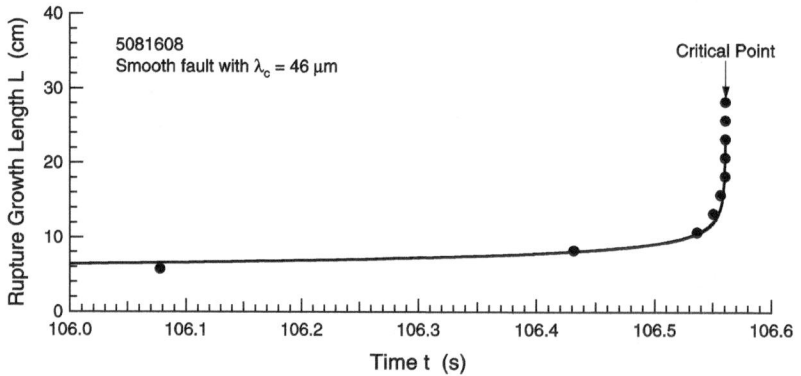

Figure 3

A plot of the rupture growth length against time for the accelerating phase of nucleation that occurred on a smooth fault with $\lambda_c = 46$ μm. Black circles denote data points, and a solid line denotes equation (10) fitted to data points.

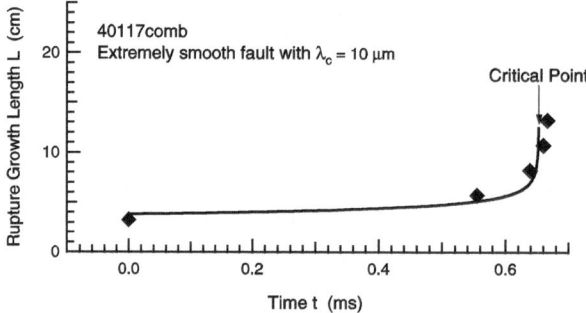

Figure 4

A plot of the rupture growth length against time for the accelerating phase of nucleation that occurred on an extremely smooth fault with $\lambda_c = 10$ μm. Black rhombuses denote data points, and a solid line denotes equation (10) fitted to data points.

quantified by $\lambda_c = 10$ μm is significantly different from that quantified by $\lambda_c = 46$ μm.

The power-law relation (10) was fitted to laboratory data plotted in Figures 3 and 4. A solid curve in Figure 3 represents relation (10) with the following values for parameters: $t_a - t_c = 6.329 \times 10^{-5}$ sec, $t_c = 0.960$ sec, and $L_c = 27$ cm. A solid curve in Figure 4 represents relation (10) with the following values for parameters: $t_a - t_c = 1.376 \times 10^{-5}$ sec, $t_c = 6.390 \times 10^{-4}$ sec, and $L_c = 7$ cm. From Figures 3 and 4, we find that there are immense differences in both the critical time t_c and the critical length L_c between these two curves. This shows that both t_c and L_c depend on the fault surface roughness characterized by the critical length λ_c, and that the nucleation time to the critical point is greatly influenced by the fault surface roughness characterized by λ_c.

In the previous experiments on frictional slip failure on a rough fault with $\lambda_c = 200$ μm (OHNAKA and SHEN, 1999), the nucleation grew stably at a steady, slow speed over the entire fault length of 29 cm, and no accelerating phase of nucleation was observed. This is because the fault length of 29 cm is too short for the rupture to grow at accelerating speeds on the rough fault characterized by $\lambda_c = 200$ μm (OHNAKA and SHEN, 1999). We can calculate an extrapolated curve of the nucleation time to the critical point for this rough fault, by using equations (10) and (11) under the assumption that $L_c = 5.6 \times 10^3 \lambda_c$, in view of equation (15) and $V_S = 2.9$ km/s (the shear wave velocity for Tsukuba granite). Figure 5 shows a calculated curve ($t_a - t_c = 2.752 \times 10^{-4}$ sec) for the rough fault with $\lambda_c = 200$ μm, and this figure can be compared with Figures 3 and 4. Scale dependence of t_c and L_c may also be corroborated from this comparison.

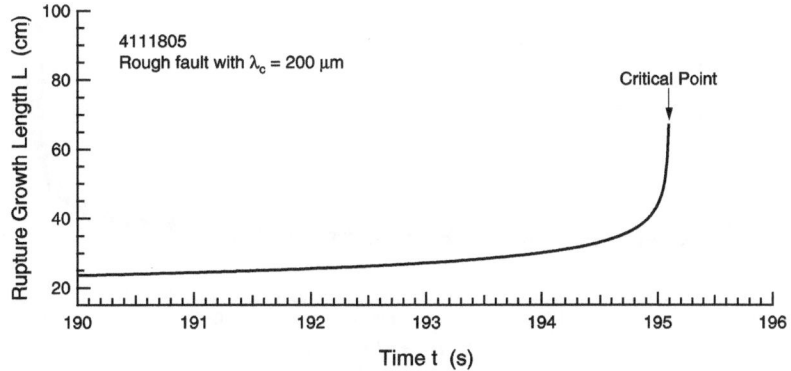

Figure 5
A plot of the rupture growth length against time for the accelerating phase of nucleation on a rough fault with $\lambda_c = 200$ μm. A solid line denotes the extrapolated curve calculated by using equations (10) and (11) under the assumption that $L_c = 5.6 \times 10^3 \lambda_c$ and $V_S = 2.9$ km/s.

The relation between $L(t)$ and t in equation (10) depends on λ_c, and hence it is scale-dependent. This scale-dependence is obvious from Figures 3, 4 and 5. To derive a universal scaling relation between $L(t)$ and t, we rewrite equation (10) as follows:

$$\frac{L(t)}{L_c} = \left(\frac{1}{1 - (t - t_c)/(t_a - t_c)}\right)^{1/(n-1)} \quad (t \le t_c) \tag{16}$$

where

$$\frac{t - t_c}{t_a - t_c} = \alpha(n - 1)\left(\frac{L_c}{\lambda_c}\right)^{n-1} \frac{t - t_c}{(\lambda_c/V_s)} \tag{17}$$

which shows that the nucleation time to the critical point, $t - t_c$, scales with λ_c/V_S. Equation (16) shows that the relation between $L(t)/L_c$ and $(t - t_c)/(t_a - t_c)$ is scale-invariant if the exponent n is scale-invariant. Equation (16) also predicts that the relation between $L(t)/L_c$ and $(t - t_c)/(t_a - t_c)$ is independent of geometric irregularity of the rupturing surfaces characterized by λ_c. This scale-independence can be checked by laboratory data which will be shown below (see Fig. 6).

The thick curve in Figure 6 represents relation (16). For comparison, laboratory data on frictional slip failure nucleation are also plotted in Figure 6 for three different fault surface roughnesses of $\lambda_c = 10$, 46, and 200 μm (shown in Figs. 3, 4 and 5). Black circles and white circles represent data points on the nucleation on the smooth fault with $\lambda_c = 46$ μm, and black rhombuses represent those on the nucleation on the extremely smooth fault with $\lambda_c = 10$ μm. The thin curve with tick bars in Figure 6 denotes the relation extrapolated from data on the nucleation on the rough fault with $\lambda_c = 200$ μm. It is confirmed from Figure 6 that scaling relation (16) unifies laboratory data.

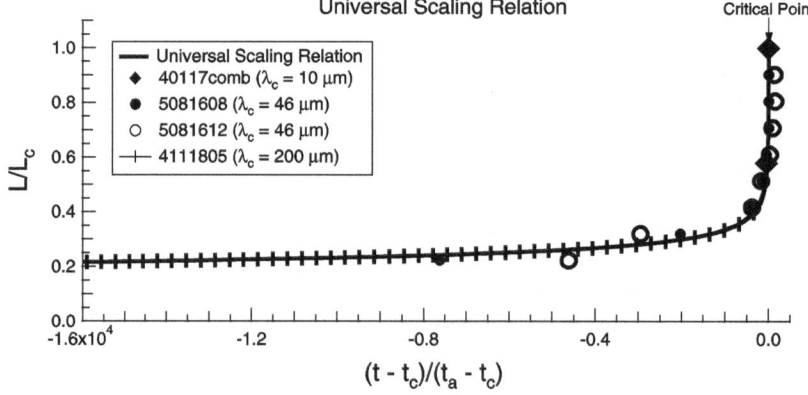

Figure 6

A plot of L/L_c against $(t - t_c)/(t_a - t_c)$. A thick solid line represents the power law relation (16) in text. Laboratory data on slip failure nucleation on faults with different surface roughnesses are also plotted for comparison with relation (16).

Discussion

It has been demonstrated that the characteristic length λ_c representing geometric irregularity (or roughness) of the rupturing surfaces plays a fundamental role in scaling scale-dependent physical quantities inherent in the rupture (OHNAKA, 2003). This poses a question regarding how large the effective characteristic length is for a real earthquake fault. Natural faults contain a wide range of characteristic length scales departed from the self-similarity. For instance, the self-similarity of natural faults is limited by the depth of seismogenic layer and fault segment size (e.g., AKI, 1992, 1996; KNOPOFF, 1996). The earthquake generation process and its eventual size can necessarily be prescribed and characterized by these macroscopic length scales (SHIMAZAKI, 1986; SCHOLZ, 1982, 1994; ROMANOWICZ, 1992; MATSU'URA and SATO, 1997; FUJII and MATSU'URA, 2000).

However, scale-dependent physical quantities such as D_c, L_c, the fracture energy G_c, and the slip acceleration $\ddot{D}_{\max}$ are controlled by a considerably smaller scale on the fault than the depth of the seismogenic layer and fault segment size (OHNAKA, 2003). A patch of high resistance to rupture growth on a fault, which may act as either "barrier" (AKI, 1979, 1984) or "asperity" (KANAMORI and STEWART, 1978; KANAMORI, 1981) against the earthquake rupture, is a candidate for providing characteristic length of such a smaller scale, departed from the self-similarity. Such a patch will be attained at portions of fault bend or stepover, at interlocking asperities on the fault surfaces, and/or at portions of cohesion healed between the mating fault surfaces.

Figure 7 shows a plot of the logarithm of D_c against the logarithm of λ_c for field data on earthquakes and laboratory data on frictional slip failure and the shear fracture of intact rock. This figure was reproduced from OHNAKA (2003). Both τ_p and λ_c are unknown for earthquakes. However, the effective characteristic length λ_c is specifically to be inferred for earthquakes for which the constitutive parameters $\Delta\tau_b$ and D_c have been estimated, if τ_p is appropriately assumed. The characteristic length λ_c for earthquakes plotted in Figure 7 has been inferred from equations (3) and (5) under the assumption that $\tau_p = 100$ MPa, using data on earthquakes taken from PAPAGEORGIOU and AKI (1983), ELLSWORTH and BEROZA (1995), and IDE and TAKEO (1997). If τ_p for an earthquake shown in Figure 7 is greater than 100 MPa, then λ_c for this particular earthquake is larger than the value plotted in Figure 7, and hence its data point shifts toward the right-hand side in the figure. If τ_p for an earthquake is less than 100 MPa, then λ_c for the earthquake is shorter than the value plotted in Figure 7, and in this case its data point shifts toward the left-hand side in the figure, with the constraint that $\Delta\tau_b \leq \tau_p$.

For laboratory data on frictional slip failure on a pre-cut fault and the shear fracture of intact rock sample, both D_c and λ_c were independently measured (see OHNAKA, 2003), and these laboratory data have also been over-plotted in Figure 7 for comparison. Solid straight lines in Figure 7 represent constitutive scaling relation

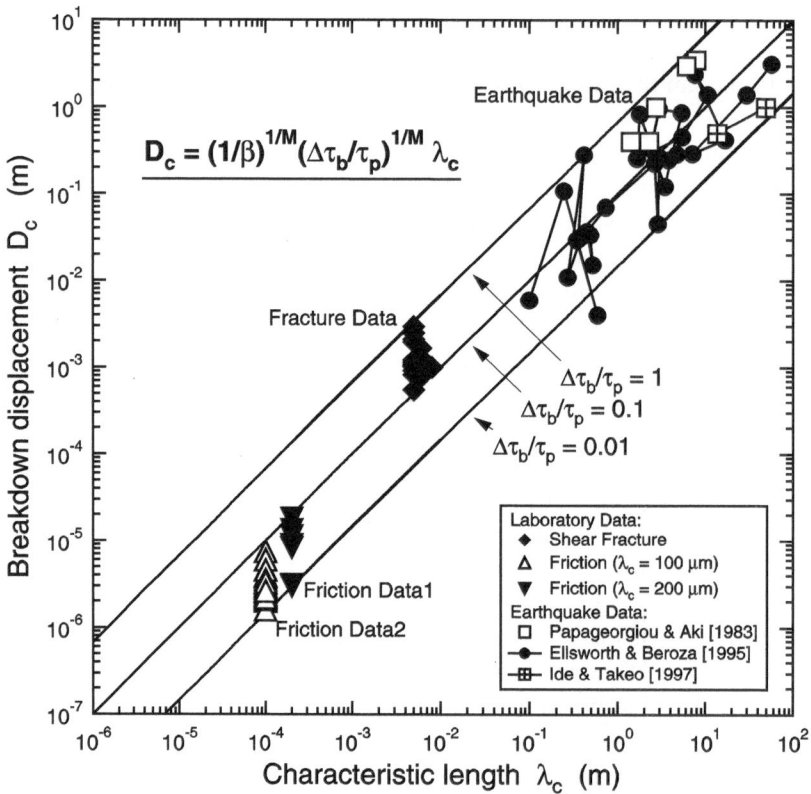

Figure 7

A plot of the breakdown displacement D_c against the characteristic length λ_c. Reproduced from OHNAKA (2003).

(3) for three cases where $\Delta\tau_b/\tau_p = 0.01$, 0.1, and 1 have been assumed. It can be seen from Figure 7 how D_c for data on frictional slip failure and shear fracture in the laboratory, and earthquake rupture in the field consistently scales with λ_c. From Figure 7, λ_c for major earthquakes can be inferred to have a value ranging from 1 to 100 m, which is substantially larger than λ_c for shear fracture and frictional slip failure of laboratory scale (OHNAKA, 2003).

It has been demonstrated in the previous section that the relation between $L(t)/L_c$ and $(t - t_c)/(t_a - t_c)$ is independent of λ_c. Indeed, equation (16) takes a mathematical form that the relation between $L(t)/L_c$ and $(t - t_c)/(t_a - t_c)$ is scale-invariant if n is scale-invariant. Although whether n is scale-invariant or not remains to be solved, the present result strongly suggests that the nucleation time to the critical point for major earthquakes is much longer than that of frictional slip failure nucleation of laboratory scale.

Let us assume here that the nucleation time to the critical point $|t - t_c|$ is given by, for instance, relation $(t - t_c)/(t_a - t_c) = -0.1 \times 10^4$ (see Fig. 6). For laboratory-scale rupture, for which λ_c has a value ranging from 10 μm to 1 mm, it can be estimated from equation (17) that the nucleation time to the critical point $|t - t_c|$ is of the order of 10 ms to 1 s, given that $\alpha = 8.87 \times 10^{-29}$, $n = 7.31$, $V_S = 2.9$ km/s, and $L_c/\lambda_c = 5 \times 10^3$. This estimate agrees with the measurement of laboratory experiments on stick-slip failure (OHNAKA and SHEN, 1999). In contrast, if it is assumed that scaling relation (16) holds for the accelerating phase of nucleation leading up to the critical point for typical, major earthquakes, and that n is scale-invariant, then it can be estimated from equation (17) that the nucleation time to the critical point $|t - t_c|$ is of the order of several tens of minutes to a few days or considerably longer, depending on seismogenic environments, because λ_c for major earthquakes is inferred to have a value ranging from 1 to 100 m. These orders of estimates give likely values for the nucleation time to the critical point of typical, major earthquakes, and are consistent with the previous conclusion that large earthquakes are in principle predictable (see OHNAKA, 2000). This may suggest that n is scale-invariant.

A power law of the form (10) or (16) has no singularity over the entire time interval $t \leq t_c$. This may be contrasted with a power law of the form called "a time-to-failure function" (e.g., BUFE and VARNES, 1993);

$$\Omega(t) = \Omega_f + \frac{K}{(t_f - t)^m}, \tag{18}$$

where $\Omega(t)$ represents strain, displacement, crack growth length, or some measurable quantity such as cumulative event count, seismic moment or Benioff strain, Ω_f, K, and m are constants, and t_f is time of failure. Though equation (18) potentially has a singularity at time of failure according to a value for the exponent m, equation (10) or (16) has no singularity. This will be a great advantage, if the law can be used for forecasting an imminent earthquake, because both its occurrence time and magnitude can be evaluated by determining t_c and L_c by curve fitting. Once L_c has been determined, the seismic moment for the predicted earthquake may be evaluated from the following relation (OHNAKA, 2000):

$$M_0 = 1 \times 10^9 (2L_c)^3 \tag{19}$$

or its corresponding magnitude M_w may be estimated from the following relation (OHNAKA and MATSU'URA, 2002):

$$\log(2L_c) = 0.5M_w + 0.03, \tag{20}$$

where L_c is measured in m. Direct measurements of the rupture growth length along the fault are not easy in the field observations. If, however, the ongoing nucleation can successfully be identified and monitored by any observational means, relation (10) or (16) may be useful for the short-term (or immediate) forecasting.

Conclusion

Since the earthquake rupture that occurs in the Earth's crust characterized by inhomogeneities is a mixed process between frictional slip failure and fracture of intact rock, it is fundamentally important to formulate the governing law for the earthquake rupture as a unifying law that governs both frictional slip failure and the shear fracture of intact rock. The governing law must also be formulated so as to scale scale-dependent physical quantities inherent in the rupture. These two requirements can be met if the governing law is formulated as a slip-dependent constitutive law, and if the effect of geometric irregularity of the rupturing surfaces is properly incorporated into the law. Both slip-dependent constitutive formulation and incorporation of the geometric irregularity into the law are the key to physical scaling of scale-dependent quantities inherent in the rupture. A power law of the form (10) or (16) has been derived to explain the accelerating phase of nucleation leading up to the critical point. The nucleation time to the critical point scales with the characteristic length λ_c which represents geometric irregularity (or roughness) of the rupturing surfaces. The present result is consistent with the previous conclusion that large earthquakes are in principle predictable (OHNAKA, 2000). A power law of the form (10) or (16) has no singularity, consequently it may be useful for the predictive purpose of an imminent, large earthquake, if the ongoing nucleation can successfully be identified and monitored by any observational means.

REFERENCES

AKI, K. (1979), *Characterization of Barriers on an Earthquake Fault*, J. Geophys. Res. *84*, 6140–6148.

AKI, K. (1984), *Asperities, Barriers, Characteristic Earthquakes and Strong Motion Prediction*, J. Geophys. Res. *89*, 5867–5872.

AKI, K. (1992), *Higher-order Interrelations between Seismogenic Structures and Earthquake Processes*, Tectonophysics *211*, 1–12.

AKI, K. (1996), *Scale-dependence in Earthquake Phenomena and its Relevance to Earthquake Prediction*, Proc. Natl. Acad. Sci. USA *93*, 3740–3747.

ANDREWS, D. J. (1976a), *Rupture Propagation with Finite Stress in Antiplane Strain*, J. Geophys. Res. *81*, 3575–3582.

ANDREWS, D. J. (1976b), *Rupture Velocity of Plane Strain Shear Cracks*, J. Geophys. Res. *81*, 5679–5687.

BEROZA, G. C. and MIKUMO, T. (1996), *Short Slip Duration in Dynamic Rupture in the Presence of Heterogeneous Fault Properties*, J. Geophys. Res. *101*, 22449–22460.

BOUCHON, M. (1997), *The State of Stress on Some Faults of the San Andreas System as Inferred from Near-field Strong Motion Data*, J. Geophys. Res. *102*, 11731–11744.

BUFE, C. G. and VARNES, D. J. (1993), *Predictive Modeling of the Seismic Cycle of the Greater San Francisco Bay Region*, J. Geophys. Res. *98*, 9871–9883.

DAY, S. M., YU, G., and. WALD, D. J. (1998), *Dynamic Stress Changes during Earthquake Rupture*, Bull. Seismol. Soc. Am. *88*, 512–522.

ELLSWORTH, W. L. and BEROZA, G. C. (1995), *Seismic Evidence for an Earthquake Nucleation Phase*, Science *268*, 851–855.

FUJII, Y. and MATSU'URA, M. (2000), *Regional Difference in Scaling Laws for Large Earthquakes and its Tectonic Implication*, Pure Appl. Geophys. *157*, 2283–2302.

IDE, S. and TAKEO, M. (1997), *Determination of Constitutive Relations of Fault Slip based on Seismic Wave Analysis*, J. Geophys. Res. *102*, 27379–27391.

KANAMORI, H. (1981) *The nature of seismic patterns before large earthquakes*. In *Earthquake Prediction: An International Review* (eds: D. W. Simpson and P. G. Richards) (AGU, Washington, DC.) pp. 1–19.

KANAMORI, H. and STEWART, G. S. (1978), *Seismological Aspects of the Guatemala Earthquake of February 4, 1976*, J. Geophys. Res. *83*, 3427–3434.

KATO, A., OHNAKA, M., and MOCHIZUKI, H. (2003), *Constitutive Properties for the Shear Failure of Intact Granite in Seismogenic Environments*, J. Geophys. Res. *108*(B1), 2060, doi:10.1029/2001JB000791.

KATO, A., YOSHIDA, S., OHNAKA, M., and MOCHIZUKI, H. (2004), *The Dependence of Constitutive Properties on Temperature and Effective Normal Stress in Seismogenic Environments*, Pure Appl. Geophys. *161*, 9/10.

KNOPOFF, L. (1996), *A Selective Phenomenology of the Seismicity of Southern California*, Proc. Natl. Acad. Sci. USA *93*, 3756–3763.

MATSU'URA, M. and SATO, T. (1997), *Loading Mechanism and Scaling Relations of Large Interplate Earthquakes*, Tectonophysics *277*, 189–198.

OHNAKA, M. (2000), *A Physical Scaling Relation between the Size of an Earthquake and its Nucleation Zone Size*, Pure Appl. Geophys. *157*, 2259–2282.

OHNAKA, M. (2003), *A Constitutive Scaling Law and a Unified Comprehension for Frictional Slip Failure, Shear Fracture of Intact Rock, and Earthquake Rupture*, J. Geophys. Res. *108*(B2), 2080, doi:10. 1029/ 2000JB000123.

OHNAKA, M. and MATSU'URA, M. (2002), *The Physics of Earthquake Generation* (University of Tokyo Press, Tokyo) *378* pp.

OHNAKA, M. and SHEN, L. -f. (1999), *Scaling of the Shear Rupture Process from Nucleation to Dynamic Propagation: Implications of Geometric Irregularity of the Rupturing Surfaces*, J. Geophys. Res. *104*, 817–844.

OHNAKA, M. and YAMASHITA, T. (1989), *A Cohesive Zone Model for Dynamic Shear Faulting Based on Experimentally Inferred Constitutive Relation and Strong Motion Source Parameters*, J. Geophys. Res. *94*, 4089–4104.

OHNAKA, M., AKATSU, M., MOCHIZUKI, H., ODEDRA, A., TAGASHIRA, F., and YAMAMOTO, Y. (1997), *A Constitutive Law for the Shear Failure of Rock under Lithospheric Conditions*, Tectonophysics *277*, 1–27.

OKUBO, P. G. and DIETERICH, J. H. (1986), *State variable fault constitutive relations for dynamic slip*. In *Earthquake Source Mechanics* (eds. S. Das, J. Boatwright, and C. H. Scholz) (AGU Geophys. Monograph *37* (Maurice Ewing Volume 6), pp. 25–35.

PAPAGEORGIOU, A. S. and AKI, K. (1983), *A Specific Barrier Model for the Quantitative Description of Inhomogeneous Faulting and the Prediction of Strong Ground Motion, Part II. Applications of the Model*, Bull. Seismol. Soc. Am. *73*, 953–978.

ROMANOWICZ, B. (1992), *Strike-slip Earthquakes on Quasi-vertical Transcurrent Faults: Inferences for General Scaling Relations*, Geophys. Res. Lett. *19*, 481–484.

SCHOLZ, C. H. (1982), *Scaling Laws for Large Earthquakes: Consequences for Physical Models*, Bull. Seismol. Soc. Am. *72*, 1–14.

SCHOLZ, C. H. (1994), *A Reappraisal of Large Earthquake Scaling*, Bull. Seismol. Soc. Am. *84*, 215–218.

SHIMAZAKI, K. (1986), *Small and Large Earthquakes: The Effects of the thickness of seismogenic layer and the free surface*. In *Earthquake Source Mechanics* (eds: S. Das, J. Boatwright, and C. H. Scholz), AGU Geophys. Monograph *37* (Maurice Ewing Volume 6), pp. 209–216.

(Received September 27, 2002, revised January 27, 2003, accepted February 10, 2003)

 To access this journal online:
http://www.birkhauser.ch

Pure appl. geophys. 161 (2004) 1931–1944
0033–4553/04/101931–14
DOI /10.1007/s00024-004-2540-3

▌Pure and Applied Geophysics

Critical Sensitivity in Driven Nonlinear Threshold Systems

Xiaohui Zhang[1], Xianghong Xu[1], Haiyin Wang[1], Mengfen Xia[1,2],
Fujiu Ke[1,3], and Yilong Bai[1]

Abstract— Rupture in heterogeneous brittle media, including earthquakes, can be regarded as complicated phenomena in driven nonlinear threshold systems. It displays catastrophe transition and sample-specificity, which results in difficulty of rupture prediction. Our numerical simulations indicate that critical sensitivity might be a common precursor of catastrophe transition and thus give a clue to catastrophe prediction. In this paper we present an analytical examination of critical sensitivity in driven nonlinear threshold systems, based on mean field approximation and damage relaxation time model. The result suggests that critical sensitivity is in reality a common feature prior to catastrophe transition in driven nonlinear threshold systems, with disordered mesoscopic heterogeneity. This result seems to be supported by rock experiments.

Key words: Critical sensitivity, catastrophe transition, sample-specificity, driven nonlinear threshold system.

1. Introduction

Damage and rupture in disordered heterogeneous brittle media, such as rocks and the earth's crust, present complexity (MEAKIN, 1991; CURRAN, 1997; CURTIN, 1997; SAHIMI and ABABI, 1993; BEN-ZION and SAMMIS 2003; JAUME, and SYKES 1999; RUNDLE et al., 2000; TIAMPO et al., 2000; BAI et al., 1994; XIA et al., 1996, 1997, 2000), especially it displays catastrophe transition and sample-specificity. Firstly, the rupture appears as a transition from globally stable accumulation of mesoscopic damage to catastrophic rupture (BAI et al., 1994; XIA et al., 2000; WEI et al., 2000). Secondly, the behavior of catastrophe transition in macroscopically identical systems exhibits macroscopic uncertainty or sample-specificity (XIA et al., 1996, 1997, 2000; WEI et al., 2000). Consequently, it is insufficient to represent catastrophe of a system

---

[1] State Key Laboratory of Nonlinear Mechanics, Institute of Mechanics, Chinese Academy Of Sciences, Beijing 100080, China.
E-mail: Zhangxh@lnm.imech.ac.cn;xxh@lnm.imech.ac.cn;why@lnm.imech.ac.cn;baiyl@lnm.imech.ac.cn
[2] Department of Physics, Peking University, Beijing 100871, China.
E-mail: xiam@lnm.imech.ac.cn
[3] Department of Applied Physics, Beijing University of Aeronautics and Astronautics, Beijing 100083, China. E-mail: kefj@lnm.imech.ac.cn

by merely its macroscopically average properties. The behavior of catastrophe may sensitively depend on the details of disordered mesoscopic heterogeneity which is usually difficult, even impossible, to be dealt with. Such complexity results in difficulty of rupture prediction (WEI *et al.*, 2000, WYSS *et al.*, 1997; GARCIMARTIN *et al.*, 1997).

To search the universal features of catastrophe transition is of genuine importance to rupture prediction. Our numerical simulation suggested that critical sensitivity might be a possible common feature prior to catastrophe in heterogeneous brittle media (XIA *et al.*2002). Critical sensitivity means that a system may become sensitive significantly as approaching its catastrophe transition point. The underlying mechanism behind critical sensitivity is the coupling effect between disordered heterogeneity on multiple scales and dynamical nonlinearity due to damage-induced stress redistribution (STEIN, 1999; XIA *et al.*, 2000, 2002; WEI, *et al.* 2000). Critical sensitivity may provide a clue to the prediction of catastrophe transition, such as material failure and great earthquakes, provided the sensitivity of the system is measurable or can be monitored.

Rupture in disordered heterogeneous brittle media can be considered as catastrophe transition in driven disordered, nonlinear threshold systems. In this paper we present an analytical approach to reveal the mechanism of critical sensitivity, based on mean field approximation and damage relaxation time model, and compare the results with experimental data on rock rupture. It is found that the critical sensitivity is a common feature of catastrophe in driven nonlinear threshold systems, which seems to be supported by the rock rupture experiments. Hence, the critical sensitivity may provide a clue to catastrophe prediction.

2. *Physical Model of Heterogeneous Brittle Media*

We consider a macroscopic system comprised of numerous interacting, nonlinear mesoscopic units. The scale of the mesoscopic units corresponds to the intrinsic characteristic scale of heterogeneous structure, and the span between macroscopic and mesoscopic scales is about 10^3–10^6 in typical cases of rupture. The mesoscopic units are identical statistically. The mesoscopic heterogeneity of material property is attributed to the quenched disorder of broken threshold of units, i.e., a predefined threshold σ_c is assigned to each unit. σ_c follows a statistical distribution function $h(\sigma_c)$, which is normalized as

$$\int_0^\infty h(\sigma_c)d\sigma_c = 1 \tag{1}$$

The system is subjected to nominal driving force σ_0, which is adopted as macroscopic, external parameter. A unit breaks as the driving force σ on it becomes higher than its threshold. When a unit breaks, it will be excluded from the distribution function. Therefore, we introduce a time-dependent distribution function of intact elements $f(\sigma_c, t)$ with initial condition

$$f(\sigma_c, t) = h(\sigma_c). \tag{2}$$

In such a system, stress redistribution induced by damage evolution plays an essential role in nonlinear dynamics. In most cases, the stress redistribution leads to stress fluctuations in stress pattern, and the coupling effect between inhomogeneous stress and the heterogeneity of the threshold of units leads to considerably more complex behavior of catastrophe. In order to take an analytical approach, we adopt a simple model based on a globally mean-field approximation. According to the mean-field approximation, the nominal driving force σ_0 is loaded uniformly on all intact units, i.e., the real driving force is determined by

$$\sigma = \frac{\sigma_0}{1 - p} \tag{3}$$

This is a widely used formula in damage mechanics, where p is damage fraction calculated from

$$p = 1 - \int_0^\infty f(\sigma_c, t) d\sigma_c. \tag{4}$$

The evolution of distribution function $f(\sigma_c, t)$ is suggested to follow an equation based on relaxation time model:

$$\frac{\partial f(\sigma_c, t)}{\partial t} = -\frac{f(\sigma_c, t)}{\tau}, \tag{5}$$

where τ is characteristic time of damage relaxation. Generally speaking, τ decreases with increasing $\frac{\sigma}{\sigma_c}$. For simplicity, we assume

$$\tau = \begin{cases} \infty, & \text{as } \sigma < \sigma_c \\ \tau_D \left(\frac{\sigma}{\sigma_c} \right)^{-q}, & \text{as } \sigma_M \geq \sigma \geq \sigma_c \ , \\ 0, & \text{as } \sigma > \sigma_M \end{cases} \tag{6}$$

Such a model implies that the damage relaxation can be characterized by three time scales: for $\sigma < \sigma_c$ the damage appears as a very slow relaxation process, and we simply assume $\tau \to \infty$; for very high driving force, the damage becomes a very fast relaxation process, and the relaxation time is nearly zero; this corresponds to the catastrophic rupture, in addition, it also means a cutoff of distribution function $h(\sigma_c)$ at $\sigma_c = \sigma_M$; and for the intermediate case, the damage relaxation can be described by a finite relaxation time which depends on $\frac{\sigma}{\sigma_c}$ and q is a positive parameter. The rupture in real heterogeneous brittle media usually presents catastrophe transition, i.e., the transition of evolution mode from globally stable accumulation of damage to catastrophic rupture. Thus, macroscopically the evolution is characterized by two time scales: a long time scale for damage accumulation and a very short time scale for catastrophe. Such a behavior can be modeled by the above-mentioned damage relaxation model with three characteristic time scales. The concern in this paper is with the common behaviors prior to $\sigma = \sigma_M$ rather than the value of σ_M.

In order to reveal the main mechanism of critical sensitivity, we adopt a simplified and ideal model which only takes the coupling between dynamical nonlinearity and mesoscopic heterogeneity into account, and a series of complicated effects are neglected. Subsequently, the model cannot describe a real system quantitatively.

3. Critical Sensitivity in the Case of Quasistatic Loading

For the case of quasistatic and monotonic loading; the damage evolution will reach equilibrium at each nominal driving force.σ_0 We denote the solution of equation (5) by $F(\sigma_c, \sigma_0)$ which is expressed by (see Fig. 1)

$$F(\sigma_c, \sigma_0) = \begin{cases} 0, & \text{for } \sigma_c < \frac{\sigma_0}{1-P}, \\ h(\sigma_c), & \text{for } \sigma_c \geq \frac{\sigma_0}{1-P}. \end{cases} \tag{7}$$

where $P = P(\sigma_0)$ is equilibrium damage fraction at σ_0 which is the solution of the following equation, which is derived from the definition (4) in the case of quasistatic and monotonic loading:

$$P(\sigma_0) = \int_0^{\frac{\sigma_0}{1-P(\sigma_0)}} h(\sigma_0) d\sigma_c \tag{8}$$

The response of the system to increasing driving force can be defined in terms of various variables, for instance, damage fraction or cumulative energy release. In this paper, it is defined by

$$R(\sigma_0) = \frac{dE(\sigma_0)}{d\sigma_0} \tag{9}$$

where $E(\sigma_0)$ is the cumulative energy release.

▶

Figure 1
Critical sensitivity in driven nonlinear threshold systems in the case of quasistatic loading based on mean-field approximation and damage relaxation model. $h(\sigma_c)$ is Weibull distribution function with modulus $m = 2$ and 5, respectively. (a) The equilibrium distribution function $F(\sigma_c, \sigma_0)$ at $\sigma_0 = \sigma_{of} - 0$, $m = 2$. (b) The equilibrium distribution function $F(\sigma_c, \sigma_0)$ at $\sigma_0 = \sigma_{of} - 0$, $m = 5$. (c) The cumulative energy release (the lower curves) and the sensitivity (the upper curves). σ_0 is the nominal driving force and the cumulative energy release E is normalized by $E_{\text{cum}} = \int_0^{\sigma_{of} - 0} R(\sigma_0) d\sigma_0$. The cumulative energy release (the lower curves) displays a catastrophe at $\sigma_0 = \sigma_{of}$. The sensitivity S (the upper curves) is defined by equation (12) and normalized by $S(\sigma_0 = 0) = m + 1$. S increases significantly prior to catastrophe, suggesting critical sensitivity.

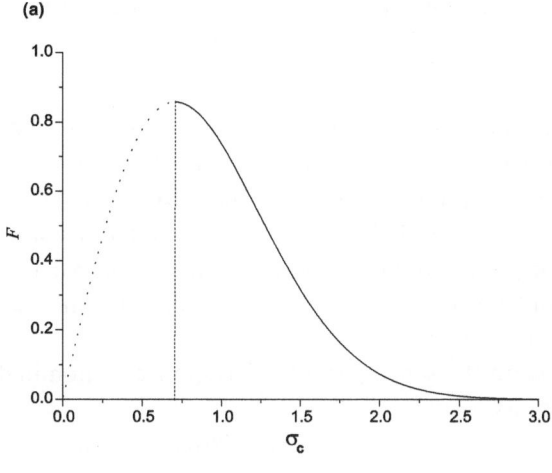

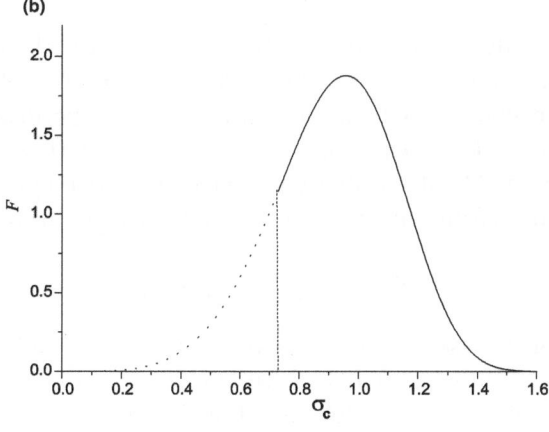

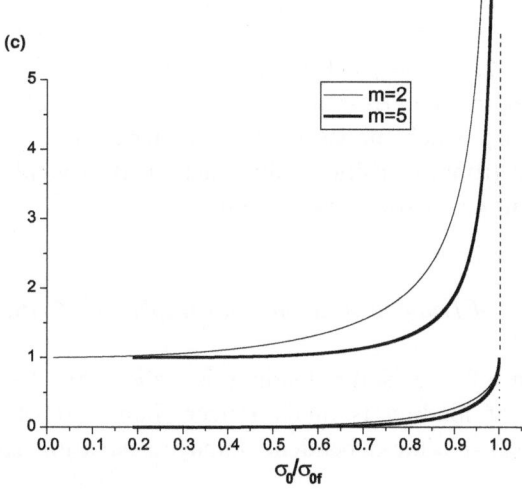

The catastrophe transition can be determined by the condition

$$\frac{dP(\sigma_0)}{d\sigma_0} = \infty \tag{10}$$

The response R displays catastrophe transition at $\sigma_0 = \sigma_{of}$ We will denote the damage fraction at catastrophe transition point by P_c

For $\sigma_0 < \sigma_{of}$ the equilibrium state of the system evolves continuously with increasing σ_0 and $P(\sigma_0) < P_c$ This is the stage of stable accumulation of damage. However, at $\sigma_0 = \sigma_{of}$ the equilibrium state jumps to global failure state (P = 1) displaying catastrophe transition, i.e., the system falls into a situation of self-sustained catastrophic failure.

In order to measure the sensitivity of the response to nominal driving force, we define the sensitivity as

$$S(\sigma_0) = \frac{\sigma_0}{R(\sigma_0)} \frac{dR(\sigma_0)}{d\sigma_0}. \tag{11}$$

A highly sensitive state is characterized by $S \gg 1$. Since there may be various definitions of the response, there also will be various definitions of the sensitivity. In particular, heterogeneous brittle media present sample specificity; the sensitivity cannot be represented by macroscopically average properties, e.g., free energy, however the definition (11) of sensitivity is available for real systems.

When the initial distribution function is assumed to be a Weibull distribution function

$$h(\sigma_c) = m\sigma_c^{m-1} \exp(-\sigma_c^m). \tag{12}$$

the cumulative energy release $E(\sigma_0)$ and sensitivity $S(\sigma_0)$ for the global mean field model are presented in Figure 1 (for a Weibull distribution function with modulus $m = 2$ and $m = 5$). Weibull distribution is a widely used function to characterize the diversity of material properties. The cumulative energy release shows that the evolution displays a catastrophe transition at $\sigma_0 = \sigma_{of} = (me)^{\frac{1}{m}}$ and $P_c = 1 - e^{-1/m}$ ($\sigma_{of} = 0.4289$ and $P_c = 0.3935$ for $m = 2$, and $\sigma_{of} = 0.5934$, $P_c = 0.1813$ for $m = 5$). In the case of $\sigma_0 \sim 0$, we derive $S \sim m + 1$, corresponding to an insensitive state. However, we can see that $S \to \infty$ as $\sigma_0 \to \sigma_{of}$ which implies that the system becomes sensitive significantly as it approaches its catastrophe transition point. Such a feature is called critical sensitivity, which is an important precursor of catastrophe.

4. *The Effects of Loading Rate on Catastrophe and Critical Sensitivity*

The assumption of quasistatic loading is valid only for the case that the characteristic time of loading is much longer than that of damage relaxation. Otherwise, the effects of time-dependent damage relaxation should be taken into account.

For the case with time-dependent nominal driving force $\sigma_0(t)$ the distribution function $f(\sigma_c, t)$ can be solved from equations (5) and (6). The energy release rate is given by

$$R(t) = \frac{N}{2k} \frac{\sigma_0^2(t)}{(1 - p(t))^2} \frac{dp(t)}{dt},$$ (13)

where N is the total of mesoscopic units and k is the stiffness of units, which is assumed to be identical for all units. We will consider the continuous limitation that $\lim_{N \to \infty} \frac{N}{k} = $ const. The sensitivity of response of the system to nominal driving force can be measured by

$$S(t) = \frac{\sigma_0(t)}{R(t)} \frac{\frac{dR(t)}{d(t)}}{\frac{d\sigma_0(t)}{dt}}$$ (14)

Now we consider the case that the nominal stress increases with time linearly:

$$\sigma_0(t) = \frac{\alpha \sigma_{of}}{\tau_D} t,$$ (15)

where σ_{of} is the catastrophe transition point in the case of quasi static loading, α is a constant. The quasi-static loading corresponds to the limitation $\alpha \to 0$.

Figure 2 illustrated the cumulative energy release and the sensitivity for the models with $\sigma_M = 4$ and $q = 0$, 1 and 2, respectively. The loading rate is characterized by $\alpha = 0.1$ and the initial threshold distribution function is adopted as Weibull distribution function with modulus $m = 3$. The cumulative energy release presents a continuous increase followed by a finite jump ΔE. This corresponds to a transition from continuous accumulation of damage to catastrophic rupture. The jump of cumulative energy release is expressed by

$$\Delta E = \frac{N}{2k} \sigma_M^2 (1 - D_M),$$ (16)

where D_M is damage fraction at $\sigma = \sigma_M$. From Figure 2 we can see that prior to catastrophic rupture, sensitivity increases significantly. Therefore, the critical sensitivity is also a precursory feature of the catastrophic rupture when the loading rate is considered. In Figure 2 the downward arrow indicates the catastrophe transition point in the case of quasistatic loading. It is found that both the catastrophic rupture and the critical sensitivity are delayed by the loading rate effect, comparable with the case of quasistatic loading. This is because the criticality of the system is determined by the coupling between the external condition and the state of the system. The appearance of catastrophe and critical sensitivity is related to the evolution of the system.

It is noted that there are differences in curves of sensitivity for the models with different q values. The higher the q value is, the faster the relaxation process becomes.

(a)

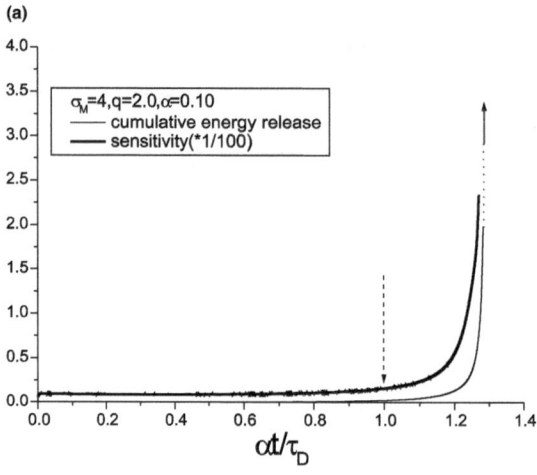

(b)

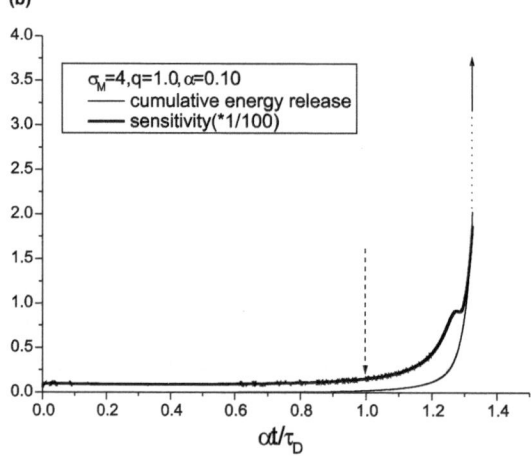

(c)

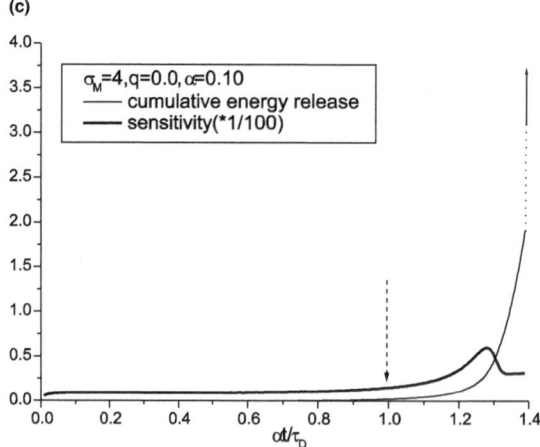

The quasistatic loading corresponds to the limitation $q \to \infty$. The characteristics of the sensitivity curve for $q = 2$, the case of faster relaxation, are similar to that for quasistatic loading, i.e., it shows monotonic increase. However, the sensitivity curve for $q = 0$, the case of slower relaxation, presents different characteristics, i.e., it shows a peak prior to catastrophic rupture. The characteristics of the sensitivity curve are determined by the coupling and competition between the effects of loading, damage relaxation and dynamical nonlinearity (redistribution of driving force). From equations (13)–(15), the sensitivity S can be expressed as

$$S = 2 + \left(\frac{2}{1-p} \frac{dp}{dt} + \left(\frac{dp}{dt} \right)^{-1} \frac{d^2p}{dt^2} \right) t. \qquad (17)$$

The factor t in the right hand of the equation derives from the linear loading, the time-dependent behavior of the damage $p(t)$ reflects the evolution of distribution function $f(\sigma_c, t)$ Figure 3 depicts the evolution of $f(\sigma_c, t)$ prior to the catastrophic rupture for various damage relaxation models, i.e., for the models with $q = 0$ and, $q = 2$ respectively. We can see that the damage relaxation process for the model with $q = 2$ is faster than that for the model wit $q = 0$.

5. Effects of Stress Fluctuations on Catastrophe and Critical Sensitivity

In mean field approximation, stress fluctuations are neglected. However, the stress fluctuations play an essential role in catastrophe (WEI, *et al* 2000). As we take the stress fluctuations into account, the problem becomes far more complicated. Numerical results revealed that the evolution also presents catastrophe transition although the catastrophe displays sample-specificity. Regardless, it is also found that the critical sensitivity is in reality the common precursory feature of catastrophe in the case with stress fluctuations, although the sensitivity manifests strong fluctuations, see reference XIA *et al* (2002).

6. Critical Sensitivity in Rock Rupture Experiments

Now we give results of rock rupture experiments YIN *et al* (2002) on samples of 1050*400*100 mm^3 (for marble and gneiss) and 1050*400*150 mm^3 (for sand-

◄

Figure 2

Cumulative energy release and sensitivity in the case of time-dependent nominal driving force $\sigma_0(t) = \alpha \frac{\sigma_{0f}}{\tau_0} t$ with $\alpha = 0.1$. $h(\sigma_c)$ is Weibull distribution function with modulus $m = 3$. (a) $\sigma_M = 4$, $q = 2$; (b) $\sigma_M = 4$, $q = 1$; (c) $\sigma_M = 4$, $q = 0$. The upward arrows indicate the catastrophic rupture and the downward arrows indicate the catastrophe transition point in the case of quasistatic loading.

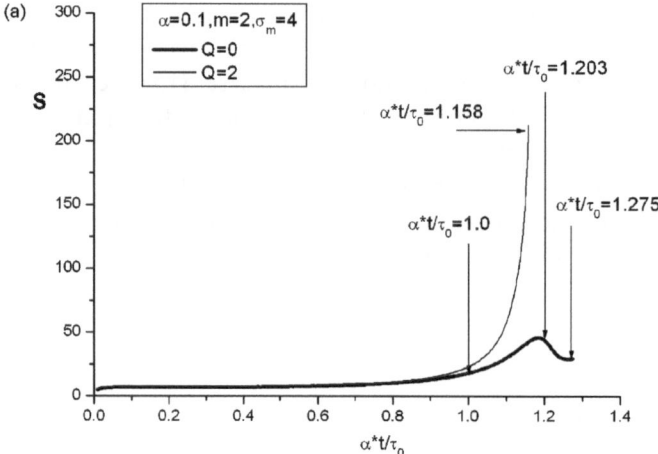

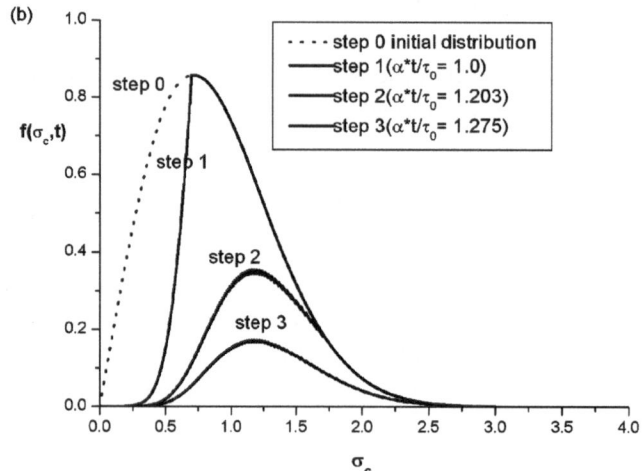

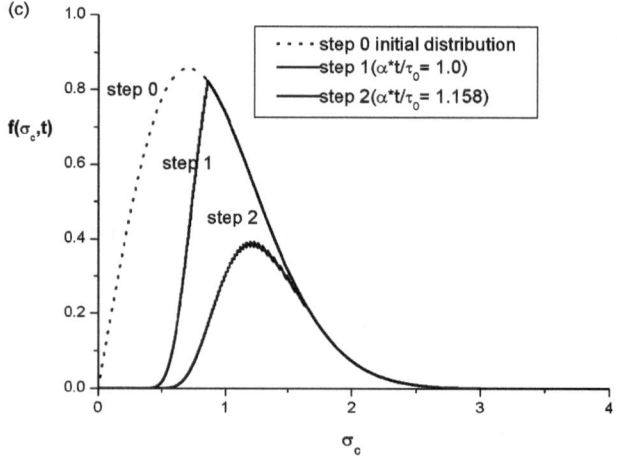

stone). The samples were compressed until rupture. The behavior shows catastrophe transition from the accumulation of mesoscopic damage to macroscopic rupture. Mesoscopic damage is characterized by energy release recorded by acoustic emission (AE). Typically, the total of recorded AE events is about $4*10^5$ in the experiments. The sensitivity of energy release to external load can be defined by

$$S = \frac{F}{R}\frac{\Delta R}{\Delta F}, \tag{18}$$

where F is the external load, R is the energy release rate, and ΔR is the increment of energy release rate corresponding to the increment of external load ΔF. The experimental results are shown in Figure 4. It is note worthy to compare the experimental results, Figure 4, with the theoretical results, Figure's 1–3. Theoretical results of sensitivity are expressed by a deterministic and smooth curve. This is related to the continuous limitation, mean-field approximation and relaxation time model. The experimental and numerical results of sensitivity display fluctuations (strong fluctuations imply high sensitivity) and sample specificity (the time series of sensitivity are different from each other for samples identical macroscopically). Such behavior is related to the fact that the mesoscopic damage events are discrete and stochastic in a real system or numerical models with stress fluctuations. Nonetheless, the rock rupture experiments seem to support the critical sensitivity: the sensitivity increases significantly prior to the catastrophic rupture point.

7. Summary

Rupture in heterogeneous brittle media is a complicated phenomenon, and rupture prediction, e.g. earthquake forecast, is a difficult problem of vital societal concern. Critical sensitivity as a universality of catastrophe transition in driven nonlinear threshold systems may help us to capture the essence of catastrophe transition. In some cases, the sensitivity might be measurable. Monitoring the sensitivity of the system may yield helpful clues to rupture prediction.

◄

Figure 3
The evolution of distribution function $f(\sigma_c, t)$ prior to the catastrophic rupture. $h(\sigma_c)$ is Weibull distribution function with modulus $m = 2$. The system is subjected to nominal driving force $\sigma_0(t) = \alpha\frac{\sigma_{of}}{\tau_0}t$ with $\alpha = 0.1$ (a) The sensitivity S for relaxation model with $\sigma_M = 4$ and $q = 0$ and 2, respectively; (b) The evolution of distribution function $f(\sigma_c, t)$ for damage relaxation model with $q = 0$; (c) The evolution of distribution function $f(\sigma_c, t)$ for damage relaxation model with $q = 2$.

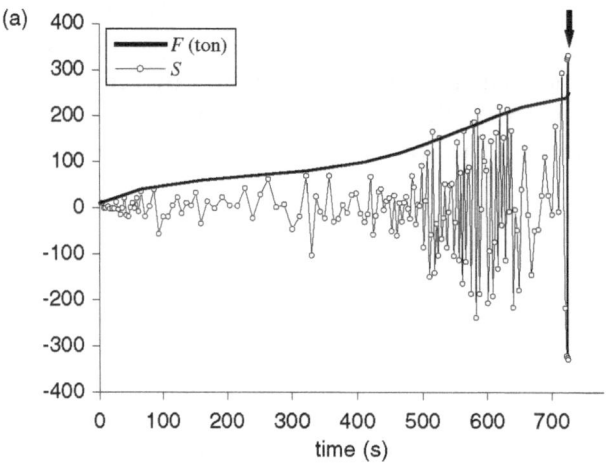

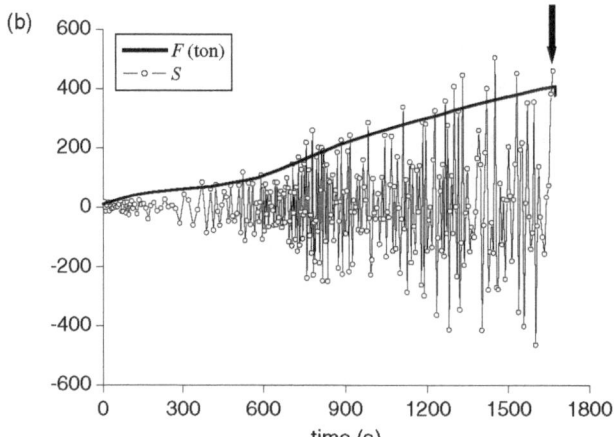

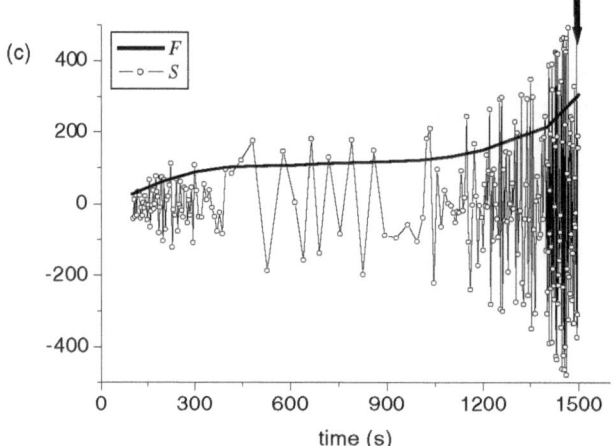

◄

Figure 4

Critical sensitivity in a rock rupture experiment. The curves are axial force $F(t)$ and the sensitivity $S(t)$ respectively, and $S(t)$ is calculated from the data of acoustic emission. The total of recorded AE events is about $4*10^5$. The arrow indicates the catastrophic rupture point. The experimental results seem to support the critical sensitivity: the sensitivity increases significantly prior to the catastrophic rupture point. (a) Gneiss; (b) Sandstone; (c) Marble.

Acknowledgement

This work is supported by the Special Funds for Major State Research Project (G2000077305) and the National Natural Science Foundation of China (No. 10372012, 10047006, 10232040, 10232050).

REFERENCES

BAI, Y. L., LU, C. S., KE, F. J., and XIA, M. F. (1994), *Evolution Induced Catastrophe*, Phys. Lett. A. *185*, 196–200.

BEN-ZION, Y. and SAMMIS, C. G. (2003), *Characterization of Fault Zones*, Pure Appl. Geophys. *160*, 677–715.

CURRAN, D. R., SEAMAN, L., and SHOCKEY, D. A. (1997), *Dynamic Failure of Solids*, Phys. Rep. *147*, 253–388.

CURTIN, W. A. (1997), *Toughening in Disordered Brittle Materials*, Phys. Rev. B *55*, 11270–11276.

GARCIMARTIN, A. GUARINO, A. BELLON, L. and CILIBERTO, S. (1997), *Statistical Properties of Fracture Precursors*, Phys. Rev. Lett. *79*, 3202–3205.

JAUME, S. C. and SYKES, L. R. (1999), *Evolving Towards a Critical Point: A Review of Accelerating Seismic Moment/Energy Release Prior to Large and Great Earthquakes*, Pure Appl. Geophys. *155*, 279–305.

MEAKIN, P. (1991), *Models for Material Failure and Deformation*, Science *252*, 226–234.

RUNDLE, J. B., KLEIN, W., TIAMPO, K. F., and GROSS, S. (2000), *Linear Pattern Dynamics in Nonlinear Threshold Systems*, Phys. Rev. E *61*, 2418–2432.

RUNDLE, J. B., KLEIN, W., TIAMPO, K. F., and GROSS, S. (2000), *Dynamics of Seismicity Patterns in Systems of Earthquake Faults, Geocomplexity and the Physics of Earthquakes*, AGU Monograph, Washington, D.C.,pp 127–146.

SAHIMI, M. and ABABI, S. (1993), *Mechanics of Disordered Solid. III. Fracture Properties*, Phys. Rev. B *47*, 713–722.

STEIN, R. S. (1999), *The Role of Stress Transfer in Earthquake Occurrence*, Nature *402*, 605–609.

TIAMPO, K. F., RUNDLE, J. B., KLEIN, W., MCGINNIS, S., and GROSS, S. J. (2000), *Observation of Systematic Variations in Non-local Seismicity Patterns from Southern California*, Geocomplexity and the Physics of Earthquakes, AGU Monograph, Washington, D. C.

WEI, Y. J., XIA, M. F., KE, F. J., YIN, X. C. and BAI, Y. L. (2000), *Evolution Induced Catastrophe and its Predictability*, Pure Appl. Geophys. *157*,1945–1957.

WYSS, M., ACEVES, R. L., PARK, S. K., GELLER, R .J., JACKSON, D. D., KAGAN, Y. Y., and MULARGIA, F. (1997), *Cannot Earthquakes Be Predicted?* Science *275*, 480–490.

XIA, M. F., KE, F. J., and BAI, Y. L. (1997), *Threshold Diversity and Trans-scale Sensitivity in a Nonlinear Evolution Model*, Phys. Lett. A *236*, 60–64.

XIA, M. F., KE, F. J., and BAI, Y. L. (2000), *Evolution Induced Catastrophe in a Nonlinear Dynamical Model of Materials Failure*, Nonlinear Dynamics *22*,205–224.

XIA, M. F., SONG, Z. Q., XU, J. B., ZHAO, K. H., and BAI, Y. L. (1996), *Sample-specific Behavior in Failure Models of Disordered Media*, Communication in Theoretical Physics. *25*, 49–54.

XIA, M. F., WEI, Y. J., KE, F. J., and BAI, Y. L. (2002), *Critical Sensitivity and Trans-scale Fluctuations in Catastrophic Rupture*, Pure Appl. Geophys. *159*, 2491–2509.

YIN, X. C., YU, H. Z., KUKSHENKO, V., XU, Z. Y., LI, Q., XIA, M. F., LI, M., PEN, K. Y., and ELIZAROV, S., *Load-Unload Response Ratio (LURR) and Accelerating Moment/Energy Release (AM/ER) and State Vector*, 3rd ACES International Workshop, May 5–10, 2002, Island of Maui, State of Hawaii, USA.

(Received September 27, 2002, revised January 27, 2003, accepted February 10, 2003)

To access this journal online:
http://www.birkhauser.ch

Pure appl. geophys. 161 (2004) 1945–1956
0033–4553/04/101945–12
DOI 10.1007/s00024-004-2541-z

© Birkhäuser Verlag, Basel, 2004

❚ **Pure and Applied Geophysics**

Intermittent Criticality and the Gutenberg-Richter Distribution

DAVID D. BOWMAN[1] and CHARLES G. SAMMIS[2]

Abstract—In recent years there has been renewed interest in observations of accelerating moment release before large earthquakes, as well as theoretical descriptions of seismicity in terms of statistical physics. Most aspects of these works are encompassed by a concept called intermittent criticality in which a region alternately approaches and retreats from a critical -point. From this perspective, the evolution of seismicity in a region is described in terms of the growth and destruction of correlation in the stress field over the course of the seismic cycle. In this paper we test the concept of intermittent criticality by investigating the temporal evolution of the Gutenberg-Richter distribution before and after two successive $M \geq 5.0$ earthquakes in western Washington State. The largest event in this distribution, M_{max}, is observed to systematically increase before each event, producing accelerating moment release, and then to subsequently decrease. Associated variations in the b-value are minimal. This is the predicted result if M_{max} is a measure of the correlation length of the regional stress field.

Key words: Seismicity, earthquake physics, earthquake stress interactions, self-organized criticality, earthquake statistics.

1. Introduction

In recent years, there have been many attempts to describe the physics of distributed regional seismicity using the framework of self-organized criticality (SOC). Self-organized criticality was originally defined on the basis of simple cellular-automaton models (e.g., BAK and TANG, 1989; SORNETTE and SORNETTE, 1989; OLAMI *et al.*, 1992). In the context of earthquakes, the most important characteristic of self-organized criticality is its inherent ability to produce power-law frequency-size statistics. This relationship, known as the Gutenberg-Richter distribution, has been known for many years to describe the frequency-magnitude statistics of earthquakes (GUTENBERG and RICHTER, 1944). BAK and TANG (1989) recognized that cellular automata models produce similar power-law size statistics, an observation that formed the basis of their claim that seismicity can be described as a self-organized critical system. Another often cited characteristic of self-organized criticality is the

---

[1] Department of Geological Sciences, California State University, Fullerton, CA 92834-6850.
E-mail: dbowman@fullerton.edu
[2] Department of Earth Sciences, University of Southern California, Los Angeles, CA 90089-0740.
E-mail: sammis@earth.usc.edu

stability of the statistics through time. Because SOC models ideally contain no tuning parameters, a system that has achieved self-organized criticality will remain in that state with constant power-law frequency-size statistics for as long as the external driving force remains constant. The apparent stationarity of the b-value of the Gutenberg-Richter distribution is one of the primary pieces of evidence supporting the model of self-organized criticality (KAGAN, 1994; JACKSON, 1996; GELLER *et al.*, 1997).

However, recent works have shown that cellular automata models of self-organized criticality break down if the model includes dissipation or strong heterogeneity. SAMMIS and SMITH (1999) demonstrated that the non-conservative automaton originally formulated by OLAMI *et al.* (1992) naturally leads to seismic cycles. In this model, some energy is removed from the grid during cascades. This energy loss may be interpreted as arising from processes in real earthquakes such as frictional heating, seismic radiation, and breaking new rock. Energy loss introduces memory into the model. Since the total energy on the grid is drastically reduced after a large event, the system must recover before it can produce another large avalanche. In this model, self-organized criticality cannot be reached on the time-scale of the seismic cycle. However, when averaged over many earthquake cycles the model reproduces a stationary power-law distribution, similar to SOC. Thus the question of determining whether or not a system is in a state of self-organized criticality depends very strongly on taking a large enough space-time window to average out fluctuations in the stress field associated with large events. Therefore, while SOC may be an important concept for understanding seismicity averaged over very long space-time domains, it is of very little use when trying to understand the space-time distribution of seismicity on a given fault network over a human time scale. An understanding of seismicity on the time scale of individual seismic cycles requires a new paradigm.

2. Intermittent Criticality

The work of SAMMIS and SMITH (1999) is representative of a new class of models which display *intermittent* criticality (see also SORNETTE and SAMMIS, 1995; SALEUR *et al.*, 1996; SAMMIS *et al.*, 1996; HEIMPEL, 1997; BOWMAN *et al.*, 1998; HUANG *et al.*, 1998; and GRASSO and SORNETTE 1998). This viewpoint is based on the hypothesis that a large regional earthquake is the end result of a process in which the stress field becomes correlated over increasingly long scale-lengths, which set the size of the largest earthquake that can be expected at any given time. The largest event possible on the fault network cannot occur until regional criticality has been achieved and stress is consequently correlated at all length scales up to the size of the region. This large event then destroys criticality on its associated network, creating a period of relative quiescence after which the process repeats by rebuilding correlation lengths towards criticality and the next large event. In contrast to self-organized criticality in which the system is always at or near criticality, intermittent criticality implies time-dependent variations in the activity during a seismic cycle.

2.1 Intermittent Criticality and Accelerating Moment Release

It is important to note that the central prediction of intermittent criticality is that large earthquakes only occur when the system is in a critical state. This large earthquake acts as a sort of "critical point" dividing the seismic cycle into a period of growing stress correlations before the great earthquake and a relatively uncorrelated stress field after. Before the large earthquake, the growing correlation length manifests itself as an increase in the frequency of intermediate-magnitude earthquakes. This is commonly referred to as the accelerating moment release model, and has been discussed by a number of authors (SYKES and JAUMÉ, 1990; BUFE and VARNES, 1993; BUFE et al., 1994; SORNETTE and SAMMIS, 1995; BOWMAN et al., 1998; BREHM and BRAILE, 1998; JAUMÉ and SYKES, 1999).

Many works have shown that accelerating moment release before large earthquakes typically occurs over a distance much larger than the size of the mainshock rupture (e.g, BOWMAN et al., 1998; BREHM and BRAILE, 1998; JAUMÉ and SYKES, 1999; ROBINSON, 2000; PAPAZACHOS and PAPAZACHOS, 2001) . KING and BOWMAN (2003) developed a simple model that relates the large region of increased activity prior to a large event to the region that is stressed by tectonic loading of the fault. In their model, the process of stress accumulation and release on the main fault perturbs the constant uniform driving stress assumed by models of self-organized criticality. This forces the system out of equilibrium and into the strongly fluctuating regime described by SMITH and SAMMIS [1999]. While KING and BOWMAN (2003) show that the size of the region experiencing accelerating moment release should scale as the size of the mainshock, there are many processes that may cause the absolute size of the regions to vary between different tectonic regimes (e.g., differences in the distribution of strength heterogeneities in the region). Also, the level of background seismicity may affect the size of the active region in the same way it affects the time duration of an aftershock sequence (see, e.g., DIETERICH, 1994).

2.2 Intermittent Criticality and Frequency-Magnitude Statistics

Intermittent criticality also makes predictions relating to the evolution of the Gutenberg-Richter distribution during the course of the earthquake cycle. The central feature of intermittent criticality is that the correlation length of the stress field controls the size of the largest event in a region (M_{max}). As the correlation length grows prior to a large earthquake, the size of the maximum expected earthquake also increases. For all magnitudes smaller than this maximum cutoff, the region is effectively at a critical state. As with self-organized criticality, this leads to power-law scaling of the frequency-magnitude distribution at the lower magnitudes. The important difference between SOC and intermittent criticality is the evolution of the correlation length. In self-organized criticality, the correlation length of the stress field is by definition constant and infinite. Thus, there should be no systematic temporal fluctuations in any parameters of the Gutenberg-Richter distribution, including the maximum magnitude of seismicity.

However, in intermittent criticality the correlation length of the stress field varies in time. As the correlation length grows before a great earthquake, M_{max} increases. This extends the Gutenberg-Richter scaling to higher magnitudes, but does not alter the underlying scaling relation. The b-value for the region will remain constant, with a small change in the a-value. Thus, traditional measurements of temporal variations in the b-value in a region will be insufficient to differentiate between self-organized criticality and intermittent criticality. A more complete description of the frequency-size statistics must also include the evolution of a and M_{max}. We will illustrate the relative importance of these parameters by analyzing seismicity before and after a series of $M \geq 5$ earthquakes in the Puget Sound region of the state of Washington, USA.

3. Seismicity of the Pacific Northwest

The data in this study were recorded by the Pacific Northwest Seismograph Network (PNSN). The PNSN, headquartered at the University of Washington, began operation in 1970, however consistent catalogs for events $M \leq 3.5$ were not available until 1980. Therefore this study will be limited to seismicity from 1980 to 2000. The dense short-period seismograph network that comprises the bulk of the PNSN covers western Washington and Oregon states from roughly 42° to 49°N latitude and from 119° to 125°W longitude.

Interpretation of seismicity in this area is complicated by the fact that the region sits astride the Cascadia subduction zone. However, as Fig. 1 shows, the vast majority of seismicity in the region occurs in the upper plate of the subduction zone. The largest earthquakes within this region are lower plate events, which can be as large as magnitude 7. In contrast, the largest instrumentally recorded events in the upper plate have been in the magnitude 5–6 range, although there is paleoseismic evidence for a large tsunamigenic crustal earthquake about 1100 years ago (BUCKNAM et al., 1992). Focal mechanisms of shallow earthquakes in the Puget Sound region indicate a consistent pattern of north-south compression in the upper plate (ZOBACK and ZOBACK, 1991). In contrast, the focal mechanisms of events in the subducting slab indicate down-dip tension (LUDWIN et al., 1991; MA and LUDWIN, 1987). This significant difference in tectonic stress orientation indicates that the plates are not mechanically coupled. Therefore, in this work we assume that the upper and lower plates are decoupled, and will only consider upper plate seismicity.

Since 1982 there have been 3 $M \geq 5$ earthquakes within 170 km of each other in the area surrounding Puget Sound. The maximum magnitude of historical upperplate seismicity in this region is in the $M \approx 5$ range. Figure 2 is a map of all seismicity $M \geq 2.5$ since 1985 in the region we shall consider, with stars denoting the epicenters of $M \geq 5.0$ events. The circles in Figure 2 delineate the critical regions before the

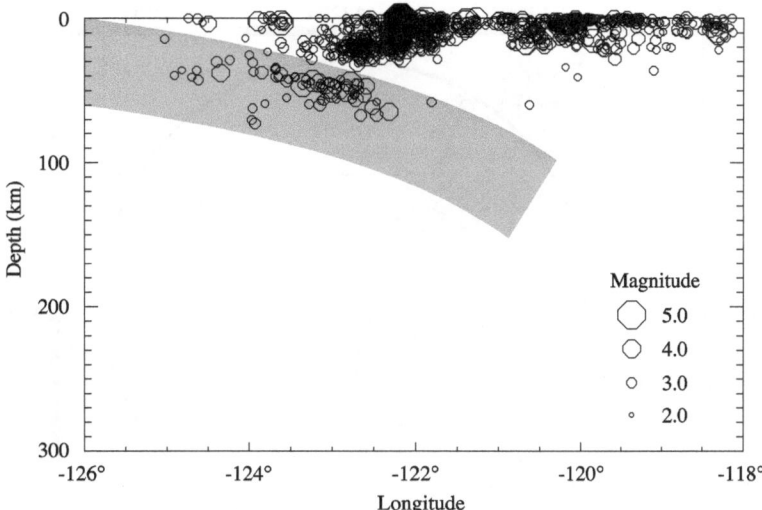

Figure 1

Crosssection of seismicity in the Pacific Northwest. Gray region is the inferred Wadati-Benioff zone. 2:1 vertical exaggeration.

1996 $M_b = 5.4$ Duvall and the 1990 $M_b = 5.0$ Deming earthquakes. These regions were found using the algorithm described by BOWMAN *et al.* (1998). Figure 3 shows the cumulative Benioff strain (square root of the seismic energy) release for both of these events. The reasons for using Benioff strain are described by a number of works (e.g., BUFE and VARNES, 1993; SORNETTE and SAMMIS, 1995; SALEUR *et al.*, 1996).

Note that the star in Figure 3d is the 1995 $M = 5.0$ Robinson's Point earthquake. While this event is large enough to be considered a "mainshock", it is significantly smaller than the 1996 Duvall event. Thus, in the context of intermittent criticality, the Robinson's Point earthquake is a large precursor to the nearby $M_b = 5.4$ Duvall event.

5. *Observed Evolution of the Gutenberg-Richter Distribution*

Ideally, a complete description of the evolution of the Gutenberg-Richter distribution through the seismic cycle would include two or more repeated events on the same fault. However, few existing seismicity catalogs are of sufficient quality for a long enough time period to measure reliable frequency-magnitude statistics (including M_{max}) over more than one earthquake cycle. In this study, we are not attempting to describe a seismic cycle in the commonly accepted sense of the term (ELLSWORTH *et al.*, 1981). Rather, we are describing regional variations in frequency-magnitude statistics due to the loading cycle on neighboring fault systems.

Figure 4 shows the frequency-magnitude statistics for all events $M \geq 2.5$ within the black circle in Figure 2 in a two-year moving window from 1981–1997. In this figure, a

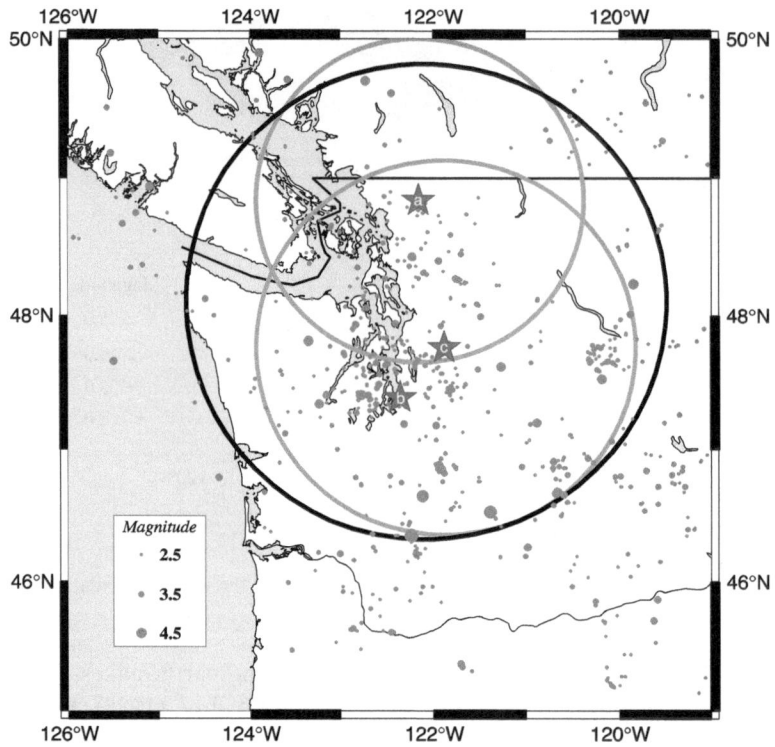

Figure 2

Seismicity of the Pacific Northwest from January 1, 1985 to November 1, 1998, $M \geq 2.5$. Stars indicate events $M \geq 5.0$. a) 1990 $m_b = 5.0$ Deming earthquake, b) 1995 $m_b = 5.0$ Robinson's Point earthquake, c) 1996 $m_b = 5.4$ Duvall earthquake. Light circles show the critical regions for the Deming and Duvall events. The dark circle encloses the larger region used to study the evolution of frequency-magnitude statistics in the time interval between these two events.

vertical section perpendicular to the time axis is the frequency-magnitude distribution for a two-year period centered at time T. The figure resembles a mountainside with prominent "valleys" at the years 1987 and 1993, and "ridges" at 1981, 1990 and 1996. Note that the b-value in this figure (the "slope of the mountainside") is relatively constant throughout the time considered here, as predicted above. The "ridge and valley" structure of the plot is a direct result of the growth and destruction of the correlation length of the regional stress field. The maximum magnitude of seismicity in the region increased prior to the two large events in 1990 and 1996, producing the pronounced ridges in the plot. The valleys are a result of the reduction in $M_{\max}$ due to the decorrelated stress field after the 1990 Deming and 1996 Duvall events. The quality of the earthquake catalogs prior to 1981 restrain us from analyzing data in this time period, however the time period from 1980–1981 saw heightened activity in the southern section of the region discussed here, including both an $M = 5.2$ event north of Elk Lake, Washington and the May 1980 eruption of Mount St. Helens.

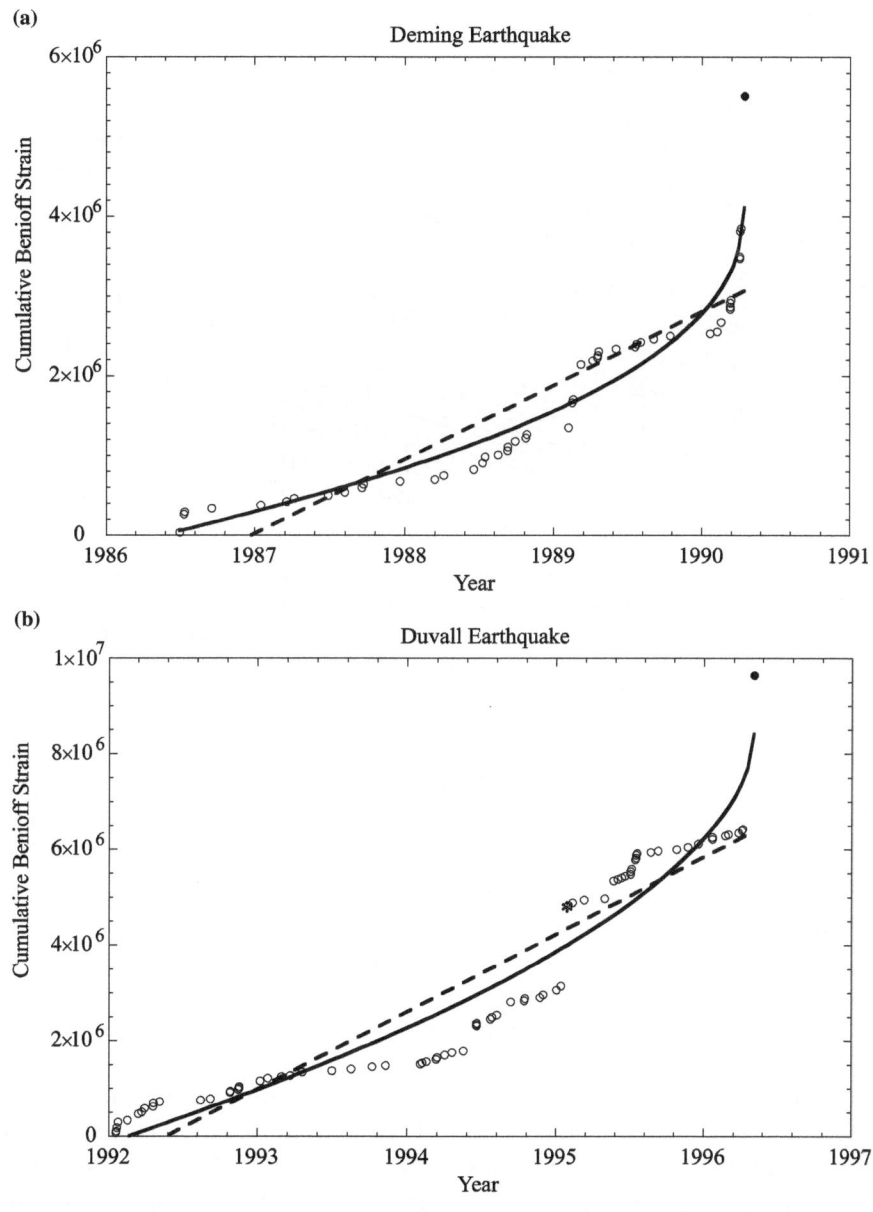

Figure 3
Cumulative Benioff strain as a function of time for the (a) 1990 $m_b = 5.0$ Deming earthquake and the (b) 1996 $m_b = 5.4$ Duvall earthquake. In both plots, the final event is denoted by a filled circle. The star in (b) is the 1995 $m_b = 5.0$ Robinson's Point earthquake. See text for details.

The evolution of the frequency-magnitude statistics over this time period can also be seen in Figure 5. From top to bottom, this figure shows the evolution of the total number of events, maximum magnitude of seismicity, and the b-value for the same

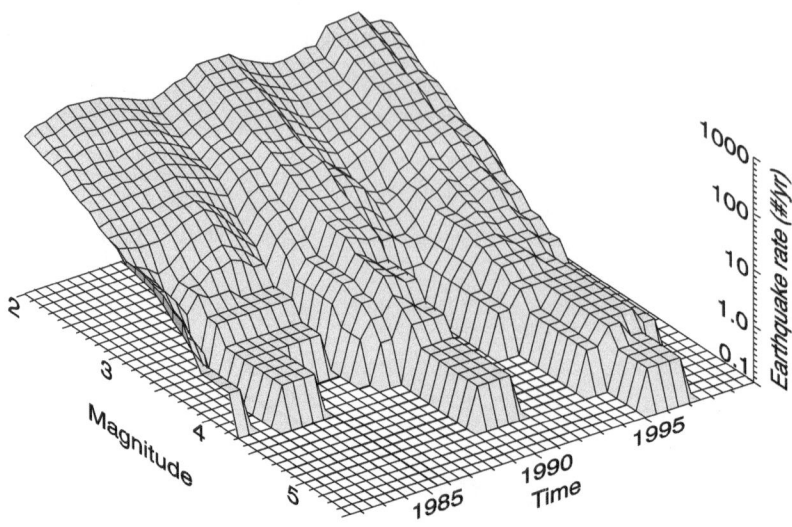

Figure 4

Frequency-magnitude statistics from 1981–2000 for the region shown in figure 3. The statistics were calculated in a two-year moving window. Note that M_{max} in the distribution slowly increases prior to 1990 and 1996. Also note that the slope (b-value) of the distribution remains relatively constant throughout the period, with fluctuations in the height (a-value) of the distribution.

two year moving window in Figure 4. Both the Deming and Robinson's Point earthquakes show a strong increase in both the maximum magnitude of seismicity and the overall level of seismicity in the five years prior to the events. The increase in the number of $M \geq 2.5$ earthquakes shows a strong correlation with the increase in M_{max}. This can be easily understood in terms of a systematic variation in the a-value in the region. The increase of a before a large earthquake creates a higher probability for intermediate-magnitude events. In a two-year moving window, this will manifest as an increase in M_{max}, while in a plot of the cumulative Benioff strain as a function of time this will be see as accelerating moment release. Following the large event, the a-value decreases, lowering M_{max}.

Note that, to first order, b is constant over the entire time period, having a mean $\{b\} = 1.13$ and a standard deviation $\sigma = \pm 0.19$. There are slight fluctuations in the b-value which are anti-correlated with M_{max} (low b-value corresponding to large M_{max}). This anti-correlation is quantified in Figure 6 which shows that the cross correlation between b and M_{max} has a strong negative peak at a time shift of 0 and weaker positive correlations at shifts of $+4$ years and -3 years. This asymmetry in the correlation function reflects the variable duration of the seismic cycle. However, note that most of the b-value data points in Figure 5 lie within one σ of the mean, so that the observed fluctuations in b-value are not statistically significant.

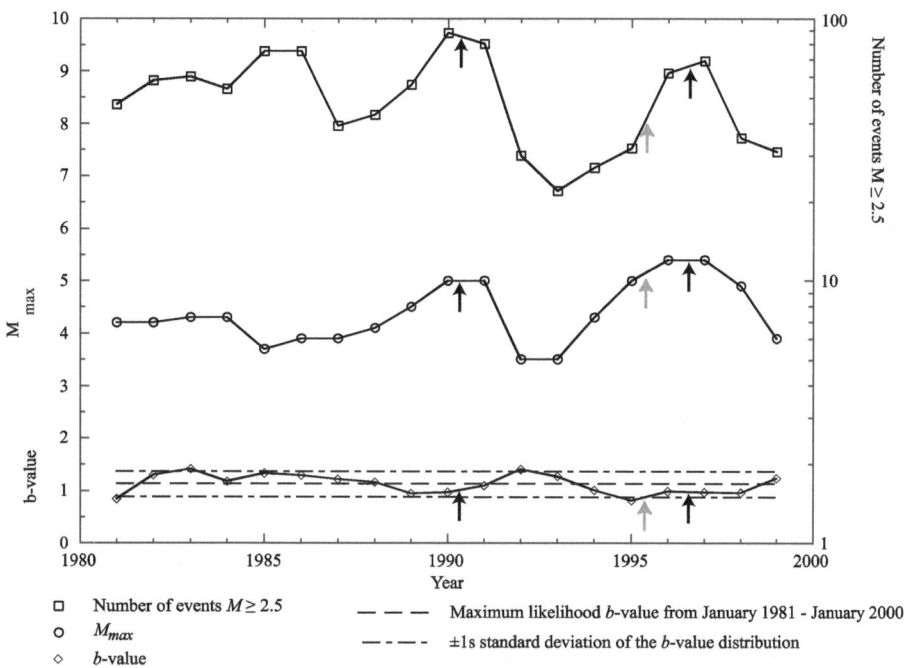

Figure 5

Variations in seismicity during the seismic cycle. The data points are from the same two-year moving windows used in Figure 6. The upper curve (square symbols) is the total number of events $M \geq 2.5$ in the region at each time step. The middle curve (round symbols) is M_{max} at each time step. The lower curve (diamond symbols) is the b-value at each time step calculated by the maximum-likelihood technique. The dashed line shows the average b-value over all time steps, and the dot-dash line shows the 1-s standard deviation of the observed b-values. The black arrows indicate the time of the 1990 Deming and 1996 Duvall earthquakes. The gray arrow is the time of the 1995 Robinson's Point event.

6. Discussion

Although the fluctuations in M_{max} found above are incompatible with the concept of self-organized criticality, they are consistent with the predictions of intermittent criticality. Of particular note are the following observations:

1) There are systematic trends in both the frequency-magnitude distribution and the cumulative Benioff strain over the course of the two earthquake cycles observed here.

2) The b-value during the study period remains roughly constant. There is minor variability in the b-value which is anti-correlated with M_{max}.

3) M_{max} grows prior to a large earthquake, and decreases following it.

4) There is an apparent variation in the a-value over the course of the seismic cycle. This variation coincides with variations in M_{max}, and may be causally related.

While it is likely that SOC systems may produce clustering of events that would produce fluctuations in the frequency-magnitude statistics, it is highly unlikely that

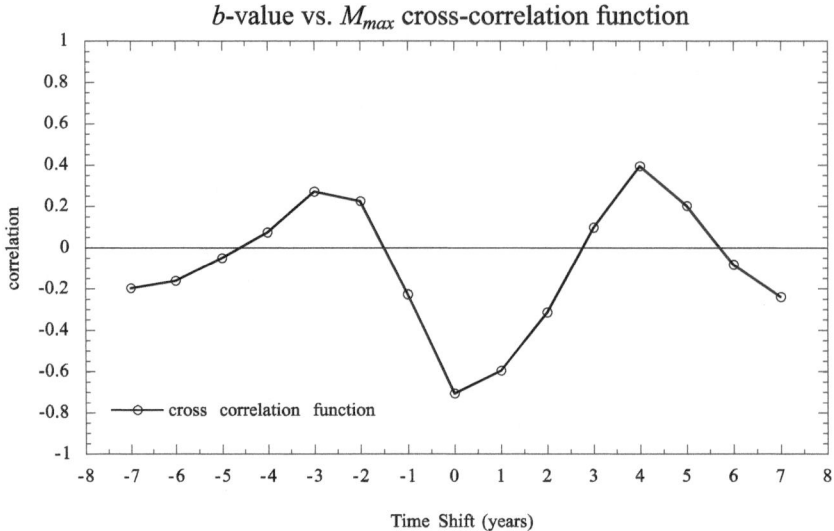

Figure 6
Correlation of cyclic variations in the temporal evolution of b-value and M_{max} in the Pacific Northwest. The strongly negative correlation at a time shift of 0 years indicates a strong anti-correlation in the cycles. The fact that a strong positive correlation cannot be found by time shifting the data is a result of the irregularity in the period of the cycles.

such clustering would display the systematic behavior noted here. However, this observation supports a basic prediction of the concept of intermittent criticality as described above.

It should be noted that the stationarity of the b-value that we describe here is contradicted by JAUMÉ and SYKES (1999). Their work asserts that accelerating moment release before large earthquakes is a direct result of a systematic decrease in the b-value. However, the changes in b-value observed by JAUMÉ and SYKES (1999) occur in the high-magnitude tail of the distribution (generally at $M > 5$). It is certainly true that an increase in M_{max} could be viewed as a change in the b-value for the largest events. However, b-value is generally thought to reflect the scaling of faults or other heterogeneities in the system (e.g., MOGI, 1962; KING, 1983). Thus, the variations of b-value in the high magnitude tail of the distribution described by JAUMÉ and SYKES (1999) can be viewed as an artifact of the more physically meaningful fluctuations in M_{max}. This is supported by the observation that a maximum-likelihood calculation of the b-value is largely insensitive to fluctuations at higher magnitudes, as shown in figure 5.

Many previous works (BUFE and VARNES, 1993; BUFE et al., 1994; SORNETTE and SAMMIS, 1995; SAMMIS et al., 1996; BOWMAN et al., 1998; BREHM and BRAILE, 1998; JAUMÉ and SYKES, 1999) have characterized accelerating seismicity before large earthquakes by fitting a power-law time to failure equation to some measure of cumulative seismic energy release (usually measured in Benioff strain) to the observed earthquake catalog. This approach is motivated by analogy to a critical point (SORNETTE and SAMMIS, 1995). However, the relatively small number of intermediate size events comprising this precursory acceleration makes the statistical significance of such analytical curve fits questionable (BOWMAN et al., 1998). In this paper we have taken the alternative approach of attempting to indirectly measure the implied correlation length by simply tracking the maximum magnitude of seismicity in a region before and after a pair of large earthquakes. This approach, which is also motivated by analogy to the behavior of a system approaching a critical point, monitors the increase of correlation length approaching the "critical event" and the corresponding decrease in the correlation length following it. As earthquake catalogs improve in quality, observable fluctuations in either the a-value or M_{max} may provide a useful tool for seismic hazard assessment.

Acknowledgements

The authors acknowledge helpful discussions with Geoffrey King and James Dolan, as well as two anonymous reviewers. This research was funded by National Science Foundation grants EAR-0105405 (CGS) and EAR-0107129 (DDB) and through Cooperative Agreements EAR-0106924 and USGS-02HQAG0008 to the Southern California Earthquake Center. This is SCEC contribution number 602.

REFERENCES

BAK, P. and TANG, C. (1989), Earthquakes as a Self-organized Critical Phenomenon, J. Geophys. Res. 94, 15,635–14,637.

BOWMAN, D. D., OUILLON, G., SAMMIS, C., G SORNETTE, A., and SORNETTE, D. (1998), An Observational Test of the Critical Earthquake Concept, J. Geophys. Res. 103, 24,359–24,372, 1998.

BREHM, D. J. and BRAILE, L. W. (1998), Intermediate-term Earthquake Prediction Using Precursory Events in the New Madrid Seismic Zone, Bull. Seismol. Soc. Am. 88, 564–580.

BUCKNAM, R. C., HEMPHIL-HALEY, E., and LEOPOLD, E. B. (1992), Abrupt uplift within the past 1700 years at southern Puget Sound, Washington, Science, 258, 1611–1614.

BUFE, C. G. and VARNES, D. J. (1993), Predictive Modelling of the Seismic Cycle of the Greater San Francisco Bay Region, J. Geophys. Res. 98, 9871–9883.

BUFE, C. G., NISHENKO, S. P., and VARNES, D. J. (1994), Seismicity Trends and potential for large earthquakes in the Alaska-Aleutian Region, Pure Appl. Geophys. 142, 83–99.

DIETERICH, J., (1994),, A constitutive Law for Rate of Earthquake Production and its Application to Earthquake Clustering, J. Geophys. Res. 99, 2601–2618.

ELLSWORTH, W. L., LINDH, A. G., PRESCOTT, W. H., and HERD, D. J. The 1906 San Francisco Earthquake and the seismic cycle. In Earthquake Prediction: An Internation. Review, Maurice Ewing Ser., vol. 4, (D.W. Simpson and P.G. Richards, eds.) (AGU, Washington, D.C., 1981) pp. 126–140.

GELLER, R. J., JACKSON, D. D., KAGAN, Y. Y., and MULARGIA, F. (1997), Earthquakes Cannot be Predicted, Science, 275, 1616–1617.

GRASSO, J. R. and SORNETTE, D. (1998), Testing Self-organized Criticality by Induced Seismicity, J. Geophys.Res., 103, 29,965–29987.

GUTENBERG, B. and RICHTER, C. F. (1994), Frequency of earthquakes in California, Bull. Seismol. Soc. Am. 34, 185–188.

HEIMPEL, M. (1997), Critical Behavior and the Evolution of Fault Strength during Earthquake Cycles, Nature, *388*, 865–868.

HUANG, SALEUR, Y., H. SAMMIS, C. G. and SORNETTE, D. (1998), *Precursors, Aftershocks, Criticality and self-organized criticality*, Europhys. Lett., *41*, 43–48.

JACKSON, D. D. (1996), *The Case for Huge Earthquakes*, Seismol. Res. Lett. *67*, 3–5.

JAUMÉ, S. C. and SYKES, L. R. (1999) *Evolving towards a Critical Point: A Review of Accelerating Moment/ energy Release prior to Large and Great Earthquakes*, Pure Appl. Geophys. *155*, 279–306.

KAGAN, Y. Y. (1994), *Observational Evidence for Earthquakes as a Nonlinear Dynamic Process*, Physica D *77*, 160–192.

KING, G. (1983), *The Accommodation of Large Strains in the Upper Lithosphere of the Earth and other Solids by Self-similar Fault Systems: The Geometrical Origin of b-value*, Pure Appl. Geophys. *121*, 761–814.

KING and BOWMAN (2003), *The evolution of regional seismicity between large earthquakes*, J. Geophys. Res. *108*, 2096, doi: 10.1029/2001JB000783.

LUDWIN, R. S., WEAVER, C. S., and CROSSON, R. S., *Seismicity of Washington and Oregon* In *Neotectonics of North America*, (D. B. Slemmons, E. R. Engdahl, M. D., eds.) 1991.

ZOBACK, and BLACKWELL, D. D. *Decade of North American Geology Map Volume 1*, pp. 77–98, Geological Society of America, Boulder, 1991.

MA, L., and LUDWIN. R. S. (1987), *Can Focal Mechanisms be Used to Separate Subduction Zone from Intraplate Earthquakes in Western Washington?*, *EOS, Trans. Am. Geophys. Un. 68*, 46.

MOGI, K. (1962), *Study of Elastic Shocks Caused by the Fracture of Heterogeneous Materials and their Relation to Earthquake Phenomena*, Bull. Earthquake Res. Inst. Univ. Tokyo, *40*, 125–73.

OLAMI, Z., FEDER, H. J. S., and CHRISTENSEN, K. (1992), *Self-organized Criticality in a Continuous, Nonconservative Cellular Automaton Modeling Earthquakes*, Phys. Rev. Lett. *68*, 1244–1247.

PAPAZACHOS, C. and PAPAZACHOS, B. (2001), *Precursory Accelerated Benioff Strain in the Aegean area*, Annali di Geofisica *44*, 461–474.

ROBINSON, R. (2000), *A Test of the Precursory Accelerating Moment Release Model on Some Recent New Zealand Earthquakes*, Geophys. J. Int. *140*, 568–576.

REASENBERG, P. A. and JONES, L. M. (1989), *Earthquake Hazard after a Mainshock in California*, Science *243*, 1173–1176.

SALEUR, H., SAMMIS, C. G., and SORNETTE, D. (1996) *Discrete Scale Invariance, Complex Fractal Dimensions, and Log-periodic Fluctuations in Seismicity*, J. Geophys. Res. *101*, 17,661–17,677.

SAMMIS, C. G., SORNETTE, D., and SALEUR, H. *Complexity and earthquake forecasting*. In *Reduction and Predictability of Natural Disasters, SFI Studies in the Sciences of Complexity*, vol. XXV, (J.B. Rundle, W. Klein, and D.L. Turcotte, eds.) , (Addison-Wesley, Reading, Mass., 1996) pp. 143–156.

SAMMIS, C. G. and SMITH, S. (1999), *Seismic Cycles and the Evolution of Stress Correlation in Cellular Automaton Models of Finite Fault Networks*, Pure Appl. Geophys. *155*, 307– 334.

SORNETTE D. and SAMMIS, C. G. (1995), *Complex Critical Exponents from Renormalization Group Theory of Earthquakes : Implications for Earthquake Predictions*, J.I. Phys 5, 607–619.

SORNETTE, A. and SORNETTE, D. (1989), *Self-organized Criticality and Earthquakes*, Europhys Lett. *9*, 197.

SYKES L. R. and JAUMÉ, S. (1990), *Seismic Activity on Neighboring Faults as a Long-term Precursor to Large Earthquakes in the San Francisco Bay Area*, Nature *348*, 595–599.

ZOBACK, M. D. and ZOBACK, M. L. *Tectonic stress field of North America and relative plate motions*. In *Neotectonics of North America* (D. B. Slemmons, E. R. Engdahl, M. D. Zoback, and D. D. Blackwell eds,) *Decade of North American Geology Map Volume 1*, (Geological Society of America, Boulder, 1991) pp. 339–366.

(Received September 27, 2002, revised April 25, 2003, accepted May 5, 2003)

To access this journal online:
http://www.birkhauser.ch

Pure appl. geophys. 161 (2004) 1957–1968
0033–4553/04/101957–12
DOI 10.1007/s00024-004-2542-1

© Birkhäuser Verlag, Basel, 2004

▌Pure and Applied Geophysics

Ergodicity in Natural Fault Systems

K. F. Tiampo[1,2], J. B. Rundle[3], W. Klein[4], and J. S. Sá Martins[5]

Abstract—Attempts to understand the physics of earthquakes over the past decade generally have focused on applying methods and theories developed based upon phase transitions, materials science, and percolation theory to a variety of numerical simulations of extended fault networks. This recent work suggests that fault systems can be interpreted as mean-field threshold systems in metastable equilibrium (Rundle *et al.*, 1995; Klein *et al.*, 1997; Ferguson *et al.*, 1999), and that these results strongly support the view that seismic activity is highly correlated across many space and time scales within large volumes of the earth's crust (Rundle *et al.*, 2000; Tiampo *et al.*, 2002). In these systems, the time averaged elastic energy of the system fluctuates around a constant value for some period of time and is punctuated by major events that reorder the system before it settles into another metastable energy well. One way to measure the stability of such a system is to check a quantity called the Thirumalai-Mountain (TM) energy metric (Thirumalai and Mountain, 1993; Klein *et al.*, 1996). In particular, using this metric, we show that the actual California fault system is ergodic in space and time for the period in question, punctuated by the occurrence of large earthquakes, and that, for individual events in the system, there are correlated regions that are a subset of the larger fault network.

Introduction

Recent work in the study of nonequilibrium systems, using computer simulations of simplified natural systems, suggests that certain equilibrium-like properties may be recovered at the appropriate spatio-temporal scales. In particular, numerical simulations of driven, mean-field systems suggest that far-from-equilibrium dissipative systems may be in a state of metastable equilibrium that can be characterized using equilibrium statistical mechanics (Rundle *et al.*, 1995; Klein *et al.*, 1996; Ferguson *et al.*, 1999; Egolf, 2000; Tiampo *et al.*, 2003). To date, however, there has been no definitive evidence that the same principles can be applied to the natural systems these

[1] CIRES, University of Colorado, Boulder, CO 80309, USA.
[2] Department of Earth Sciences, University of Western Ontario, London, ONT, Canada.
E-mail: ktiampo@uwo.ca.
[3] Center for Computational Science and Engineering, University of California, Davis, CA 95616,USA. E-mail: rundle@physics.ucdavis.edu
[4] Department of Physics, Boston University, Boston, MA 02215, USA and Center for Nonlinear Science, Los Alamos National Laboratory, Los Alamos, NM 87545, USA. E-mail: klein@buphy.edu
[5] Instituto de Fisica, Universidade Federal Fluminense, Av. Litoranea s/n, Boa Viagem, Niteroi 24210-340, RJ, Brazil. E-mail: jssm@if.uff.br

simulations are expected to reproduce. Here we show that the natural earthquake fault system displays ergodicity in its activity in order to demonstrate that it, like the slider block simulations models, is a mean-field system in a state of metastable equilibrium. In addition, we apply this concept to varying region sizes in order to determine the correlated region for several large historic events in southern California.

Background

Threshold systems are known to be some of the most important nonlinear, self-organizing systems in nature, including networks of earthquake faults, neural networks, superconductors and semiconductors, and the World Wide Web, as well as many political, social, and ecological systems. All of these systems have dynamics that are strongly correlated in space and time, and all typically display a multiplicity of spatial and temporal scales. In particular, if the range of interactions between elements of the system is long and weak, so that the dynamics can be understood as mean-field, fluctuations tend to be suppressed and the system may approach a stationary state (KAC *et al.*, 1961; GASPARD *et al.*, 1998). In addition, Boltzmann fluctuations, which are an important property of equilibrium systems, were directly observed in driven mean-field slider block simulations (RUNDLE *et al.*, 1995; MOREIN *et al.*, 1997; KLEIN *et al.*, 2000; MAIN *et al.*, 2000; RUNDLE *et al.*, 2002). Recently, FERGUSON (1999) demonstrated that a fluctuation metric (THIRUMALAI et al., 1989; THIRUMALAI and MOUNTAIN, 1993) originally developed to test for the presence of ergodic behavior in equilibrium systems showed that driven mean-field slider block models could also be considered to demonstrate ergodic behavior over finite intervals of time. Finally, direct observations of Gaussian fluctuations and detailed balance in transition probabilities suggest ergodic behavior in a driven system of mean-field coupled map lattices (EGOLF, 2000).

The critical question is whether these conclusions from model driven systems can be extended to natural driven systems. In particular, we might consider as candidates the class of driven threshold systems, of which the slider block models are one case. Earthquake fault systems are one example of such a complex nonlinear driven threshold system. As in other threshold systems, interactions among a spatial network of fault segments are mediated by means of a potential that, in this case, allows stresses to be redistributed to other segments following slip on any particular segment. For faults embedded in a linear elastic host, this potential is a stress Green's function whose exact form is calculated from the equations of linear elasticity, once the geometry of the fault system is specified. A persistent driving force, arising from plate tectonic motions, increases stress on the fault segments (RUNDLE et al., 1995; MOREIN *et al.* 1997; KLEIN *et al.*, 2000).

Once the stresses reach a threshold characterizing the limit of stability of the fault, a sudden slip event occurs. The slipping segment can also trigger slip at other locations on the fault surface whose stress levels are near the failure threshold as the event begins. In this manner, earthquakes result from the interactions and nonlinear nature of the stress thresholds (HUANG et al., 1998; RUNDLE et al., 1999; KLEIN et al., 2000).

Finally, in the mean-field regime, as the interaction length becomes large, leading to a damping of the fluctuations, a mean-field spinodal appears that is the classical limit of stability of a spatially extended system (LIEBOWITZ, 1979; GOPAL and DURIAN, 1995). Examined in this limit, driven threshold systems appear to be locally ergodic, and display equilibrium behavior when driven at a uniform rate. Following the initial discovery that driven mean-field slider block systems with microscopic noise display equilibrium properties (RUNDLE et al., 1995), other studies have confirmed local ergodicity, the existence of Boltzmann fluctuations in both these and other mean- or near mean-field systems, and the appearance of an energy landscape, similar to other equilibrium systems (KLEIN et al., 1996; FERGUSON et al., 1999; MAIN et al., 2000; RUNDLE et al., 2002). Thus the origin of the physics of scaling, critical phenomena and nucleation appears to lie, at least in part, in the ergodic properties of these mean-field systems.

Note that the spatial and temporal firing patterns of such driven threshold systems are complex and often difficult to understand and interpret from a deterministic perspective, as these patterns are emergent processes that develop from the obscure underlying structures, parameters, and dynamics of a multidimensional nonlinear system (NIJHOUT, 1997). For example, in the earth there is no means at present to measure the stress and strain at every point in an earthquake fault system, or the constitutive parameters that characterize this heterogeneous medium and its dynamics.

However, the seismicity, the firing patterns that are the surface expression or proxy for the dynamical state of the underlying fault system, can be located in both space and time with considerable accuracy (BAKUN and MCEVILLY, 1984; SIEH et al., 1989; HILL et al., 1990). If this natural system is also, as simulations suggest, a mean-field threshold system in metastable equilibrium (KLEIN et al., 1997), then the time averaged elastic energy of the system fluctuates around a constant value for some period of time. These periods are punctuated by major events that reorder the system before it settles into another metastable energy well. Here we wish to compare the ergodic properties of the simulated and natural systems in terms of their energy release, an easily measured quantity. We employ a quantity called the Thirumalai-Mountain (TM) fluctuation metric in order to make this comparison (THIRUMALAI et al., 1989; THIRUMALAI and MOUNTAIN, 1993; KLEIN et al., 1996; FERGUSON et al., 1999).

Method

The TM metric measures effective ergodicity, the difference between the time average of the quantity E_j at each site, and the ensemble average of that average over the entire system, because the derivation of the fundamental equations requires the equivalence of the time and ensemble averages of the *E*-fluctuating quantities. The fundamental idea is that of statistical symmetry, in which the *N* oscillators, particles, cells, or spins in the system are statistically identical, in terms of its averaged properties (THIRUMALAI *et al.*, 1989; THIRUMALAI and MOUNTAIN, 1993). While, in general, a system is ergodic for infinite averaging times, if the actual measurement time scales are finite, but long, all regions of phase space are sampled with equal likelihood, and the system is effectively ergodic (THIRUMALAI *et al.*, 1989). Statistically, ergodicity means that, over a large enough representative sample in time and space, the spatial and temporal averages are constant. Ergodicity is a behavior that is limited to equilibrium states, and implies stationarity as well. Note that if such a system is ergodic, it is also in metastable equilibrium and can be analyzed as such.

Again, a system is said to be effectively ergodic if, for a given time interval, the system has equivalent time-averaged and ensemble-averaged properties (THIRUMALAI *et al.*, 1989; THIRUMALAI and MOUNTAIN, 1993). The fluctuation metric $\Omega_e(t)$, proposed by Thirumalai and Mountain is

$$\Omega_e(t) = \frac{1}{N} \sum_{i=1}^{N} [\varepsilon_i(t) - \bar{\varepsilon}(t)]^2, \tag{1}$$

$$\varepsilon_i(t) = \frac{1}{t} \int_0^t E_i(t') dt' \tag{2}$$

where is the time average of a particular individual property, $E_i(t)$, and

$$\bar{\varepsilon}(t) = \frac{1}{N} \sum_i^N \varepsilon_i(t) \tag{3}$$

the ensemble average over the entire system. If the system is effectively ergodic at long times, $\Omega_e(t) = D/t$, where *D* is a constant that measures the rate of ergodic convergence (THIRUMALAI *et al.*, 1989; THIRUMALAI and MOUNTAIN, 1993). A more physical way of thinking about the fluctuation metric is to consider that the oscillator has been mapped into a Brownian particle, and equation (1) is therefore an expression of the equivalence between the time averages of particle behavior, and the ensemble averages over particle states after a series of Brownian increments.

In slider block models used to replicate the behavior of earthquake fault networks, as the interaction range increases, the system approaches mean-field limit behavior. If, as a result, these slider block models are in metastable equilibrium, they can be analyzed using the methods and principles of statistical mechanics. FERGUSON (1999) applied the TM metric to slider block numerical simulations in order to show that the system was ergodic at external velocities, V, that approach $V = 0$, and retrieved a linear relationship between the inverse TM metric for slider block energy and time, denoting effective ergodicity as defined above (KLEIN *et al.*, 1996; FERGUSON *et al.*, 1999).

Figure 1 shows a plot of the inverse TM metric not for energy, but for numbers of events in a similar slider block model, with a 128 × 128 member lattice, a loader plate velocity approaching zero, and the addition of a small amount of precursory slip (RUNDLE *et al.*, 2002). In this calculation of the TM metric, $E_i(t) \equiv R_i(t)$, the number of events greater than a certain magnitude. Here we use a magnitude cutoff that corresponds to magnitude 3.0, as in the actual data below. Note the linear relationship between the inverse TM metric and time, denoting effective ergodicity as defined above. There is also an initial transient phase in addition to the linear sections, punctuated by the occurrence of larger events.

The question that remains to be answered is whether the same applies to natural earthquake fault networks. Interactions in a natural driven system should be mean-field if the results on ergodicity are to hold. Since slider block models were originally conceived as models of earthquake faults, it is therefore logical to investigate the presence or absence of ergodic behavior in systems of earthquake faults. We proceed to test this hypothesis using the TM fluctuation metric.

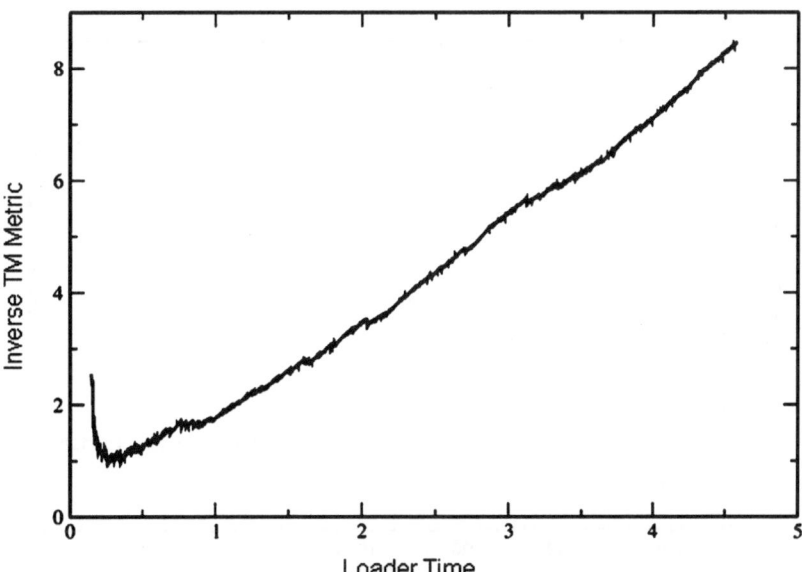

Figure 1

The inverse TM metric for seismicity numbers plotted versus loader time, in a slider block model with precursory slip.

Here we apply the TM metric to the surface expression of the underlying energy landscape, the seismicity in a regional fault system, southern California. For application to the earthquake fault system, the number of earthquakes of a particular magnitude or greater can be expressed as a function of the seismic energy release (KANAMORI, 1977; TURCOTTE, 1997). Therefore, in this calculation of the TM metric, $E_i(t) \equiv R_i(t)$, the number of events greater than magnitude three in central and southern California in each year.

The seismicity data employed in our analysis are taken from existing observations in California between the years 1932 and the present. Using only the subset of this data at locations **x** in southern California and covering the period from January 1, 1932 through December 31, 2001, we compute the TM metric for southern California seismicity over the region 32° to 40° latitude, −115° to −125° longitude. Note that we use only events with magnitude $M \geq 3$ to ensure completeness of the catalog, and the catalog is in no way declustered.

Results

In Figure 2, we plot the inverse TM fluctuation metric for the variance in the number of events in southern California over time, from 1932 through 2001. Note, again, the linear relationship between the inverse TM metric and time. The natural fault system appears to be effectively ergodic for relatively long periods of time,

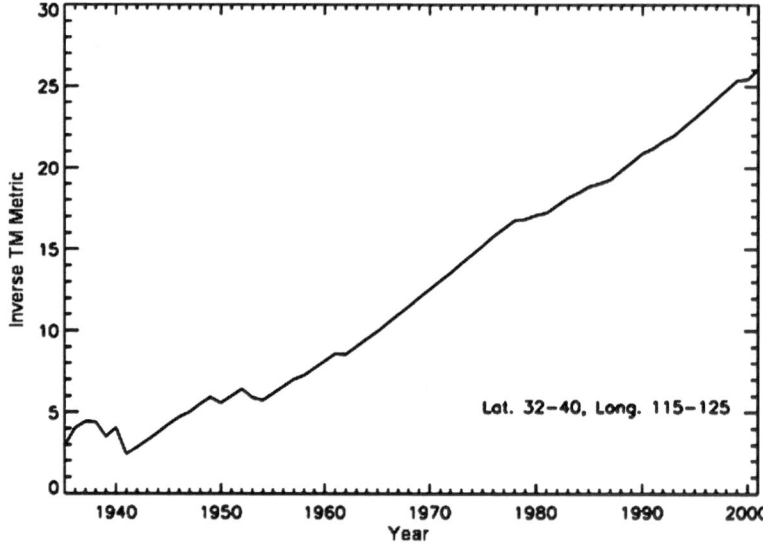

Figure 2
The inverse TM metric for seismicity numbers plotted versus time, for southern California. Note the periods of ergodic behavior, punctuated by the occurrence of large events.

punctuated by the occurrence of large earthquakes, such as the Kern County event of 1952, the Imperial Valley earthquake in 1979, the Landers sequence of 1991, and the Hector Mine earthquake in 1999 (see Fig. 3). We interpret these events in terms of a physical picture where the fault system resides for long periods in an ergodic, local energy minimum on a complex energy landscape. Occasionally, however, the nonlinear dynamics leads to an earthquake, which has the effect of causing the system to suddenly depart from its local energy minimum, and to migrate to a new local minimum, where it again resides in an effectively ergodic state.

The question over what region sizes are applicable for this ergodic state, whether this is constant for the entire region, and if they are related to either large events or fault structure. We decided to study three different events in the central and southern California region. The Coalinga earthquake of 1983, a 6.4 M_W thrust event, the Landers sequence of 1992, a magnitude 7.3 M_W right-lateral strike- slip event in the Mojave desert, and the 1994 Northridge earthquake, a 6.7 M_W blind thrust event in the Los Angeles basin (see Fig. 3).

Figure 4 shows the same Thirumalai-Mountain metric calculation as above for three different region sizes; each centered on the epicenter of the Coalinga earthquake. The region sizes are approximately square. The first, shown at the top, is one-half of a degree in latitude and longitude to a side. The second, shown in the middle, is one degree in latitude and longitude to a side, and the third, on the bottom, is two degrees in latitude and longitude to a side. The slope of the line does not become linear until the region size is 2 degrees in latitude and longitude to a side. Note that the earthquake itself has very little effect on the slope of the line.

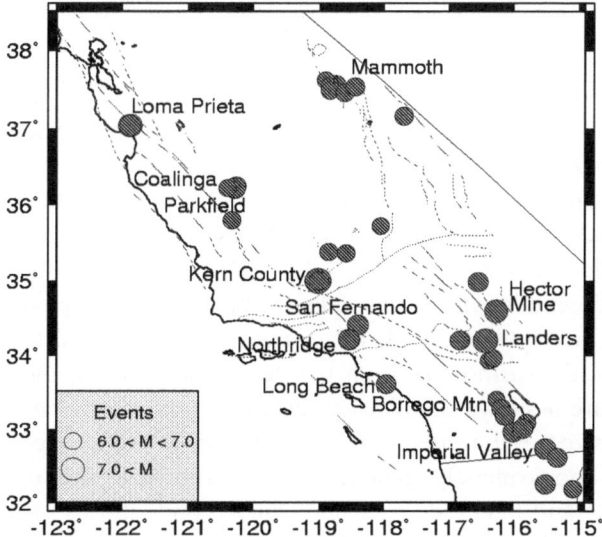

Figure 3
Location of historic earthquakes in southern California.

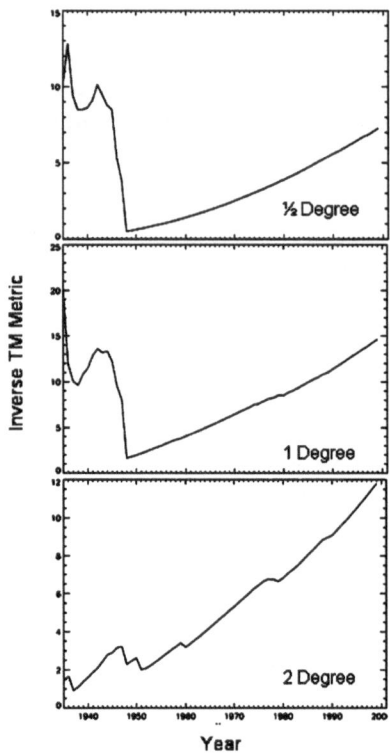

Figure 4
The inverse Thirumalai-Mountain metric for three increasing region sizes, each centered on the Coalinga epicenter. The top is for a box of ½ a latitude/longitude degree on a side, the middle is for a box 1 degree to a side, and the bottom is for 2 degrees to a side.

Figure 5 also shows the Thirumalai-Mountain metric calculation for three different region sizes, each centered on the epicenter of the Northridge earthquake. The region sizes are approximately square. The first, shown at the top, is one-half of a degree in latitude and longitude to a side. The second, shown in the middle, is one degree in latitude and longitude to a side, and the third, on the bottom, is 1-1/2 degrees in latitude and longitude to a side. In this case, the slope of the line becomes linear when the box size reaches 1–1/2 degrees in latitude and longitude to a side. Note that, again, the earthquake itself has very little effect on the slope of the line.

Finally, Figure 6 shows the Thirumalai-Mountain metric calculation for three different region sizes, each centered on the epicenter of the Landers earthquake. The region sizes are approximately square. The first, shown at the top, is one-half of a degree in latitude and longitude to a side. The second, shown in the middle, is one degree in latitude and longitude to a side, and the third, on the bottom, is two degrees in latitude and longitude to a side. The slope of the line again does not become linear

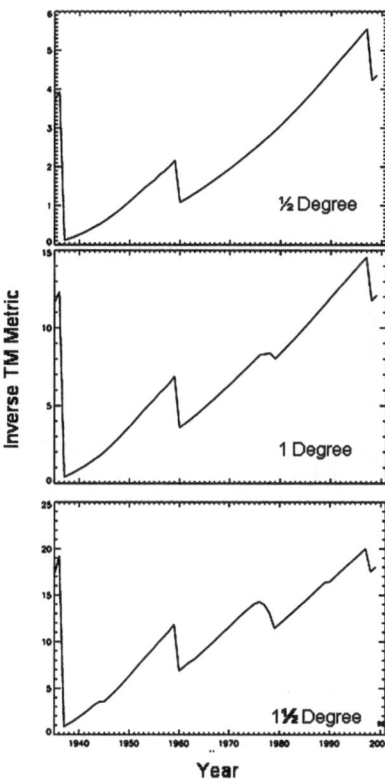

Figure 5

The inverse Thirumalai-Mountain metric for three increasing region sizes, each centered on the Northridge epicenter. The top is for a box of ½ a latitude/longitude degree on a side, the middle is for a box 1 degree to a side, and the bottom is for a box 1-1/2 degrees to a side.

until the region size is 2 degrees in latitude and longitude to a side. Note that the earthquake itself has very little effect on the slope of the line.

There are several interesting points that should be mentioned here. The first is that this region size does not appear to be related to the size of the event $\frac{1}{M}$ both the largest and the smallest event do not become ergodic until they reach two degrees to a side. This suggests that there is some other factor that is entering into the growth of correlated regions, perhaps the underlying fault structure itself. The second point it that there appears to be certain events that disrupt the ergodic structure for every region, such as the 1952 Kern County earthquake and the 1979 Imperial Valley event. On the other hand, each region appears to have other earthquakes that primarily influence only that area, such as the 1983 Coalinga earthquake for the Landers earthquake shown in Figure 6. Finally, it is evidence that there are smaller, correlated regions in the fault network where the system is effectively ergodic, but with a different temporal signature than the entire system.

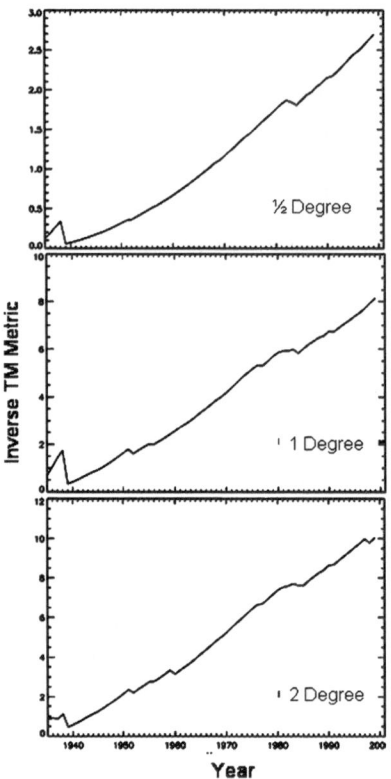

Figure 6

The inverse Thirumalai-Mountain metric for three increasing region sizes, each centered on the Landers epicenter. The top is for a box of ½ a latitude/longitude degree on a side, the middle is for a box 1 degree to a side, and the bottom is for a box 2 degrees to a side.

Conclusions

In summary, the observable properties of self-organizing, strongly correlated, near mean-field driven threshold systems arise from the underlying, locally ergodic dynamics. We employ the well-known Thirumalia-Mountain fluctuation metric to identify the presence of one equilibrium property in the dynamics of the natural earthquake fault system in California, ergodicity, using data from existing seismic monitoring networks, as numerical simulations have demonstrated that the dynamics of driven mean-field systems of interacting slider blocks and coupled map lattices display strong evidence of ergodic behavior. Our results suggest that this natural system is effectively ergodic, in metastable wells for periods of time on the order of several decades, and, therefore, mean-field as in the numerical simulations that are used to study these systems, displaying critical point behavior with correlations over a range of spatial and temporal scales. While the identification of ergodicity in this system is based upon the study of a particular phase space, it has been shown that the number of events is a proxy for both seismic energy release and stress release, related by empirical constants that are a function of the regional fault system (KANAMORI, 1977; TURCOTTE, 1997). As a result, this particular phase space description should be fundamental, with universal properties that do not depend on particular regional characteristics.

This also suggests that many of the observed properties of the natural system, such as Gutenberg-Richter scaling, large correlation lengths, and the classification of earthquakes as nonclassical nucleation events, can be understood as manifestations of an effectively ergodic, nonlinear system. As the dynamical system evolves, migration to a new energy minimum occurs in association with the occurrence of a large earthquake. In addition, it appears that large, correlated regions where the fault system is effectively ergodic can be identified around the location of several of the major events in southern California. Finally, this supports the previous work using numerical simulations and validates their use in the study of natural, driven threshold systems in general and earthquake fault systems in particular.

Acknowledgments

Research by KFT was funded by The Southern California Earquake Center. SCEC is funded by NSF Agreement EAR - 8920136 and USGS Cooperative Agreements 14-18-0001-A0899 and 1434-HQ-97AG01718. The SCES Contribution number for this paper is 703. Research by JBR was funded by USDOE/OBES grant DE-FG03-95ER14499 (theory), and by NASA grant NAG5-5168 (simulations). This research was supported by the Southern California Earthquake Center. SCEC is funded by NSF Cooperative Agreement EAR-8920136 and USGS Cooperative Agreements 14-08-0001-A0899 and 1434-HQ97AG01718. The SCEC contribution number for this paper is 703.Research by WK was supported by USDOE/OBES grant DE-FG02-95ER14498 and W-7405-ENG-6 at LANL. WK would also like to acknowledge the hospitality and support of CNLS at LANL.

REFERENCES

BAKUN, W. H. and MCEVILLY, T. V. (1984), *Recurrence Models and Parkfield, California, Earthquakes*, J. Geophys. Res. *89*, 3051–3058.

EGOLF, D. (2000), *From Nonequilibrium Chaos to Statistical Mechanics*, Science *287*, 101–104.

FERGUSON, C. D., KLEIN, W., and RUNDLE, J. B. (1999), *Spinodals, Scaling, and Ergodicity in a Threshold Model with Long-range Stress Transfer*, Phys. Rev. E *60*, 1359–1373.

GASPARD, P., BRIGGS, M. E., FRANCIS, M. K. *et al.*, (1998), *Experimental Evidence for Microscopic Chaos*, Nature *394*, 865–868.

GOPAL, A. D. and DURIAN, D. J. (1995), *Nonlinear Bubble Dynamics in a Slowly Driven Foam*, Phys. Rev. Lett. *75*, 2610–2613.

HILL, D., EATON, J. P., and JONES, L. M., *Seismicity, 1980–86*. In *The San Andreas Fault System, USGS Prof. Paper 1515* (U. S. GPO, 1990) pp. 115–152.

HOPFIELD, J. J. (1982), *Neural Networks and Physical Systems with Emergent Collective Computational Abilities*, Proc. Nat. Acad. Sci. U.S.A. *79*, 2554–2558.

HUANG, Y., SALEUR, H., SAMMIS, C., and SORNETTE, D. (1998), *Precursors, Aftershocks, Criticality and Self-organized Criticality*, Europhys. Lett. *41*, 43–49.

KAC, M., UHLENBECK, G. E., and HEMMER, P. C. (1963), *On the van der Waals Theory of the Vapor-liquid Equilibrium. I. Discussion of a One-dimensional Model*, J. Math. Phys. *4*, 216–228.

KANAMORI, H. (1977), *Energy-release in Great Earthquakes*, J. Geophys. Res. *82*, 2981–2987.

KLEIN, W., ANGHEL, M., FERGUSON, C. D., RUNDLE, J. B., and SÁ MARTINS, J. S., *Statistical analysis of a model for earthquake faults with long-range stress transfer. In Geocomplexity and the Physics of Earthquakes*, Geophysical Monograph Ser. *120* (J. B. Rundle, D. L. Turcotte, W. Klein, eds.) (AGU, 2000) pp. 43–71.

KLEIN, W., FERGUSON, C., and RUNDLE, J. B., Spinodals *and scaling in slider-block models. In Reduction and Predictability of Natural Disasters, SFI Series in the Science of Complexity, XXV* (J. B. Rundle, D. L. Turcotte, W. Klein, eds.) (Addison-Wesley, Reading, MA, 1996), 223-242.

KLEIN, W., RUNDLE, J. B., and FERGUSON, C. D. (1997), *Scaling and Nucleation in Models of Earthquake Faults*, Phys. Rev. Lett. *78*, 3793–3796.

LEIBOWITZ, J. L., *Towards a Rigorous Molecular Theory of Metastability. In Studies in Statistical Mechanics: Fluctuation Phenomena* (Penrose, O., ed.) (North-Holland 1979), pp 295–340.

MAIN, I. G., O'BRIEN, G., and HENDERSON, J. R. (2000), *Statistical Mechanics of Earthquake: Comparison of Distribution Exponents for Source Area and Potential Energy and the Dynamic Emergence of Log-periodic Energy Quanta*, J. Geophys. Res. *105*, 6105–6126.

MOREIN, G., TURCOTTE, D. L., GABRIELOV, A. (1997), *On the Statistical Mechanics of Distributed Seismicity*, Geophys. J. Int. *131*, 552–558.

NIJHOUT, H. F., *Pattern formation and biological systems. In Pattern Formation in the Physical and Biological Sciences, Lecture Notes V, SFI* (Addison Wesley, 1997) pp 269-298.

RUNDLE, J. B., KLEIN, W., and GROSS, S. (1999), *Physical Basis for Statistical Patterns in Complex Earthquake Populations: Models, Predictions and Tests*, Pure Appl. Geophs. *155*, 575–607.

RUNDLE, J. B., KLEIN, W., GROSS, S., and TURCOTTE, D. L. (1995), *Boltzmann Fluctuations in Numerical Simulations of Nonequilibrium Lattice Threshold Systems*, Phys. Rev. Lett. *75*, 1658–1661.

RUNDLE, J. B., KLEIN, W., TIAMPO, K. F., and SÁ MARTINS, J. S. (2002), Self-organization in Leaky Threshold Systems: The Influence of Near-mean Field Dynamics and its Implications for Earthquakes, Neurobiology, and Forecasting, Proc. Nat. Acad. Sci. U.S.A., Suppl. 1, *99*, 2514–2521.

SIEH, K., STUIVER, M., and BRILLINGER, D. (1989), *A More Precise Chronology of Earthquakes Produced by the San Andreas Fault in Southern California*, J. Geophys. Res. *94*, 603–623.

THIRUMALAI, D. and MOUNTAIN, R. D. (1993), *Activated Dynamics, Loss of Ergodicity, and Transport in Supercooled Liquids*, Phys. Rev. E. *47*, 479–489.

THIRUMALAI, D., MOUNTAIN, R. D., and KIRKPATRICK, T. R. (1989), *Ergodic Behavior in Supercooled Liquids and in Glasses*, Phys. Rev. A *39*, 3563–3574.

TIAMPO, K. F., RUNDLE, J. B., GROSS, S. J., MCGINNIS, S., and KLEIN, W. (2002), J. Geophys. Res. *107*, 2354, doi:10.1029/2001JB000562.

TIAMPO, K. F., RUNDLE, J. B., KLEIN W., SÁ MARTINS, J. S., and FERGUSON, C. D., Phys. Rev. Lett. *91*, 238501.

TURCOTTE, D. L., *Fractals and Chaos in Geology and Geophysics*, 2nd ed. (Cambridge University Press, 1997).

(Received September 27, 2002, revised February 28, 2003, accepted March 7, 2003)

 To access this journal online:
http://www.birkhauser.ch

Pure appl. geophys. 161 (2004) 1969–1978
0033–4553/04/101969–10
DOI 10.1007/s00024-004-2543-0

© Birkhäuser Verlag, Basel, 2004

▌Pure and Applied Geophysics

Focal Mechanism Dependence of a Few Seismic Phenomena and its Implications for the Physics of Earthquakes

Z. L. Wu[1,2], Y. G. Wan[2], and G. W. Zhou[2]

Abstract—We studied the scaling of reduced energy versus seismic moment, the Coulomb stress triggering, and the property of clustering for global shallow earthquakes with different focal mechanisms. Clear focal mechanism dependence can be observed for these three phenomena. For strike-slip earthquakes, reduced energy increases slightly, nearly constant with the increase of seismic moment, while for thrust and normal earthquakes, reduced energy decreases with seismic moment. The level of reduced energy for thrust and normal earthquakes is several times smaller than that for strike-slip earthquakes. Thrust earthquakes exhibit clear stress triggering effect, strike-slip earthquakes possess almost no such a property, while normal faulting type earthquakes lie in between. Thrust earthquakes are more likely to occur in 'pairs' or 'groups', while strike-slip earthquakes are more likely to be 'single'. These three phenomena are related to each other in the physics of earthquakes.

Key words: Reduced energy, apparent stress, Coulomb stress triggering, earthquake clustering, focal mechanism.

Introduction

Numerous earthquake phenomena have been recognized as focal mechanism dependent, such as frequency-magnitude relation (FRÖHLICH and DAVIS, 1993), tidal triggering of earthquakes (TSURUOKA *et al.*, 1995; TANAKA *et al.*, 2002), level of apparent stress (CHOY and BOATWRIGHT, 1995; PEREZ-CAMPOS and BEROZA, 2001), scaling of earthquake parameters (STOCK and SMITH, 2000), and rate of foreshocks (REASENBERG, 1999), reflecting the physics of earthquake preparation and occurrence. Here we discuss the focal mechanism dependence of some of the newly observed earthquake phenomena. These phenomena are observed using 'modern' earthquake catalogues, which are produced on the basis of the analysis and inversion of digital/broadband seismograms and provide more source parameters in describing the physical properties of earthquakes. Some of such phenomena could also be observed using traditional earthquake catalogues, but using 'modern' earthquake

[1] Center for Earth System Science, Graduate School, Chinese Academy of Sciences, 100039 Beijing, China.
[2] Institute of Geophysics, China Earthquake Administration, 100081 Beijing, China.

catalogues, these phenomena can be studied more systematically, more straightforward, and in more detail.

We discuss the focal mechanism dependence of three earthquake phenomena: The scaling of reduced energy or apparent stress, the static stress triggering, and the probability for strong aftershocks or paired/grouped shocks. In the study the Harvard CMT (DZIEWONSKI *et al.*, 1981) catalogue and the NEIC broadband radiated energy (BOATWRIGHT and CHOY, 1986) catalogue are used. The study focuses on global shallow earthquakes. Definition of focal mechanism types follows FRÖHLICH (1992). It is interesting that the focal mechanism dependences of these three phenomena seem to be correlated to each other, and have clear significance in the physics of earthquakes.

1. Scaling of Reduced Energy

Reduced energy is the ratio of radiated energy by seismic moment (KANAMORI and HEATON, 2000). In seismology, reduced energy times the rigidity of the source medium is called 'apparent stress' (WYSS and BRUNE, 1968). By considering the efficiency of seismic wave radiation, apparent stress can be related to the average stress level causing the earthquake slip, a useful working assumption having been tested by several experiments and field observations (e.g., MCGARR, 1999). Comparison of apparent stress can obtain the relative difference in the ambient stress level (CHOY and BOATWRIGHT, 1995).

Scaling of reduced energy or apparent stress versus seismic moment is a controversial issue in seismology. While some evidence shows that apparent stress keeps constant with the changing of seismic moment (IDE and BEROZA, 2001), other studies show that there is an increase of apparent stress with seismic moment (IZUTANI and KANAMORI, 2001). In solving this problem, it seems interesting to investigate whether or not the scaling of reduced energy or apparent stress has focal mechanism dependence.

We use the radiated energy estimation of NEIC and the seismic moment from the Harvard CMT catalogue to study the scaling of reduced energy for different focal mechanisms. We use the data from January 1987 to December 2000, which are more extensive than the data set used before (CHOY and BOATWRIGHT, 1995; Wu, 2001). In the previous work detailed analysis was provided for the properties of the data set and the uncertainties of the result. Figure 1 shows, for thrust, strike-slip, and normal focal mechanisms, respectively, the reduced energy versus seismic moment for global shallow earthquakes. Two characteristics can be observed in Figure 1. One is that the level of reduced energy, and thus the level of ambient stress is different for different types of focal mechanisms: Similar to previous results (CHOY and BOATWRIGHT, 1995; PEREZ-CAMPOS and BEROZA, 2001), strike-slip earthquakes have a higher stress level, about a few times that for thrust and normal events. Another characteristic is that the scaling of reduced energy versus seismic moment is different for different

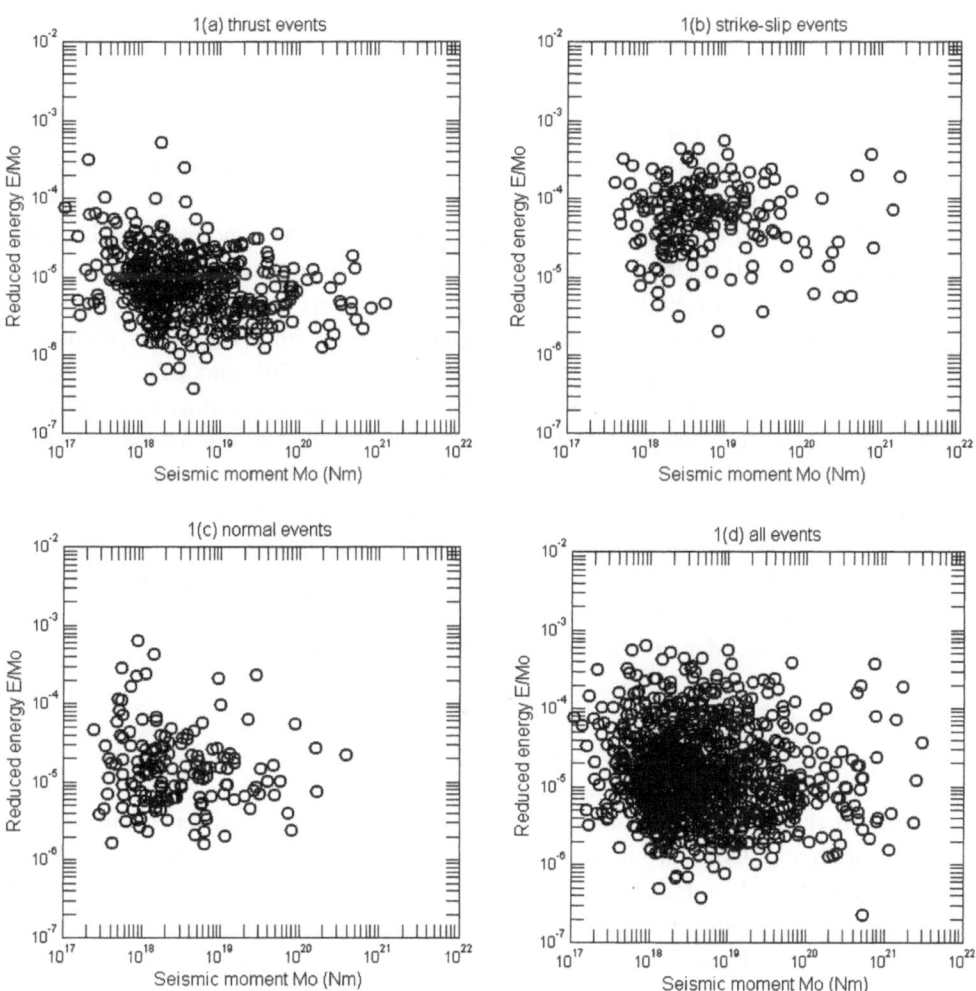

Figure 1(a)

Reduced energy versus seismic moment for global shallow earthquakes from 1987 to 2000: Thrust events. Reduced energy is dimensionless because the dimensions of seismic moment and radiated energy are all the same. If the average rigidity of source medium is taken for each earthquake, then this figure also shows the scaling of apparent stress.

Figure 1(b)

Reduced energy versus seismic moment for global shallow earthquakes from 1987 to 2000: Strike-slip events.

Figure 1(c)

Reduced energy versus seismic moment for global shallow earthquakes from 1987 to 2000: Normal events.

Figure 1(d)

Reduced energy versus seismic moment for global shallow earthquakes from 1987 to 2000: All events. Note that in this figure, not only thrust, strike-slip, and normal events, but also the so-called 'odd' events which belong to none of the former three types (FRÖHLICH, 1992), are included.

focal mechanisms: For thrust and normal earthquakes, reduced energy decreases with seismic moment, while for strike-slip earthquakes, reduced energy is almost constant, with a weak trend of increase. This result implies that, as to the controversy on the scaling of reduced energy or apparent stress, both sides may be correct. When considering strike-slip earthquakes only, a weak trend of increase of reduced energy with seismic moment can be seen; when considering small regions such as America this trend is more evident (WU, 2001), which is similar to the conclusion of IZUTANI and KANAMORI (2001). On the other hand, when earthquakes with different focal mechanisms are mixed together, apparently reduced energy will remain unchanged with seismic moment, with a relatively longer range of uncertainty, as shown in Figure 1(d), which is corresponding with the argument of IDE and BEROZA (2001). Our result shows that for thrust and normal type earthquakes, reduced energy may decrease with seismic moment, which has not been discussed in the previous results.

It should be noted that in the regime of elasticity, in principle it is impossible to measure the absolute stress level. The estimate of ambient stress level is generally through a set of theoretical/empirical assumptions such as the evaluation of stress level based on apparent stress through the efficiency of seismic wave radiation. In this sense the comparison of 'stress level', as well as the discussions based on such a comparison, is model dependent by its nature.

2. Static Stress Triggering

When an earthquake (referred to as a 'source earthquake' hereafter) occurs, it can cause the change of Coulomb failure stress (CFS) on the seismogenic fault of another earthquake (referred to as a 'target earthquake' hereafter) and thus can change the probability of occurrence of the 'target earthquake' (e.g., HARRIS, 1998). If ΔCFS is larger than a threshold value (currently adopted as 0.01 MPa), then the 'target earthquake' is said to be 'encouraged', or 'triggered' by the 'source earthquake'. In contrast, if ΔCFS is less than a threshold value (currently adopted as −0.01 MPa), then the 'target earthquake' is said to be 'discouraged', or in the 'stress shadow' produced by the 'source earthquake'. One of the controversial issues is whether or not such a small change of CFS (typically of the order of 10^{-2} MPa) is sufficient to change the preparation process of the 'target earthquake'.

To study this problem we use the Harvard CMT catalogue of global shallow earthquakes from 1977 to 1998. In such a catalogue there are many 'earthquake pairs', which are a group of two successive earthquakes occurring near each other in space and time, with similar focal mechanisms. Utilization of this concept is hinted by the study of 'earthquake doublets' (KAGAN and JACKSON, 1999). That study proposed that for large earthquakes worldwide there are 'earthquake doublets', which are two successive earthquakes in proximity to each other with the temporal interval between them considerably less than their recurrence period as defined by the geodynamical loading rate. The twin earthquakes in an 'earthquake doublet' are tectonically connected to each other and have similar focal mechanisms. In this study we expand this concept to intermediate-size earthquakes. The 'earthquake pairs' studied here include both the originally proposed 'earthquake doublets' and other clustered earthquakes such as those in an aftershock sequence.

We address the problem of whether or not the occurrence of the first earthquake in the 'earthquake pair' (that is, the 'source earthquake') 'encourages' the occurrence of the second one (that is, the 'target earthquake') by changing the Coulomb failure stress (CFS). In the calculations the nodal plane which is nearer the vector linking the two centroids of the two earthquakes is chosen as the assumed earthquake faulting plane, a concept similar to that proposed by KAGAN and JACKSON (1999) in studying earthquake 'doublets'. We eliminate the vague cases in which it is hard to determine which nodal plane is nearer the vector linking the two centroids of the two earthquakes. We also removed the 'earthquake pairs' in which the centroid depths are not well constrained (i.e., 'fixed' to 15 km or 33 km, and so on). This geometrical treatment is only valid in a statistical or overall sense, which is a limitation of this assumption. However, we consider just the overall effect; in this case such an assumption can be used in the comparison between different classes of earthquakes. As an analogue to this overall concept, there is a similar one in the physics of earthquakes when considering the physical significance of hypocenter and centroid (SCHOLZ, 1990): In a general sense the hypocenter of an earthquake can be understood as the starting point of the seismic rupture, the centroid can be understood as the centroid of the entire earthquake rupture area, and the vector from the hypocenter to the centroid can be considered as pointing to the direction of rupture propagation. This understanding is valid in a statistical or overall sense. For individual earthquakes, however, the situation is more complex.

We use the empirical relations among seismic moment, magnitude, rupture length, rupture width, and average slip as collected previously. This treatment is equivalent to that of a simple rectangular geometry of rupture area and a simple homogeneous slip distribution for each earthquake. Once again because what we consider is an overall effect, and because for all earthquakes the same treatment is taken equally, such a simplification, although it may affect the result of individual 'earthquake pairs', will not affect the overall statistics.

The CMT catalogue provides standard errors of the moment tensor inversion as well as the centroid location, and the empirical relationships also give the standard errors of the linear aggregation. These two errors can be transferred into the errors of the CFS result by simple error-transfer calculation, providing an estimate of the uncertainty of the CFS calculation. We eliminate the vague cases in which the absolute value of the CFS change is less than the level of uncertainty. We use the algorithm of OKADA (1992) in the calculation of the values of ΔCFS. Considering the complexity of the physics related to the effective friction coefficient, in the calculation, we take a range of this parameter from 0.2 to 0.8. It is ascertained

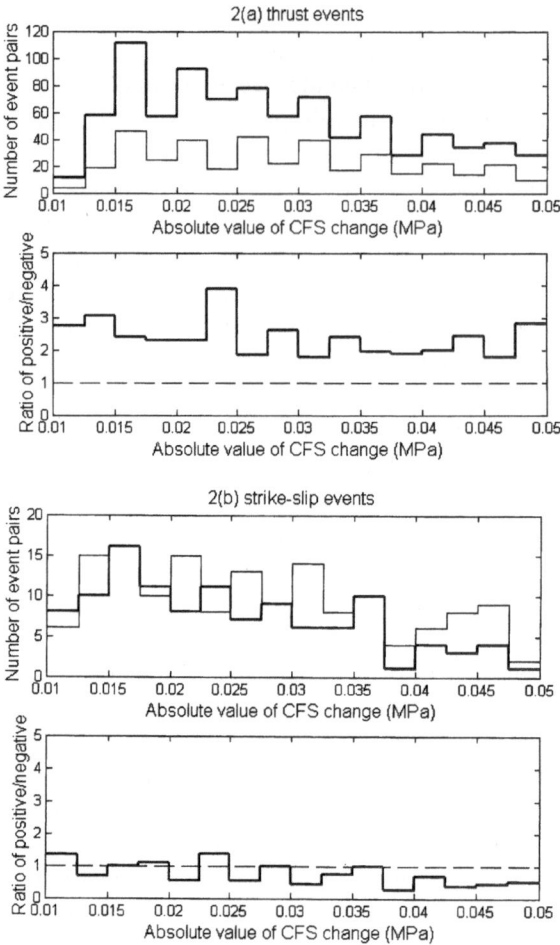

Figure 2(a)

Number of earthquake pairs versus CFS changes of the 'target earthquake' produced by the 'source earthquake': Thrust events. In the figure the thick line represents the result of positive ΔCFS values, while the thin line represents the result of negative ΔCFS values. The bottom figure is the ratio of the number of events with positive ΔCFS values over that of events with negative ΔCFS values, in which the dashed line shows the reference line representing that the ratio is unity. See text for details.

Figure 2(b)

Number of earthquake pairs versus CFS changes of the 'target earthquake' produced by the 'source earthquake': Strike-slip events.

that such a change does not overly affect the general tendency of change. The physics underlying this is that the focal mechanisms of the 'source earthquake' and the 'target earthquake' are similar to each other. In Figure 2 we show only the result for the effective friction coefficient being 0.4. The top figure indicates the numbers of 'earthquake pairs' with positive ΔCFS (thick line) and negative ΔCFS (thin line), respectively. In the bottom figure a solid line denotes the ratio of the number of

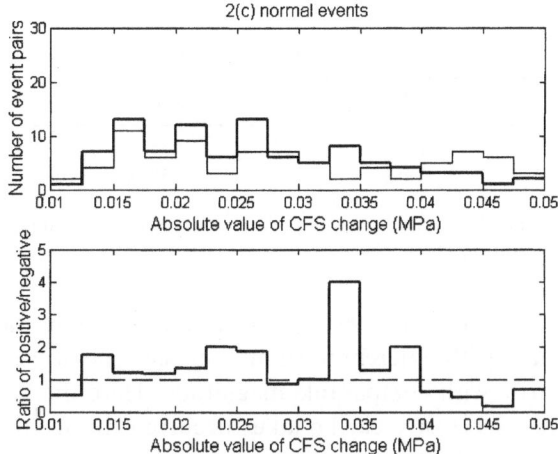

Figure 2(c)

Number of earthquake pairs versus CFS changes of the 'target earthquake' produced by the 'source earthquake': Normal events.

events with positive ΔCFS over the number of events with negative ΔCFS, while the dashed line is the reference line showing that the ratio equals unity. In the figure only the results for the 'earthquake pairs' with the distance between the centroids of the two earthquakes less than 100 km have been shown. We also investigated the cases in which the distance between the centroids of the two earthquakes is less than 50 km, 150 km, and other values, and obtained similar results.

We divide the earthquakes into three groups: thrust, strike-slip, and normal, respectively, in which the grouping is made according to the focal mechanism of the 'target earthquake'. Due to the overall similarity between the focal mechanisms of the 'source earthquake' and the 'target earthquake', in such a classification we can also use the focal mechanism of the 'source earthquake'. However, we are interested in the effect of stress triggering on different types of earthquakes, therefore, using the focal mechanism of the 'target earthquake' in the classification is more physical, although the conclusions for the two cases are similar to each other. Figure 2 demonstrates that the stress triggering effect has a clear focal mechanism dependence: For thrust type earthquakes more evident triggering effect can be observed; strike-slip earthquakes seem less in support of the stress triggering hypothesis; normal faulting earthquakes lie in between.

3. Likelihood for Strong Aftershocks or Paired/Grouped Shocks

Related to the stress triggering effect is the clustering property, i.e., whether or not earthquakes tend to be in 'pairs' or 'groups'. Using the Harvard CMT catalogue

Table 1

'Paired' or 'grouped' earthquakes for different focal mechanisms

	1977–98	1977–87	1988–98
Thrust events	84%	89%	80%
Strike-slip events	14%	25%	9%

Note: In the table the ratio is the number of earthquake pairs over the total number of earthquakes for a specific type of focal mechanism.

of global seismicity from 1977 to 1998, we conducted a simple statistics to illustrate how many earthquakes with different focal mechanism types are followed by another event occurring nearby with comparable magnitude. Here the term 'paired' and/or 'grouped' means that successive earthquakes occurred during three years within a spatial range of 200 km. If the preceding earthquake is larger than the following one(s), then the following earthquake is just a strong aftershock in a classical sense. Due to the limitation on the number of samples, the comparison is mainly between thrust and strike-slip earthquakes, whereas for normal type events no definite conclusions can be drawn. The result shows that for shallow earthquakes with seismic moment larger than 10^{19}Nm, there are only 14% of strike-slip earthquakes which occurred in 'pairs' or 'groups', with the earthquake pairs numbering 83. As a comparison, the 'paired' or 'grouped' thrust events are 84% of all the thrust events, with the earthquake pairs numbering 159. Changing the time window produces the results shown in Table 1, which reflects the fluctuation of the statistics.

Using the NEIC broadband radiated energy catalogue from 1987 to 1998 for shallow earthquakes larger than magnitude 5.5, we also discovered that only 24% of strike-slip events occurred in 'pairs' or 'groups'. As a comparison, the 'paired' or 'grouped' thrust events are 71% of all of the thrust ones. Here we change the criteria for the 'paired' or 'grouped' earthquakes slightly, to see whether or not the conclusion is robust against the arbitrariness of the definition. In using the NEIC energy data, because moderate events are also included, the spatial/temporal range is changed into 100 km and two years, with another criterion added that the difference between magnitudes of different events within a 'pair' or 'group' is less than 1.0. Although the values are different for different cases, the tendency always holds that strike-slip earthquakes tend to have fewer 'succeeding' shocks, and thrust earthquakes are more likely to occur in 'pairs' or 'groups'.

4. Discussion and Conclusions

From the results as described above, it can be seen that the scaling of reduced energy or apparent stress, the static stress triggering effect, and the clustering property of earthquakes are all focal mechanism dependent. For strike-slip earthquakes, reduced energy increases slightly with seismic moment; for thrust and normal earthquakes, reduced energy decreases with seismic moment. The level of reduced energy for strike-slip earthquakes is higher than that for thrust and normal earthquakes. Thrust earthquakes exhibit more clear stress triggering effect; strike-slip earthquakes seem unsupportive of the stress triggering hypothesis; normal earthquakes lie in between. Thrust earthquakes are more likely to be in 'pairs' or 'groups', whereas strike-slip earthquakes are more likely to be 'single'.

The focal mechanism dependence of the scaling of reduced energy, the static stress triggering effect, and the clustering properties may be correlated to each other, and the physics seems straightforward: For strike-slip events with higher ambient stress level, the same perturbation by ΔCFS plays a less significant role; for thrust earthquakes, in contrast, the ambient stress level is lower, so the same perturbation by ΔCFS has a more important effect. As to normal type events which are just in between, one characteristic is that, seen from the comparison between Figures 1(a), (b) and (c), there are fewer large normal events in the catalogue, as shown by the narrower horizontal span of data points in Figure 1(c). Another characteristic is that, according to the scaling of apparent stress versus seismic moment as shown in Figure 1(c), the smaller the normal event, the higher the apparent stress level. As a result, the average stress level for normal events is between those of thrust and strike-slip ones. It is understandable, therefore, that the position of normal earthquakes is between strike-slip and thrust earthquakes. For similar reason, compared to strike-slip earthquakes, thrust earthquakes are more likely to affect each other, and thus are more likely to occur in clusters.

Acknowledgements

We thank Prof. Yin Xiangchu and Dr. Zhang Yongxian for their assistance in participating in the ACES activities. This work is supported by Project NSFC 40274013, MOST 2001CB711000, and MOST 2001BA601B02.

REFERENCES

BOATWRIGHT, J. and CHOY, G. L. (1986), *Teleseismic Estimates of the Energy Radiated by Shallow Earthquakes*, J. Geophys. Res. *91*, 2095–2112.

CHOY, G. L. and BOATWRIGHT, J. L. (1995), *Global Patterns of Radiated Seismic Energy and Apparent Stress*, J. Geophys. Res. *100*, 18205–18228.

DZIEWONSKI, A. M., CHOU, T.-A., and WOODHOUSE, J. H. (1981), *Determination of Earthquake Source Parameters from Waveform Data for Studies of Global and Regional Seismicity*, J. Geophys. Res. *86*, 2825–2852.

FRÖHLICH, C. (1992), *Triangle Diagrams: Ternary Graphs to Display Similarity and Diversity of Earthquake Focal Mechanisms*, Phys. Earth Planet. Inter. *75*, 193–198.

FRÖHLICH, C. and DAVIS, S. D. (1993), *Teleseismic b Values; or, Much Ado About 1.0*, J. Geophys. Res. *98*, 631–644.

HARRIS, R. A. (1998), *Introduction to Special Section: Stress Triggers, Stress Shadows, and Implications for Seismic Hazard*, J. Geophys. Res. *103*, 24347–24358.

IDE, S. and BEROZA, G. C. (2001), *Does Apparent Stress Vary with Earthquake Size?* Geophy, Res. Lett. *28*, 3349–3352.

IZUTANI, Y. and KANAMORI, H. (2001), *Scale-dependence of Seismic Energy-to-moment Ratio for Strike-slip Earthquakes in Japan*, Geophys. Res. Lett. 28, 4007–4010.

KAGAN, Y. Y. and JACKSON, D. D. (1999), *Worldwide Doublets of Large Shallow earthquakes*, Bull. Seism. Soc. Am. *89*, 1147–1155.

KANAMORI, H. and HEATON, T. H. (2000), *Microscopic and Macroscopic Physics of Earthquakes, In: GeoComplexity and the Physics of Earthquakes* (Rundle, J. B., Turcotte, D. L. and Klein, W. eds.) AGU, Washington, D. C. 147–163.

McGARR, A. (1999), *On Relating Apparent Stress to the Stress Causing Earthquake Fault Slip*, J. Geophys. Res. *104*, 3003–3011.

OKADA, Y. (1992), *Internal Deformation due to Shear and Tensile Faults in a Half Space*, Bull. Seism. Soc. Am. *82*, 1018–1040.

PEREZ-CAMPOS, X. and BEROZA, G. C. (2001), *An Apparent Mechanism Dependence of Radiated Seismic Energy*, J. Geophys. Res. *106*, 11127–11136.

REASENBERG, P. A. (1999), *Foreshock Occurrence before Large Earthquakes*, J. Geophys. Res. *104*, 4755–4768.

SCHOLZ, C. H, *The Mechanics of Earthquakes and Faulting* (Cambridge: Cambridge Univ. Press 1990) 179.

STOCK, C. and SMITH, E. G. C. (2000), *Evidence for Different Scaling of Earthquake Source Parameters for Large Earthquakes Depending on Faulting Mechanism*, Geophys. J. Int. *143*, 157–162.

TANAKA, S., OHTAKE, M., and SATO, H. (2002), *Evidence for Tidal Triggering of Earthquakes as Revealed from Statistical Analysis of Global Data*, J. Geophys. Res. *107*, B10, 2211, doi: 10.1029/2001JB001577.

TSURUOKA, H., OHTAKE, M., and SATO, H. (1995), *Statistical Test of the Tidal Triggering of Earthquakes: Contribution of the Ocean Tide Loading Effect*, Geophys. J. Int. *122*,183–194.

WU, Z. L. (2001), *Scaling of Apparent Stress from Broadband Radiated Energy Catalogue and Seismic Moment Catalogue and Its Focal Mechanism Dependence*, Earth Planets Space *53*, 943–948.

WYSS, M. and BRUNE, J. N. (1968), *Seismic Moment, Stress, and Source Dimensions for Earthquakes in the California-Nevada Region*, J. Geophys. Res. *73*, 4681–4694.

(Received September 27, 2002, revised April 25, 2003, accepted May 5, 2003)

To access this journal online:
http://www.birkhauser.ch

Pure appl. geophys. 161 (2004) 1979–1989
0033–4553/04/101979–11
DOI 10.1007/s00024-004-2544-2

© Birkhäuser Verlag, Basel, 2004

▌Pure and Applied Geophysics

Continuum Fractal Mechanics of the Earth's Crust

ARCADY V. DYSKIN[1]

Abstract— In the cases when the Earth's crust possesses self-similar structure its mechanical behaviour can be modelled by a continuous sequence of continua each determined by the size of the averaging volume element. It is shown that tensorial properties and integral state variables scale by power laws with exponents common for all components of the tensors. Thus the scaling is always isotropic with anisotropy accounted for by the prefactors. As an example, scaling laws for effective moduli of the Earth's crust with self-similar cracking are derived for the cases of isotropic distribution of disk-like cracks and two mutually orthogonal sets of 2-D cracks. Real systems are not self-similar therefore the proposed approach is based on their approximation by self-similar systems. A necessary condition is formulated for such an approximation.

Key words: Continuum fractal mechanics, multi-scale modelling, effective characteristics, self-similar approximation.

1. Introduction

Modelling geological phenomena using the concept of fractals and self-similarity has attracted considerable interest (e.g. SADOVSKIY, 1983; SCHOLZ, 1990; BARTON and ZOBACK, 1992; TURCOTTE, 1993; DUBOIS, 1998) because it offers a rational method of dealing with such a highly irregular structure as the Earth's crust. The modelling is based on determining power- scaling laws pertinent to different aspects of the Earth's crust behaviour. When the mechanics of fractal objects is considered, two essential problems arise.

First, these highly irregular objects no longer permit the introduction of a scale at which they can be treated using the methods of continuum mechanics. Even the basic notions, such as stress and strain cannot be introduced. For instance, the stress renormalisation accounting for the fractal scaling of elementary area leads to non-traditional units of stress (CARPINTERI, 1994) that can, in principle, depend on the position within the material. On the other hand, the methods of discontinuous mechanics (like the discrete element method) become computationally prohibitive when multi-scale objects have to be modelled. Thus, a rational method is needed that

[1]School of Civil and Resource Engineering, The University of Western Australia, 35 Stirling Hwy, Crawley, WA, 6009, Australia. E-mail: adyskin@cyllene.uwa.edu.au

can reconcile the irregular nature of fractal objects with the highly developed machinery of continuum mechanics.

Second, the fractal description is based on the idea that the object is self-similar. This is a very strong property leading to the conclusion that all functions which are functions of the scale must only be power functions. In reality however, the natural dependencies are only approximated by the power functions. Then the conditions have to be formulated in which the power-law approximation is consistent with the notion of self-similarity.

The present paper considers these two problems.

2. Multi-scale Modelling vs. Continuum Fractal Modelling of Self-similar Objects

Traditional Continuum Mechanics is based on the introduction of infinitesimal lengths such as ds and, subsequently, areas and volumes. Geological materials are discontinuous, as all other materials, nevertheless, in the cases when the characteristic length of discontinuities, l is considerably smaller than the problem scale, L, one can represent the real material by a continuum. This is done by introducing an intermediate quantity H ($l << H << L$) which plays the role of an infinitesimal length of the continuum description from which the (representative) volume elements are constructed enabling the introduction of conventional continuum mechanics quantities such as stress and strain (e.g., DYSKIN et al., 1992; NEMAT-NASSER and HORI, 1993). In many cases however, geomaterials posses discontinuities with dimensions covering a wide range scales (e.g., SADOVSKIY, 1983) such that a unique definition of the volume element size H is not possible. If these scales are distinctively different, it was suggested to use a set of continua each characterised by its own yardstick H_i ($i = 1,...n$ where n is the number of continua). This approach leads to a so-called *multiscale modelling* (DYSKIN et al. 1992). The left part of Figure 1 illustrates the main idea of the multi-scale modelling in application to a material with multi-scale sets of discontinuities.

The main obstacle for applying the multi-scale modelling to a fractal material is the absence of characteristic scales: the material can for instance possess discontinuities of all sizes. Then it is no longer possible to model it by a finite set of continua. In order to overcome this obstacle consider a material, say a volume of the Earth's crust with fractal structure, choose a scale, H, and remove all discontinuities of the size H and greater. Thus we obtain a material with truncated structure and, at the same time specify a scale at which the truncated material can be modelled as a continuum. Therefore, the actual computations can be conducted for an object that is not fractal but rather a continuum that models the material with certain structure of discontinuities of the sizes smaller than H. By varying H one can model the self-similar fractal objects by a continuous set of continua each of them being characterized by its own yardstick, H, specifying the scale. Each continuum is to

Multiscale modelling

Fractal continuum modelling

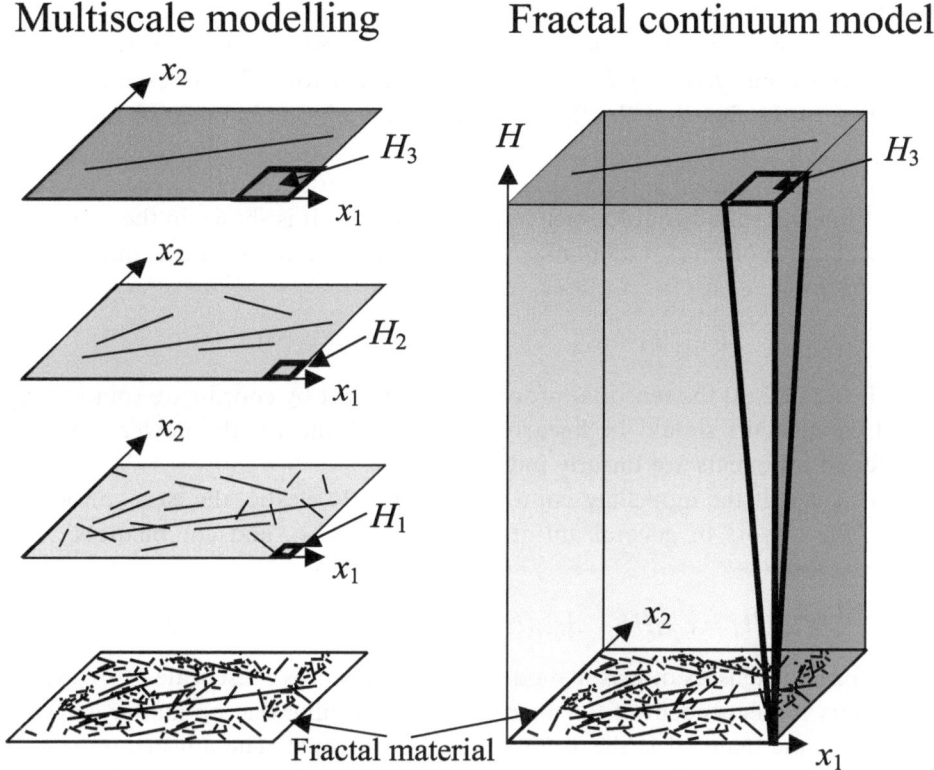

Figure 1
Multi-scale modelling vs. continuous fractal modelling. Multi-scale modelling involves setting a finite set of continuum models representing the material at a fixed number of scales determined by the representative volume element sizes H_1, H_2, ... (eg., DYSKIN et al., 1992). The continuous fractal modelling involves a continuous set of models such that a new dimension (scale H) is added to the conventional three or two spatial dimensions (only two dimensions are shown in the diagrams) and the time (RODIONOV et al., 1989).

model the fractal material at the scale H in the sense that the volume elements of size H cut from both the original material and the modelling continuum respond somewhat equally to uniform loading. The yardstick H determines the resolution: no features with all dimensions smaller than H are viewable in the H-continuum (DYSKIN, 2001). Thus the H-continuum replaces the original material with the one possessing modified microstructure in which only those microstructural elements are present that have characteristics sizes less than H. This leads to continuum fractal modelling (the right part of Fig. 1) in which a model is produced that possesses an additional dimension, the scale (RODIONOV et al., 1989), on top of conventional dimensions (four in the general case—three spatial dimensions and time).

When a fractal (self-similar) object is modelled by a continuum of continua, the overall characteristics of each continuum (e.g., effective moduli) become functions of

H. Usually these characteristics have the same sign for all continua, so according to the general theorem (e.g., BARENBLATT and BOTVINA, 1980) they must be represented by power functions, $f(H) = f^* H^\alpha$, where f^* is a prefactor. This is essentially the consequence of the fact that the fractal (self-similar) objects have no characteristic length.

Of a particular interest is the scaling of overall tensorial quantities (e.g., effective moduli or integral state variables such as average stress). It is shown in the Appendix that the scaling exponents for all non-zero components of a tensor must coincide, i.e., for any tensor

$$f_{ijk...}(H) = f^*_{ijk...} H^{\alpha_{ijk...}}, \quad \alpha_{ijk...} = \alpha = \text{const.} \tag{1}$$

This is because (a) the tensorial property implies that by coordinate rotations all tensorial components should be linearly transformed and (b) the power functions with different exponents are linearly independent.

In particular, if the modelling continua are linearly elastic, the case considered hereafter, the tensors of general anisotropic moduli, C_{ijkl}, and compliances, A_{ijkl}, should scale as

$$C_{ijkl}(H) = c_{ijkl} H^\alpha, \quad A_{ijkl}(H) = a_{ijkl} H^\beta, \quad i, j, k, l = 1, 2, 3. \tag{2}$$

Therefore, the tensors of elastic moduli and compliances must scale *isotropically*. This property is independent of the microstructure meaning that no matter what the anisotropy of the material is the scaling must be isotropic. The anisotropy is only accounted for by the prefactors, c_{ijkl} and a_{ijkl}. The particular values of α, β, c_{ijkl} and a_{ijkl} depend on the material structure. Examples of the Earth's crust with self-similar crack distributions are considered in the following section.

3. Scaling Laws for Elastic Moduli in the Earth's Crust with Self-similar Distributions of Cracks

Consider a material containing a self-similar distribution of cracks which is represented by a power distribution function

$$f(R) = wR^{-4} \tag{3}$$

normalised as follows

$$\int_{R_{\min}}^{R_{\max}} R^3 f(R) \, dR = v_t. \tag{3a}$$

Here, w is the concentration factor that ensures a specified total concentration v_t. This distribution has a remarkable property no other self-similar distribution possesses, which makes it possible to suggest an accurate procedure of computing the effective moduli as power functions of the crack size. Consider the probability, $P(n)$, that in a vicinity of a crack of size R, i.e., a region of size proportional to R_2, there are cracks of smaller sizes, say from R/n to R, where $n > 1$. This probability is equal to $P(n) \sim w(n^3-1)$, i.e., it does not depend upon the crack size, R. Since eq. (3) represents real distributions only asymptotically as $R_{max}/R_{min} \to \infty$, i.e., as $w \to 0$ ($v_{t\,=}$ const.), for any n the value of w can be chosen sufficiently small to make this probability negligible for any crack size.

This property can be interpreted in the sense that for any crack the probability to find nearby a crack of a similar size is negligible; only cracks of greatly different sizes can be found there. Mechanically it means that the interaction between the cracks of similar sizes can be neglected; only the interaction between cracks of very different sizes is to be taken into account. This suggests the use of the SALGANIK's (1973) differential self-consistent method for calculating the effective characteristics.

According to the method, the compliance increment ΔA_{ij}, $i,j = 1,\ldots,6$, at each scale is determined by the contribution of non-interacting cracks of the corresponding scale considered in an effective continuum representing all cracks of smaller scales. The contribution is proportional to the concentration of this group of cracks, wdH/H. Therefore

$$A_{ij}(H + dH) = A_{ij}(H) + wS_{ij}(A_{11}, \ldots, A_{66})dH/H$$

$$C_{ij}(H + dH) = C_{ij}(H) - w\Lambda_{ij}(C_{11}, \ldots, C_{66})dH/H, \qquad (4)$$

where S_{ij}, Λ_{ij} are homogeneous functions of the first degree specific for the geometry and distribution of parameters of the cracks.

Substituting (2) into (4) gives the scaling equations

$$\beta a_{ij} = wS_{ij}(a_{11}, a_{12}, \ldots a_{66}), \alpha c_{ij} = w\Lambda_{ij}(c_{11}, c_{12}, \ldots c_{66}), \quad i,j = 1\ldots 6. \qquad (5)$$

Below these equations are solved for two specific cases.

3.1. Isotropically Oriented Disk-like Cracks

In the case of randomly oriented disk-like cracks, the material is isotropic. Thus, the components of Λ_{ij} can be extracted from the expressions for effective Young's modulus, E, and Poisson's ratio, v, written for the case of non-interacting cracks (e.g., SALGANIK, 1973):

$$E = E_m \left[1 - \frac{16}{45}(10 - 3v_m)\frac{1 - v_m^2}{2 - v_m}v \right], v = v_m \left[1 - \frac{16}{15}(3 - v_m)\frac{1 - v_m^2}{2 - v_m}v \right], \qquad (6)$$

where v is the crack concentration, E_m and v_m are the Young's modulus and Poisson's ratio of the material. The coefficients at v in the square brackets in (6) play the role of components of the function Λ_{ij}. After substituting them into the second equation in (5) one has the following equations for the exponent, α, and the prefactors e and v:

$$\alpha e = -\frac{16}{45}(10 - 3v)\frac{1 - v^2}{2 - v}we, \quad \alpha v = -\frac{16}{15}(3 - v)\frac{1 - v^2}{2 - v}wv. \tag{7}$$

Since the Poisson's ratio is bounded, the second equation of (7) offers two solutions: either $\alpha = 0$ or $v = 0$. The first solution being substituted into the first equation of (7) leads to a trivial case of $e = 0$. The second solution after substituting into the first equation of (7) gives the following scaling law

$$v = 0, \quad E = eH^\alpha, \quad \alpha = -16w/9. \tag{8}$$

In this scaling law the prefactor e is undeterminable and should be found independently, for instance from the measurements.

The exponent in the obtained scaling law determines the rate with which the Young's modulus reduces with the increase of the scale. The exponent is proportional to the concentration factor w which reflects crack concentration at each scale.

It should be noted that the obtained exponent has no relevance to the fractal dimension $D = 3$ for the case of cracks, since the cracks have zero internal volume.

3.2. Plane with Two Mutually Orthogonal Sets of Cracks

Consider now a 2-D problem for a plane with two mutually orthogonal sets of cracks (Figure 2a). The self-similar distribution function in 2-D has the form

$$f(l) = \omega l^{-3}, \quad \int_{l_{min}}^{l_{max}} l^2 f(l)\, dl = \Omega_t, \tag{9}$$

where, ω is the concentration factor ensuring the specified total concentration Ω_t, l is the crack length. It is assumed that the set of cracks perpendicular to the x_i axis is characterised by the distribution $\omega_i l^{-3}$ such that the total distribution is characterised by the concentration factor $\omega = \omega_1 + \omega_2$.

We consider VAVAKIN and SALGANIK's (1978) solution for the effective compliances for an orthotropic plate with a set of non-interacting cracks which is aligned to one of the symmetry axes of the material and generalise it to two sets of non-interacting cracks by superimposing the contributions of each set of cracks. Using the method outlined above the scaling equations can be obtained:

$$\beta a_{ii} = \frac{\pi}{4}\omega_i\sqrt{a_{ii}(a_{66} + 2\sqrt{a_{11}a_{22}})}, \quad i = 1, 2$$

$$\beta a_{66} = \frac{\pi}{4}\omega_2\sqrt{a_{11}(a_{66} + 2\sqrt{a_{11}a_{22}})} + \frac{\pi}{4}\omega_1\sqrt{a_{22}(a_{66} + 2\sqrt{a_{11}a_{22}})}. \tag{10}$$

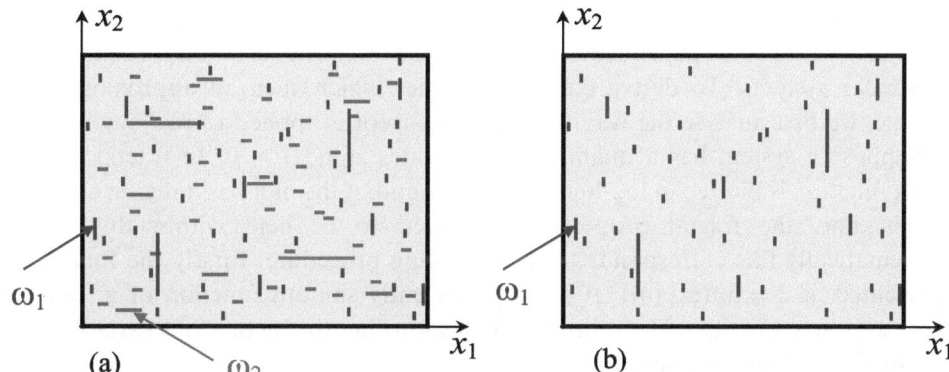

Figure 2
A plate with two mutually orthogonal sets of cracks with the concentration factors ω_1 and ω_2 respectively (a) and a limiting case of the plate with a single set of parallel cracks (b).

These equations have a unique solution that produces the following scaling laws:

$$A_{ij} = a_{ij}H^\beta, \quad \beta = (\pi/2)\sqrt{\omega_1\omega_2},$$
$$a_{22} = a_{11}(\omega_2/\omega_1)^2, \quad a_{66} = 2a_{11}(\omega_2/\omega_1), \quad a_{12} = 0. \tag{11}$$

For the case of a single set of cracks, i.e., when the concentration of the other set vanishes, say $\omega_2 \to 0$, (Figure 2b), the exponent and all compliances except a_{11} vanish. The material becomes completely rigid in the direction x_2. This can easily be understood if one considers that in fractal materials the total crack concentration is infinite implying the infinite total compliancy. What the scaling law determines is the way the moduli change in transition from one scale to another. Relative to such moduli the material without cracks as well as the material in the directions not affected by cracks (in the direction x_2 in Figure 2b) become infinitely rigid. In real situations we do not have true fractal materials as there are always lower and upper cutoffs. The scaling laws (8), (11) can still offer a good approximation provided that the material in question can be approximated by a self-similar one (see the following section for conditions of such an approximation). In this case the complete rigidity will simply mean a very high modulus as compared to the values of other moduli.

4. Self-similar Approximation

The fact that all components of the tensorial functions of H must scale with the same exponent suggests that the property of self-similarity is very restrictive, the restrictions reaching far beyond the imposition of power laws. In particular, the specific form of scaling laws for tensorial quantities (scaling isotropy) implies that not all systems which show dependencies close to the power ones can be approximated by self-similar systems. To derive conditions under which such an approximation is possible, we first analyse the way the fractal concept is applied to real systems.

Suppose a system has a quantity which scales as $f(H) > 0$. In fractal analysis such a function is plotted in log-log coordinates and, if the plot has a region close to a straight line, the fractal property is declared to be held within this region. Mathematically this corresponds to the following procedure. Firstly the function is represented as $\xi = \ln f(\exp(\eta))$. If it is a sufficiently smooth function of η, one can linearise it in a vicinity of $\eta_0 = \ln H_0$. Then returning to the original variables, one obtains the self-similar approximation:

$$f(H) \approx f(H_0)\left(\frac{H}{H_0}\right)^{\alpha}, \quad \alpha = \frac{f'(H_0)H_0}{f(H_0)}, \quad f(H_0) \neq 0. \tag{12}$$

Thus, one can always approximate any function by a power one at list in a sufficiently narrow vicinity of H_0. Obviously, the formula can be generalised to the case $f(H) < 0$.

Real systems shall of course be characterised by a set of scale-dependent properties. If there is a subset of properties with the same units $f_1(H), f_2(H), \ldots$ then by summation, multiplication and multiplication by a dimensionless factor new properties can be derived (for instance operations of linear combination correspond to rotations of the tensor of elastic moduli). Such quantities which permit summation and multiplication will be called *compatible*. (All other functions on the set of compatible quantities can be reduced to, possibly infinite, sequences of summations and multiplications.) Therefore, the self-similar approximation to be meaningful should preserve these operations. This means that the selfsimilar approximation of the product or the sum of two functions in a vicinity of the same point, H_0, should be equal to the product or the sum of the corresponding approximations, respectively. We shall check these properties now. (The case of multiplication by a dimensionless factor is trivial.) Suppose

$$f_i(H) \approx f_i(H_0)\left(\frac{H}{H_0}\right)^{\alpha_i}, \quad \alpha_i = \frac{f_i'(H_0)H_0}{f_i(H_0)}. \tag{13}$$

Then for the multiplication we have the necessary property:

$$f_1(H) \cdot f_2(H) \approx f_1(H_0)f_2(H_0)\left(\frac{H}{H_0}\right)^{\alpha_1+\alpha_2} \tag{14}$$

However, for the summation a similar property

$$f_1(H) + f_2(H) \approx [f_1(H_0) + f_2(H_0)]\left(\frac{H}{H_0}\right)^{\alpha} \tag{15}$$

only holds when

$$\alpha_1 = \alpha_2 \quad \text{or} \quad \frac{f_1'(H_0)}{f_1(H_0)} = \frac{f_2'(H_0)}{f_2(H_0)}. \tag{16}$$

Equation (16) establishes a condition of equivalence between the summation of functions and their self-similar approximations. This is a *necessary* condition of the self-similar modelling: only systems described by sets of *compatible* quantities equivalent in the sense of (16) allow self-similar approximation.

5. Conclusions

The proposed continuum fractal modelling of mechanical behaviour of the Earth's crust with self-similar structure is based on representing the object as a continuum of continua of different scales, which is essentially a generalisation of multi-scale modelling. In the continuum fractal modelling the scaling of mechanical properties, i.e., transition from one continuum to another is described by power laws. The tensorial mechanical properties scale by power laws with exponents common for all components of the tensors.

Real systems are not self-similar, so one can only talk about their approximation by the self-similar ones. Such an approximation is based on representing all scale-dependent properties in log-log coordinates and linearising them in a vicinity of a certain scale. This provides a local (in terms of the scale) power-law approximation. When the real system is described by a set of quantities, each of them should be approximated in a vicinity of the same scale. However, the self-similar approximation is only possible when all compatible quantities (the ones which permit summation and multiplication) have power-law approximations with the same exponent.

Acknowledgement

The author is grateful to the Australian Research Council for the financial support (Large grant A00104937).

Appendix

Scaling Laws for Tensors

The fact that the components of a tensor must change with the coordinate rotation in accordance with a certain rule imposes further restrictions on the scaling laws. Consider, for the sake of simplicity a Cartesian coordinate set (x_1, x_2, x_3).

Suppose a tensor $T_{ijk...}(H) \propto H^{\alpha}_{ijk...}$ where $i, j, k... = 1...3$. When the coordinate set is rotated with a matrix r_{ij}, the new components of the tensor assume the form

$$T'_{ijk...} = r_{il}r_{jm}r_{kn}\ldots T_{lmn...}, \quad T'_{ijk...} \propto H^{\beta_{ijk}}, \quad T_{lmn...} \propto H^{\alpha_{lmn}} \tag{A1}$$

Here summation over repeated indices is presumed.

The tensorial equation (A1) establishes a linear relationship between power functions. However, power functions with different exponents are linearly independent (ACHIESER, 1956). Consequently, the following proposition can be formulated.

Proposition. *For a tensor that depends on scale, H, as the only variable of length, all non-zero components must scale with the same exponent, i.e.,*
Either

$$T_{ijk...} = 0 \quad \text{or} \quad T_{ijk...} \propto H^{\alpha}, \tag{A2}$$

where α is a constant common for all $i, j, k...$

Proof. Suppose the exponents in (A1) assume p different values, $\alpha_1, ... \alpha_p$. Then by grouping the terms with the same powers in (A1) one gets

$$\sum_{q=1}^{p} C_q H^{\alpha_q} = 0, \tag{A3}$$

where C_q are the sums of the corresponding terms $r_{il}r_{jm}r_{kn}\ldots T_{lmn...}$; one of these sums contains (or just consists of) the component $-T'_{ijk...}$. Because the power functions with different exponents are linearly independent,

$$C_q = 0, \quad q = 1, \ldots p. \tag{A4}$$

When another coordinate rotation, r, is chosen, only terms $-T'_{ijk...}$ could, in principle, change the power and either migrate to another group or form a new one. Since the number of groups, p, is finite, while the number of rotations r is infinite (even continuum), one can always find a group that does not contain terms $-T'_{ijk...}$ for a sufficiently large number of different rotations. Thus, one arrives to a homogeneous system of equations

$$r_{il}r_{jm}r_{kn}\ldots T_{lmn...} = 0, \tag{A5}$$

where the summation is presumed over the values of indices $l, m, n,...$ that belong to the group in question. Since different components of the rotation matrix are linearly independent, $T_{lmn...} = 0$.

Continuing in this way, one can eliminate all groups except the one containing $-T'_{ijk...}$. Therefore, all nonlinear components of the tensor will have the same power, the one that corresponds to this last group. This finalises the proof.

References

ACHIESER, N. I., *Theory of Approximation* (Frederic Unger Publishing Co, New York 1956).

BARENBLATT, G. I. and BOTVINA, L. R., *Application of the similarity method to damage calculation and fatigue crack growth studies*, In *Defects and Fracture* (eds. Sih. G.C. and Zorski, H.) (Martinus Nijhoff Publishers 1980) pp. 71–79.

BARTON, C. A. and ZOBACK, M. D. (1992), *Self-similar Distribution and Properties of Macroscopic Fractures at Depth in Crystalline Rock in the Cajon Pass Scientific Drill Hole*, J. Geophys. Res. *97*B, 5181–5200.

CARPINTERI, A. (1994), *Scaling Laws and Renormalization Groups for Strength and Toughness of Disordered Materials*, Int. J. Solids Struct. *31*, 291–302.

DUBOIS, J., *Non-Linear Dynamics in Geophysics* (John Wiley and Sons, Chichester, New York, Weinheim, Brisbane, Singapore, Toronto 1998).

DYSKIN, A. V., *Stress-strain calculations in materials with self-similar or fractal microstructure*. In *Computational Mechanics – New Frontiers for New Millennium* (eds. Valiapan. S. and Khalili. N.) (World Scientific 2001), *2*, pp. 1173–1178.

DYSKIN, A. V., *Mechanics of fractal materials*. In *Proc. of the IUTAM Symposium on Analytical and Computational Fracture Mechanics of Non-homogeneous Materials* (ed. Karihaloo. B.) (Kluwer Academic Press 2002) pp. 73–82.

DYSKIN, A. V., SALGANIK, R. L., and USTINOV, K. B., *Multi-scale geomechanical modelling*. In *Proc. of Western Australian Conference of Mining Geomechanics* (eds. Szwedziki. T., Baird., G. R., and Little., T. N.) (Curtin University, WASM, Kalgoorlie, Western Australia 1992) pp. 235–246.

NEMAT-NASSER, S. and HORI, M., *Micromechanics: Overall Properties of Heterogeneous Materials* (North-Holland-Amsterdam, London, New York, Tokyo 1993).

RODIONOV, V. N., SIZOV, I. A., and KOCHARYAN, G. G., *The discrete properties of geophysical medium. In Modelling of Natural Objects in Geomechanics* (Nauka, Moscow 1989) pp. 14–18 (in Russian).

SADOVSKIY, M. A. (1983), *Distribution of Preferential Sizes in Solids*. Transactions USSR Academy of Sciences, Earth Science Series *269*. 1, 8–11.

SALGANIK, R. L, (1973), *Mechanics of Bodies with Many Cracks*. Mech. Sol. *8*, 135–143.

SCHOLZ, C. H., *The Mechanics of Earthquakes and Faulting* (Cambridge University Press, Cambridge, New York, Port Chester, Melbourne, Sydney 1990).

TURCOTTE, D. L., *Fractals and Chaos in Geology and Geophysics* (Cambridge University Press 1993).

VAVAKIN, A. S. and SALGANIK, R. L. (1978), *Effective Elastic Characteristics of Bodies with Isolated Cracks, Cavities, and Rigid Nonhomogeneities.*, Mech. Sol. *13*, 87–97.

(Received September 27, 2002, revised February 28, 2003, accepted March 7, 2003)

To access this journal online:
http://www.birkhauser.ch

Pure appl. geophys. 161 (2004) 1991–2003
0033–4553/04/101991–13
DOI 10.1007/s00024-004-2545-y

▌Pure and Applied Geophysics

Using Eigenpattern Analysis to Constrain Seasonal Signals in Southern California

K. F. Tiampo[1,2], J. B. Rundle[3], W. Klein[4], Y. Ben-Zion[5], and S. McGinnis[1]

Abstract — Earthquake fault systems are now thought to be an example of a complex nonlinear system (Bak, *et al.*, 1987; Rundle and Klein, 1995). The spatial and temporal complexity of this system translates into a similar complexity in the surface expression of the underlying physics, including deformation and seismicity. Here we show that a new pattern dynamic methodology can be used to define a unique, finite set of deformation patterns for the Southern California Integrated GPS Network (SCIGN). Similar in nature to the empirical orthogonal functions historically employed in the analysis of atmospheric and oceanographic phenomena (Preisendorfer, 1988), the method derives the eigenvalues and eigenstates from the diagonalization of the correlation matrix using a Karhunen-Loeve expansion (KLE) (Fukunaga, 1970; Rundle *et al.*, 2000; Tiampo *et al.*, 2002). This KLE technique may be used to determine the important modes in both time and space for the southern California GPS data, modes that potentially include such time-dependent signals as plate velocities, viscoelasticity, and seasonal effects. Here we attempt to characterize several of the seasonal vertical signals on various spatial scales. These, in turn, can be used to better model geophysical signals of interest such as coseismic deformation, viscoelastic effects, and creep, as well as provide data assimilation and model verification for large-scale numerical simulations of southern California.

Introduction

Data assimilation is the process by which observations are incorporated into models to set their parameters, and to tune them in real time as new data become available. The result of the data assimilation process is a framework that is maximally consistent with the observed data, producing a model that is useful in ensemble forecasting. The idea is that the state of the model follows an evolutionary path through state space as time progresses, and that observations can be used to

[1] CIRES, University of Colorado, Boulder, CO 80309, USA and

[2] Dept. of Earth Sciences, University of Western Ontario, London, ONT, Canada.
E-mail: ktiampo@uwo.ca

[3] Center for Computational Science and Engineering, University of California, Davis, CA 95616, USA. E-mail: rundle@physics.ucdavis.edu

[4] Department of Physics, Boston University, Boston, MA 02215, USA and Center for Nonlinear Science, Los Alamos National Laboratory, Los Alamos, NM 87545, USA. E-mail: klein@buphy.edu

[5] Department of Earth Sciences, University of Southern California, Los Angels, CA, USA.
E-mail: benzion@terra.usc.edu

periodically adjust model parameters. One of the observations that potentially can be used to constrain large-scale models of the fault network in southern California is the horizontal and vertical deformation measurements produced by SCIGN.

The ability to incorporate SCIGN data into these models is limited by the difficulty in differentiating between the large number of different geophysical signals in the aggregate geodetic measurement. For example, the relatively large seasonal effects, including those with both annual and semiannual signals, can mask the small interseismic tectonic crustal deformation that this concentrated GPS network was deployed to identify and quantify (The SCIGN Project Report to NSF, 1998; DONG *et al.*, 2002; WATSON *et al.*, 2002).

Here we show that the decomposition of the SCIGN data, based upon the underlying correlations, into its spatial eigenvectors and associated temporal signature, can separate out the various modes based upon the important spatial and temporal scales. This technique, called a Karhunen-Loeve expansion (KLE) analysis, is similar to the empirical orthogonal function (EOF) analysis frequently employed in the atmospheric sciences (PREISENDORFER, 1988), with the primary difference being that the EOF analysis differentiates the modes based upon the covariance in the data, while the KLE analysis decomposes them based upon their correlations. The advantage is that the source identification and modeling is clearer, based upon the spatial wavelength and the simpler temporal signals that result from their isolation from other modes. We illustrate that principle in this paper by examining the seasonal vertical modes.

KLE Analysis

Pattern evolution and prediction in nonlinear systems is complicated by nonlinear interactions and noise, but understanding such patterns, which are simply the surface expression of the underlying dynamics, is critical to understanding and perhaps characterizing the physics which control the system. RUNDLE *et al.* (2000), proposed a method of decomposing seismicity, one set of complex spatial and temporal patterns that are the surface expression of the obscure dynamics of the physical system, into the orthonormal pattern eigenstates. This decomposition implicitly assumes that one is dealing with a process that is both Markov and stationary in time. The procedure involves constructing a correlation operator, $C(x_i, x_j)$, for the sites that contains the spatial relationship of slip events over time. $C(x_i, x_j)$ is decomposed into the orthonormal spatial eigenmodes for the nonlinear threshold system, e_j, and their associated time series, $a_j(t)$ (TIAMPO *et al.*, 2002). These spatial-temporal pattern states can be used to reconstruct the primary modes of the system, with or without noise, and quantify their relative magnitude and importance. In addition, these primary modes can be used to characterize the underlying dynamics and the physical parameters such as stress levels and interactions that control the observable patterns of events. ANGHEL et al. (2001, 2003), apply a similar methodology to modeled deformation data in order to capture the coherent structures and their interactions. Here we apply this technique to the Southern California Integrated GPS Network (SCIGN) data in order to determine the principal modes of deformation for the southern California fault system.

Similar to the empirical orthogonal function (EOF) technique developed by PREISENDORFER(1988) for the atmospheric sciences, the Karhunen-Loeve expansion is obtained from the p time series that record the deformation history at particular locations in space. Each time series, $y(x_s, t_i) = y_i^s$, $s = 1, \dots p$, consists of n time steps, $i = 1, \dots n$. The goal is to construct a time series for each of a large number of locations for a given short period of time. If, for example, the time interval was decimated into units of days, the result could be a time series of 365 time steps for every year of data, with values of deformation for that location at each time step. These time series are incorporated into a matrix, T, consisting of time series of the same measurement for p different locations, i.e.,

$$T = [\bar{y}_1, \bar{y}_2, \dots \bar{y}_p] = \begin{bmatrix} y_1^1 & y_1^2 & \cdots & y_1^p \\ y_2^1 & y_2^2 & \cdots & y_2^p \\ \vdots & \vdots & \ddots & \vdots \\ y_n^1 & y_n^2 & \cdots & y_n^p \end{bmatrix}. \tag{1}$$

For analysis of SCIGN data, the values in the matrix T consist of deformation measurements, horizontal or vertical. The covariance matrix, $S(x_i, x_j)$, for these events is formed by multiplying T by T^T, where S is a p x p real, symmetric matrix. The covariance matrix, $S(x_i, x_j)$, is converted to a correlation operator, $C(x_i, x_j)$, by dividing each element of $S(x_i, x_j)$, by the variance of each time series, $y(x_i, t)$ and $y(x_j, t)$,

$$\sigma_p = \sqrt{\frac{1}{n} \sum_{k=1}^{n} (y_k^p)^2}, \tag{2}$$

and

$$C = \begin{bmatrix} \frac{s_{11}}{\sigma_1 \sigma_1} & \frac{s_{12}}{\sigma_1 \sigma_2} & \cdots & \frac{s_{1p}}{\sigma_1 \sigma_p} \\ \frac{s_{21}}{\sigma_2 \sigma_1} & \frac{s_{22}}{\sigma_2 \sigma_2} & \cdots & \frac{s_{2p}}{\sigma_2 \sigma_p} \\ \vdots & \vdots & \ddots & \vdots \\ \frac{s_{p1}}{\sigma_p \sigma_1} & \frac{s_{p2}}{\sigma_p \sigma_2} & \cdots & \frac{s_{pp}}{\sigma_p \sigma_p} \end{bmatrix}. \tag{3}$$

This equal-time correlation operator, $C(x_i, x_j)$, is decomposed into its eigenvalues and eigenvectors in two parts. The first employs the trireduction technique to reduce the matrix $\mathbf{C}$ to a symmetric tridiagonal matrix, using a Householder reduction. The second part employs a ql algorithm to find the eigenvalues, λ_j^2, and eigenvectors, e_j, of the tridiagonal matrix (PRESS et al., 1992). These eigenvectors, or eigenstates, are orthonormal basis vectors arranged in order of decreasing variance that reflect the spatial relationship of events in time. If one divides the corresponding eigenvalues, λ_j^2, by the sum of the eigenvalues, the result is that percent of the correlation accounted for by that particular mode. We then reconstruct the time series associated with each location for each eigenstate by projecting the initial data back onto these basis vectors in what is called a principal component analysis (PCA) (PREISENDOR-FER, 1988). These time-dependent expansion coefficients, $a_j(t)$, which represent temporal eigenvectors, are reconstructed by multiplying the original data matrix by the eigenvectors, i.e.,

$$ a_j(t_i) = \vec{e}^T \cdot T = \sum_{s=1}^{p} e_j y_i^s, \tag{4} $$

where $j, s = 1, \ldots p$ and $i = 1, \ldots n$. This eigenstate decomposition technique produces the orthonormal spatial eigenmodes for this nonlinear threshold system, e_j, and the associated principal component time series, $a_j(t)$. These principal component time series represent the signal associated with each particular eigenmode over time. For purposes of clarity, the spatial eigenvectors are designated KLE modes and the associated time series Principal Component (PC) vectors.

The KL expansion, a linear decomposition technique in which a dynamical system is decomposed into a complete set of orthonormal subspaces, has been applied to a number of other complex nonlinear systems over the last fifty years, including the ocean-atmosphere interface, turbulence, meteorology, biometrics, statistics, deformation, and seismicity (PREISENDORFER, 1988; SAVAGE, 1988; PENLAND, 1989; VAUTARD and GHIL, 1989; FUKUNAGA, 1970; POSADAS *et al.*, 1993; PENLAND and SARDESHMUKH, 1995; TIAMPO *et al.*, 2002). Here we use this technique to decompose daily time series constructed from SCIGN data.

Data

The Southern California Integrated GPS Network (SCIGN) is a regional network of GPS stations installed in and around the Los Angeles basin for the purpose of measuring the response of the large, complex southern California fault system to regional strains, to identify localized, unknown fault features and sources such as blind thrust faults, to estimate earthquake potential, and to quantify the physical parameters of the fault system itself. Important work related to this continuous array includes the calculation of station velocities (The SCIGN Project Report to NSF, 1998), the estimation of coseismic and postseismic displacements relating to both the 1992 Landers earthquake and the 1994 Northridge earthquake and their aftershocks (BOCK *et al.*, 1997; DONNELLAN and WEBB 1998; WDOWINSKI *et al.*, 1997), and the analysis of various error or noise sources (ZHANG *et al.*, 1997;

DONG *et al.*, 2002). However, much still remains to be accomplished toward the quantization of fault parameters and fault mechanics, and the identification and detailing of those local faults or events which are integral to the system but currently unidentified, such as blind thrusts. In addition, these stations have rarely been used to study time-dependent motions such as viscoelastic response to local or regional strain. The large, complex nature of the fault network, coupled with the often obscure underlying dynamics, precludes the simple analysis of its surficial expression, whether it is seismicity or deformation.

SCIGN is a network of continuous GPS stations used to monitor deformation in southern California. The first stations were installed in 1991, and the network was expanded rapidly after the occurrence of the 1994 Northridge earthquake. Today there are over 250 stations throughout southern California, many of which are concentrated in the Los Angeles basin. Data for this study was preprocessed, using two different data analyses methods, SCIGN 1.0 and 2.0, downloaded directly from the SCIGN website (www.scign.org). SCIGN 1.0 has repeatabilities of 3.7 mm latitude, 5.5 mm longitude, and 10.3 mm vertical. SCIGN 2.0 has repeatabilities of 1.2 mm latitude, 1.3 mm longitude, and 4.4 mm vertical. We broke the decompositions down into pre- and post-1998, as that marks the break between SCIGN 1.0 and SCIGN 2.0. The KLE method was applied to both the SCIGN 1.0 vertical data and the latitude-longitude (horizontal) data, for the time period 1993–1997, inclusive. analysis of the data beginning 1 January 1998 included only the SCIGN 2.0 data, ending in mid-2000. This same analysis, pre- and post-1998, vertical and horizontal, was performed for both the entire data set, consisting of approximately 200 stations in 2000, and just the Los Angeles (LA) basin.

Results

Here we investigate the separation of vertical motions from horizontal motions, leading to the identification of modes whose surface expression has a large vertical component, such as seasonal environmental patterns or coseismic response. Identification of these signals, and the separation of these modes based on their different spatial and temporal scales, not only allows them to be modeled more effectively, but also allows for the investigation of other modes, without the interaction of these effects. A partial list would include plate motions, creep events or 'slow' earthquakes, blind thrust faults, viscoelastic response, and local variations in strain rate direction.

Here an analysis has been performed on vertical motions, with data obtained directly from the SCIGN web site, as described above. Shown in Figure 1 is the time series for the first vertical KL mode for all stations, post-1998. The color scale has been normalized to the maximum, so that red areas are correlated with red and anticorrelated with blue, while blue regions are correlated with each other. Note the

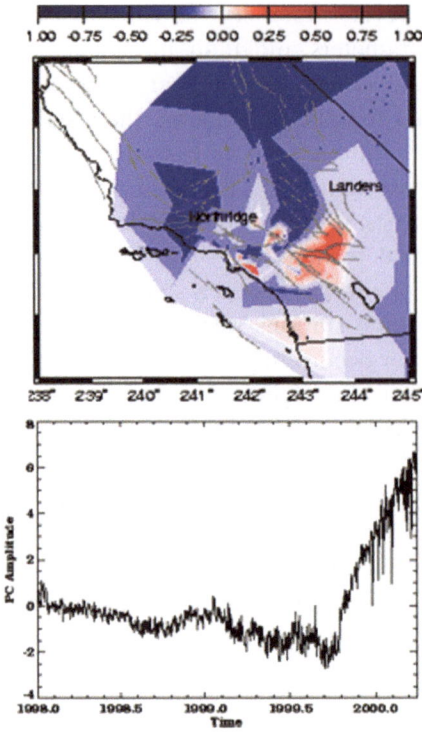

Figure 1

KL Mode 1, for vertical SCIGN 2.0 data, after 1998. At the top is shown the spatial eigenmode, where the color bar represents normalized correlation, running from −1.0 (blue) through white (0.0) to 1.0 (red). At the bottom is shown the PCA signal for this eigenpattern, from 1998 through the spring of 2000.

correlation through the Mojave Desert, which incorporates both the area of the Landers sequence of 1992 and the Hector Mine event of 1999. This corresponds to a jump in the associated time series, also shown in the attached figure, at the time of the Hector Mine event in the fall of 1999. Unexpected results include the correlations just inshore of the 1933 Long Beach earthquake, and the correlated increase in vertical motion near the location of the 1899 Cajon Pass event.

In Figure 2 is shown the first KL mode for vertical deformation from SCIGN 1.0. Notice that this mode characterizes the system-wide correlations in the GPS network, stronger in the western part of the state than the east, and that this differs from the first vertical mode for SCIGN 2.0. This is a result of the change in processing between SCIGN 1.0 and SCIGN 2.0, where the conversion from a global to a regional reference frame decreased the effects of region-wide signals. As a result, if there were an effect that might vary slightly in size over the region, but whose temporal signal was effectively the same, it would be reduced in the SCIGN 2.0 analysis as a result of the regional reference frame (ZHANG *et al.*, 1997; ZUMBERGE *et al.*, 1997; DONG *et al.*, 1998).

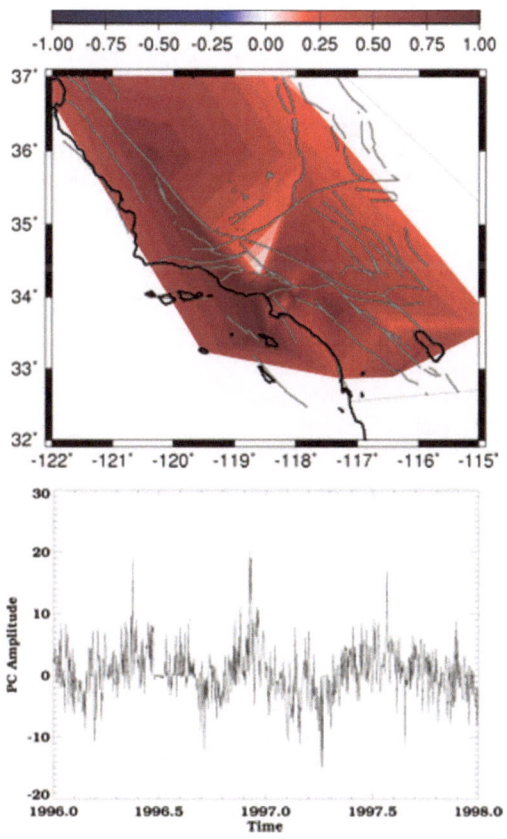

Figure 2

KL mode 1 for vertical deformation, 1996–1998, all of southern California. Top shows the spatial eigenvector, with the same color scale, and on the bottom is the associated temporal signal, 1996–1998.

Recent work has attempted to quantify regional seasonal effects as a variety of signals, such as thermal noise, groundwater, snow mass, precipitation, and tidal or nontidal ocean mass (GROSS and LARSON, 1998; BAWDEN *et al.,* 2001; VAN DAM *et al.,* 2001; DONG *et al.,* 2002; WATSON *et al.,* 2002). Here we can see from the temporal signature of the separated mode that it is a more complicated combination of a variety of signals.

The first assumption in modeling this signal is that it is an aggregate of several seasonal signals. One component of this is likely related to the temperature, although the minimum in the deformation signal is offset from the minimum local temperature, which occurs in January (California Weather Databases, www.ipm.ucdavis.edu/WEATHER). As a result, we began by modeling the thermoelastic strain in a half-space (BERGER, 1975; BEN-ZION and LEARY, 1986). This model consists of an elastically decoupled layer overlying a uniform elastic half-

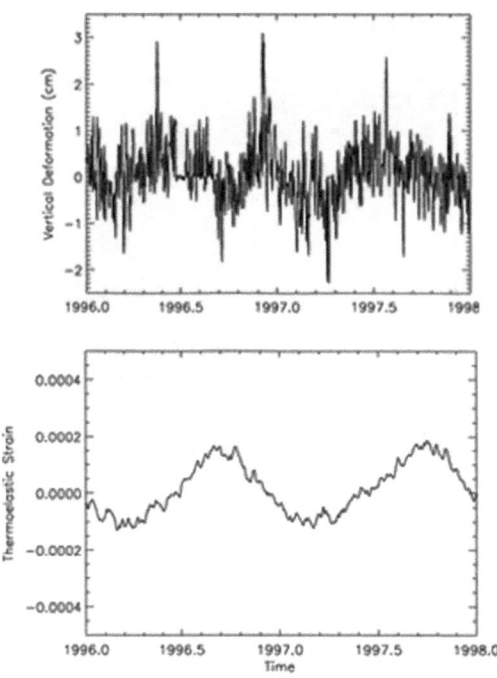

Figure 3
Top—vertical deformation at SCIGN station CLAR, 1996–1998, computed from KL mode 1, shown in
Figure 2. Bottom — thermoelastic strain at CLAR for the same time period.

space. This decoupled layer of unconsolidated material delays, filters, and attenuates
the thermal source field, a stationary temperature wave. It can be shown then that the
horizontal and vertical strain at the surface, ε_{xx} and ε_{yy}, are

$$\varepsilon_{xx} \approx 2(1 + \sigma)\frac{k}{\gamma}\beta T_0 e^{i(\omega t + kx)}, \tag{5}$$

$$\varepsilon_{yy} \approx \frac{(1 + \sigma)}{(1 - \sigma)}\beta T_0 e^{i(\omega t + kx)}, \tag{6}$$

and

$$\gamma \approx (1 + i)[\omega/2\kappa]^{1/2}, \tag{7}$$

where the surface temperature, $T = T_0 e^{i(\omega t + kx)}$, ω is the frequency, k is the
wavenumber (10^{-3} m^{-1}, β is the coefficient of linear thermal expansion (10^{-5} °C^{-1}), κ
is the thermal diffusivity (8.64×10^{-2} m^2/day), σ is Poisson's ratio (1/3), and γ is
approximately 0.3 m^{-1} for periods of one year.

For this work, we calculated the vertical deformation (see Fig. 3, top), in cms, at
CLAR, a SCIGN station located on the eastern side of the LA basin, from the first

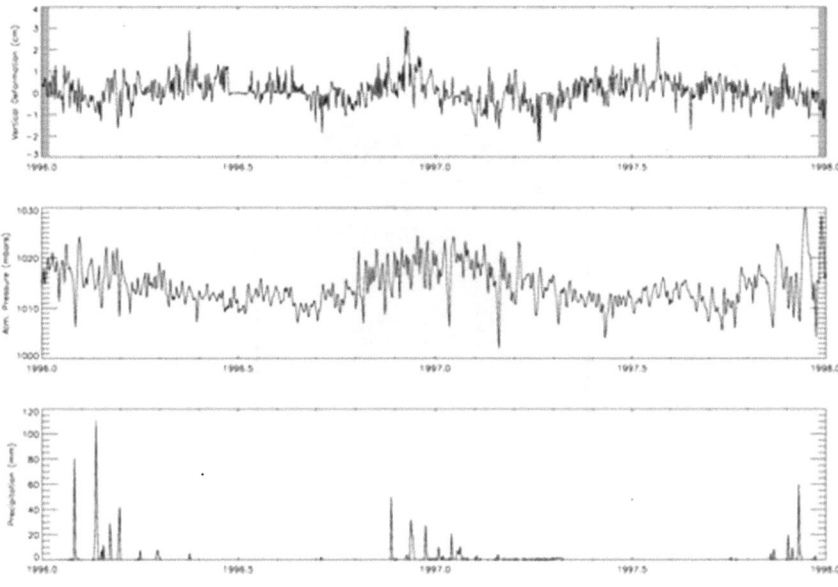

Figure 4

Top—vertical deformation at SCIGN station CLAR, 1996–1998, computed from KL mode 1, as shown in Figure 3. Middle—barometric pressure at CLAR (mbars) for the same time period. Bottom—precipitation at CLAR (mm), again for the same time period.

KL mode (Fig. 2). We also used the algorithmic steps and computer programs of BEN-ZION and LEARY (1986), and the temperature data available from the California Weather Database for Claremont, CA to calculate the horizontal strains at CLAR. The model results were then converted to vertical strain (Fig. 3, bottom) with the above equations. The predicted strain provides a good fit for the delay in the deformation minimum, but there are clearly additional signals causing the maximum deformation in the late spring and the deformation low in the late fall. The restriction on these mechanisms remains that they must be correlated region-wide, and that they are probably interrelated in some way.

Figure 4 illustrates two potential sources of the remaining seasonal signal. At the top, again, is shown the deformation at station CLAR, computed for this mode. The middle shows a plot of the atmospheric pressure near CLAR, from the NOAA weather database (ingrid.ldeo.columbia.edu/SOURCES/.NOAA/), which is correlated, in this region, with temperature (GROSS and LARSON, 1998). On the bottom is shown a plot of the local precipitation for the same time period. Note that the highs in barometric pressure are generally correlated with lows in the deformation, except where they appear to be modulated by heavy precipitation. WATSON et al. (2002) found that increases in the GPS vertical deformation in southern California were preceded by precipitation highs, with a time lag of several months and a magnitude on the order of that seen here. Finally, the variation in the pattern of vertical

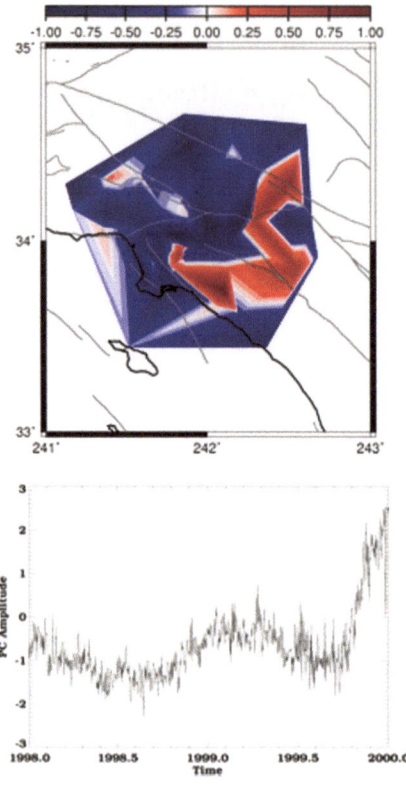

Figure 5

The first vertical KLE mode, 1998–2000, for only the LA basin. Again, the top shows the spatial eigenmode, while the bottom shows the temporal signal.

deformation between 1996 and 1997 may be accounted for by the onset of the 1997–1998 El Niño, which can be seen in the plot of barometric pressure (MCPHADEN, 1999; DONG *et al.*, 2002). Of particular relevance here is the link between the two mechanisms that is demonstrated by their appearance together in this particular spatial and temporal mode, suggesting that the cause of the signal is a modulation of the incipient groundwater signal by the variation in atmospheric pressure on groundwater elevation.

The next step in this analysis from the perspective of data assimilation is to model the aggregate of these signals, tuning the parameters that directly affect the magnitude and frequency of this mode at each station, such as the horizontal wavelength, thermoelastic depth, and soil moisture, using an adaptive search procedure.

What this particular analysis provides is the ability to separate out signals based on both spatial and temporal correlations. While the above example is for a regionally correlated spatiotemporal signal, there are others with correlations on a smaller scale. In Figure 5 is shown the first vertical KL mode for an analysis of stations in the LA basin only. This signal is very similar, both spatially and temporally, to the seasonal groundwater signal identified by BAWDEN et al. (2001) and WATSON et al. (2002) including the sharp discontinuity at the Newport-Inglewood fault zone.

Conclusions

Our results illustrate that the application of this correlation operator technique to coherent GPS data, specifically the SCIGN data array of southern California, can be used to characterize the important deformation modes based on the spatial and temporal scales of the signals and their sources. It should be noted that, while this technique assumes a linearization of the superimposed modes, any nonlinear coupling across the modes will be important in those modes in which either the spatial or temporal scales are of the same order, and should be considered accordingly. In this case, the analysis provides the ability to separate out the regionally correlated seasonal signal from those seasonal signals with smaller spatial scales, and allows us to model them separately from each other and from smaller tectonic deformation.

Quantifying small crustal deformation, such as interseismic signals, using continuous GPS is a difficult task due to the relatively large magnitude of a variety of other signals. The separation of these signals allows for the individual analysis and modeling of the tectonic signals. In addition, the quantification of the spatial and temporal eigenmodes of the tectonic signals will assist in the identification of the appropriate tectonic sources for those signals, as was done for the seasonal signals above. Finally, in order to assimilate the continuous GPS data into models and to better constrain the associated parameters, it is imperative that the underlying tectonic signal be separated from other components. This technique provides a mechanism for removing those modes which are not related to the tectonics of the fault system, or that have not been incorporated into the models as yet.

Acknowledgements

Research by KFT was funded by NGT5-30025 (NASA). Research by JBR was funded by USDOE/OBES grant DE-FG03-95ER14499 (theory), and by NASA grant NAG5-5168 (simulations). Research by WK was supported by USDOE/OBES grant DE-FG02-95ER14498 and W-7405-ENG-6 at LANL. WK would also like to acknowledge the hospitality and support of CNLS at LANL. KFT would also like to acknowledge the work of an anonymous reviewer, and useful discussions with Dr. Roger Bilham, of CIRES, University of Colorado at Boulder.

REFERENCES

ANGHEL, M. and BEN-ZION, Y. (2001), *Nonlinear System Identification and Forecasting of Earthquake Fault Dynamics using Artificial Neural Networks*, EOS Trans., AGU *82*, F571.

ANGHEL, M., BEN-ZION, Y., and MARTINEZ, R. R. (2003), *Dynamical System Analysis and Forecasting of Deformation Produced by an Earthquake Fault*, Pure Appl. Geophys., in press.

BAK, P., TANG, C., and WEISENFIELD, K. (1987), *Self-organized Criticality: An Explanation of the 1/f Noise*, Phys. Rev. Lett. *59*, 381–384.

BAWDEN, G. W., THATCHER, W., STEIN, R. S., HUDNUT, K. W., and PELTZER, G. (2001), *Tectonic Contraction Across Los Angeles After Removal of Groundwater Pumping Effects*, Nature *412*, 812–815.

BEN-ZION, Y. and LEARY, P. (1986), *Thermoelastic Strain in a Half-space Covered by Unconsolidated Material*, Bull. Seismol. Soc. Am. *76*, 1447–1460.

BERGER, J. (1975), *A Note on Thermoelastic Strains and Tilts*, J. Geophys. Res. *80*, 274–277.

BOCK, Y., WDOWINSKI, S., FANG, P., ZHANG, J., WILLIAMS, S., JOHNSON, H., BEHR, J., GENRICH, J., DEAN, J., VAN DOMSELAAR, M., AGNEW, D., WYATT, F., STARK, K., ORAL, B., HUDNUT, K., KING, R., HERRING, T., DINARDO, S., YOUNG, W., JACKSON, D., and GURTNER, W. (1997), *Southern California Permanent GPS Geodetic Array: Continuous Measurements of Crustal Deformation between the 1992 Landers and 1994 Northridge Earthquakes*, J. Geophys. Res. *102*, 18,013–18,033.

CALIFORNIA WEATHER DATABASES, www.ipm.ucdavis.edu.

DONG, D., HERRING, T. A., and KING, R. A. (1998), *Estimating Regional Deformation from a Combination of Space and Terrestrial Geodetic Data*, J. Geod. *72*, 200–214.

DONG, D., FANG, P., BOCK, Y., CHENG, M. K., and MIYAZAKI, S. (2002), *Anatomy of Apparent Seasonal Variations from GPS-derived Site Position Time Series*, J. Geophys. Res. *107*, doi:10.1029/2001JB000573.

DONNELLAN, A. and WEBB, F. H. (1998), *Geodetic Observations of the M 5.1 January 29, 1994 Northridge Aftershock*, Geophys. Res. Lett. *25*, 667–670.

FUKUNAGA, K., *Introduction to Statistical Pattern Recognition* (Academic Press 1970).

GARCIA, A. and PENLAND, C. (1991), *Fluctuating Hydrodynamics and Principal Oscillation Pattern Analysis*, J. Stat. Phys. *64*, 1121–1132.

GROSS, S. and LARSON, K. (1998), *Annual Signals in Continuous GPS Time Series*, Eos Trans. AGU *79*, F182.

MCPHADEN, M. J. (1999), *Genesis and Evolution of the 1997–98 El Niño*, Science *283*, 950–954.

MOGHADDAM, B., WAHID, W., and PENTLAND, A. (1998), *Beyond Eigenfaces: Probabilistic Matching for Face Recognition*, Third IEEE Intl. Conf. on Automatic Face and Gesture Recognition, pp 1–6.

PENLAND, C. (1989), *Random Forcing and Forecasting Using Principal Oscillation Pattern Analysis*, Mon. Wea. Rev. *117*, 2165–2185.

PENLAND, C., and SARDESHMUKH, P.D. (1999), *The Optimal Growth of Tropical Sea Surface Temperature Anomalies*, J. Climate *8*, 1999–2024.

POSADAS, A. M., VIDAL, F., DE MIGUEL, F., ALGUACIL, G., PEÑA, J., IBAÑEZ, J. M., and MORALES, J. (1993), *Spatial-temporal Analysis of a Seismic Series Using the Principal Components Method. The Antequera Series (Spain), 1989*, J. Geophys. Res. *98*, 1923–1932.

PREISENDORFER, R. W., *Principle Component Analysis in Meteorology and Oceanography* (ed. C. D. Mobley) Develop. Atm. Sci. *17* (Elsevier, 1988).

PRESS, W., TEUKOLOSKY, S., VETTERING, W., and FLANNERY, B. *Numerical Recipes in C: The Art of Scientific Computing* (Cambridge University, 1992).

RUNDLE, J. B. and KLEIN, W. (1995), *New Ideas About the Physics of Earthquakes*, Rev. Geophys. Space Phys. Suppl. (July) *283*, 283–286.

RUNDLE, J. B., KLEIN, W., TIAMPO, K. F., and GROSS, S. (2000), *Linear Pattern Dynamics of Nonlinear Threshold Systems*, Phys. Rev. E *61*, 2418–2432.

SAVAGE, J. C. (1988), *Principal Component Analysis of Geodetically Measured Deformation In Long Valley Caldera, Eastern California, 1983–1987*, J. Geophys. Res. *93*, 13,297–13,305.

The SCIGN Project Report to NSF (SCEC, 1998).

TIAMPO, K. F., RUNDLE, J. B., GROSS, S. J., and KLEIN, W. (2002), *Eigenpattern in Southern California Seismicity*, J. Geophys. Res. *107*, doi:10.1029/2001JB000562, 2354.

VAN DAM, T., WAHR, J., MILLY, P. C. D., SHMAKIN, A. B., BLEWITT, G., LAVALLEE, D., and LARSON, K. (2001), *Crustal Displacements due to Continental Water Loading*, Geophys. Res. Lett. *28*, 651–654.

VAUTARD, R. and GHIL, M. (1989), *Singular Spectrum Analysis in Nonlinear Dynamics, with Applications to Paleodynamic Time Series*, Physica D *35*, 395–424.

WATSON, K. M., BOCK, Y., and SANDWELL, D. T. (2002), *Satellite Interferometric Observations of Displacements Associated with Seasonal Groundwater*, J. Geophys. Res. *107*, doi:10.1029/2001JB000470.

WDOWINSKI, S., BOCK, Y., ZHANG, J., and FANG, P. (1997), *Southern California Permanent GPS Geodetic Array: Spatial Filtering of Daily Positions for Estimating Coseismic and Postseismic Displacements Induced by the 1992 Landers Earthquake*, J. Geophys. Res. *102*, 18,057–18,070.

ZHANG, J., BOCK, Y., JOHNSON, H., FANG, P., GENRICH, J., WILLIAMS, S., WDOWINSKI S., and BEHR, J. (1997), *Southern California Permanent GPS Geodetic Array: Error Analysis of Daily Position Estimates and Site Velocities*, J. Geophys. Res. *102*, 18,035–18,055.

ZUMBERGE, J., HEFLIN, M. B., JEFFERSON, D. C., WATKINS, M. M., and WEBB, F. H. (1997), *Precise Point Positioning for the Efficient and Robust Analysis of GPS Data from Large Networks*, J. Geophys. Res. *102*, 5005–5017.

(Received September 27, 2002, revised February 28, 2003, accepted March 7, 2003)

To access this journal online:
http://www.birkhauser.ch

Pure appl. geophys. 161 (2004) 2005–2019
0033–4553/04/102005–15
DOI 10.1007/s00024-004-2546-x

❙ **Pure and Applied Geophysics**

Accelerating Precursory Activity within a Class of Earthquake Analogue Automata

Dion Weatherley[1] and Peter Mora[2]

Abstract — A statistical fractal automaton model is described which displays two modes of dynamical behaviour. The first mode, termed *recurrent criticality*, is characterised by quasi-periodic, characteristic events that are preceded by accelerating precursory activity. The second mode is more reminiscent of SOC automata in which large events are not preceded by an acceleration in activity. Extending upon previous studies of statistical fractal automata, a redistribution law is introduced which incorporates two model parameters: a dissipation factor and a stress transfer ratio. Results from a parameter space investigation indicate that a straight line through parameter space marks a transition from recurrent criticality to unpredictable dynamics. Recurrent criticality only occurs for models within one corner of the parameter space. The location of the transition displays a simple dependence upon the fractal correlation dimension of the cell strength distribution. Analysis of stress field evolution indicates that recurrent criticality occurs in models with significant long-range stress correlations. A constant rate of activity is associated with a decorrelated stress field.

Key words: Critical point hypothesis, cellular automata, accelerating moment release.

Introduction

The physical origins of two empirical relations governing seismicity have been elucidated in recent years, due to the infusion of concepts from statistical physics. The first such relation is the Gutenberg-Richter relation governing the size-scaling of earthquakes (Gutenberg and Richter, 1954). Power-law event scaling of this form has been identified as the signature of extended dynamical systems operating at or near a Critical Point. A broad class of such systems naturally self-organise towards a Critical Point, regardless of the initial conditions, where these systems remain (Bak *et al.*, 1988). These systems are termed Self-Organised Critical (SOC). Comparison of the archetypical SOC model, the sandpile automaton, with a simplified mechanical model for earthquakes, the Burridge-Knopoff model (Burridge and Knopoff, 1967), led Bak and Tang (1989) to propose the Self-Organised Crust Hypothesis: the Earth's crust is perpetually in a state of SOC and the Gutenberg-Richter relation is the temporal signature of this state.

[1] QUAKES, The University of Queensland, Brisbane, 4072, Australia. E-mail: dion@quakes.uq.edu.au
[2] QUAKES, The University of Queensland, Brisbane, 4072, Australia. E-mail: mora@quakes.uq.edu.au

The second empirical relation to find explanation in statistical physics is the Bufe-Varnes relation (BUFE and VARNES, 1993) which characterises an observed acceleration in the rate of seismic energy release prior to a number of historical large or great earthquakes (see JAUMÉ and SYKES, 1999 for a review). Employing Renormalisation Group models for catastrophic failure of hierarchical fibre-bundles, SORNETTE and SAMMIS (1995) identified accelerating energy release as the signature of systems approaching a Critical Point (catastrophic failure).

Numerical models for catastrophic failure were unable to explain repeating cycles of accelerating energy release because such models terminated with catastrophic failure. These models did not allow for healing of previously failed fibres. On the other hand, SOC models may be employed to simulate long histories of seismicity because failed elements are assumed to heal instantaneously and hence, may fail multiple times during a given simulation. However, SOC models do not generally display accelerating energy release once the self-organised critical state is reached; SOC is a statistically stationary state.

SAMMIS and SMITH (1998) made progress in explaining both the Gutenberg-Richter and Bufe-Varnes relations within a common framework, via numerical studies of a hybrid fibre-bundle/SOC automaton involving a geometrical fractal hierarchy of cells. Sammis and Smith found that recurrent cycles of accelerating energy release culminating in events which fail the largest cell may occur if such large events fail a significant portion of the model region and dissipate sufficient energy from the system to perturb the system away from the SOC state. The dynamics displayed by such models consists of intervals in which the system self-organises into the critical state, punctuated by large events which perturb the system away from the critical state. Hereafter we refer to such behaviour as *recurrent criticality*.

While encouraging, the results of SAMMIS and SMITH (1998) were qualitative in nature and it was unclear whether recurrent criticality is a universal class of dynamical behaviour (displayed by a broad range of models). One would like to explore the effect of varying the size of the largest events and the amount of energy dissipated in a given event. However, the geometrical fractal automaton may not readily be employed for this task. An alternative class of models more condusive to this task is examined in this paper. This class is a generalisation of two previously examined models (WEATHERLEY et al., 2000) one of which was found to display recurrent criticality. The automata employ a statistical fractal distribution of cell strengths rather than a geometrical fractal distribution of cell sizes. In the models examined here, the maximum size of events and the amount of energy dissipated in a given event, are governed by tuning parameters.

Numerical evidence from a parameter space study of these models indicates that recurrent criticality is displayed by models residing within a portion of the parameter space. Such models have an over-abundance of larger events, relative to the power-law distribution of small events and the mean energy of the models displays large, saw-tooth fluctuations. A significant percentage of the largest events is preceded by intervals of accelerating energy release. A line of models displaying power-law scaling for all event sizes, marks the boundary in parameter space between models displaying recurrent criticality and models that display a constant rate of energy release.

A significant result is the relationship between the degree of correlations in the stress field and the occurrence of recurrent criticality. Recurrent criticality models contain broad regions of highly correlated stress. Cells within these regions concurrently accumulate stress and fail in the same event. Conversely, models outside the recurrent criticality regime do not contain broad correlated regions; the stress field is decorrelated over long distances. Varying the correlation dimension of the fractal strength distribution systematically alters the location of the boundary of power-law scaling simulations. Increasing the long-range correlations in the strength distribution increases the size of the recurrent criticality regime, strengthening the claim that long-range correlations are an important factor determining whether recurrent criticality is observed.

The paper is organised as follows: The first section contains a description of the models and a brief discussion of the effect of varying the tuning parameters. The following section presents numerical results from a parameter space investigation of the models. The role of stress field correlations is discussed prior to a summary of the pertinent results in the concluding section.

Model Description

The models consist of a rectangular grid of 128×128 cells, each of which is assigned a constant, scalar strength and a variable, scalar stress. The strengths of the cells are pre-defined according to a statistical fractal distribution that is generated using a Fourier filtering technique (TURCOTTE, 1997). Initially, the stresses of all cells are set to zero. External loading is achieved by periodically incrementing the stress of all cells by an amount $\Delta\sigma$, computed dynamically as the minimum stress increment necessary to fail at least one cell. A cell fails when its stress equals or exceeds the strength of that cell.

The two statistical fractal automata examined by WEATHERLEY *et al.* (2000) differed only in the redistribution laws employed to transfer stress from failed cells to the nearest neighbours. Recurrent criticality was identified in a model which transferred all of the stress of failed cells only to the nearest neighbours which had not previously failed in the current rupture cascade. SOC-like dynamics was identified in the other model, in which a proportion of the stress of failed cells was transferred to previously failed neighbours.

This difference motivated the derivation of a new redistribution law in which the ratio of stress transferred to previously failed neighbours compared with the stress transferred to previously unfailed neighbours, is given by a model parameter (κ)

termed the stress transfer ratio. In addition to this new redistribution law, circular (rather than dissipative) boundaries are employed in the new models. A dissipation factor (γ) controls the amount of stress dissipated from failed cells prior to redistribution of stress to the four nearest neighbours.

If a failed cell supports a stress of σ_f, and n_b of its four nearest neighbours have previously failed since the last stress increment was applied, the stress transferred to previously unbroken or previously broken nearest neighbours is given by:

$$\Delta\sigma_u = \frac{\gamma\sigma_i}{[4 - (1 - \kappa)n_b]}, \tag{1}$$

$$\Delta\sigma_b = \frac{\kappa\gamma\sigma_i}{[4 - (1 - \kappa)n_b]}, \tag{2}$$

where $\Delta\sigma_u$ is the amount of stress transferred to unbroken neighbours and $\Delta\sigma_b$ is the stress transferred to broken neighbours. Note that by definition, $\frac{\Delta\sigma_b}{\Delta\sigma_u} = \kappa$, the stress transfer ratio.

It is worthwhile to discuss the properties of models with particular values of these two model parameters. For the case $\kappa = 0$, no stress is transferred to previously failed cells. This defines a crack-like automaton and results in very large stress concentrations along the boundary of expanding ruptures and no stress remains within the failed region. Conversely, $\kappa = 1$ corresponds to the stress redistribution mechanism of the original sand-pile automaton. All neighbours receive an equal amount of stress regardless of whether those neighbours have previously failed. Decreasing κ has the effect of increasing the maximum size of events, due to the increasing amount of stress that is concentrated along rupture fronts.

A dissipation factor of $\gamma = 0$ implies that no stress is transferred to neighbours. All of the stress of failed cells is dissipated from the model. There is no coupling between cells and consequently cells fail independently, i.e., no rupture cascades occur for $\gamma = 0$. A dissipation factor of $\gamma = 1$ stipulates that no stress is dissipated from failed cells. In a model employing periodic boundary conditions, this would result in unbounded increase in the stress accumulated within the model. Such a situation is unphysical and hence all simulations described below employ dissipation factors $\gamma \leq 0.95$. Preliminary tests showed that 5% dissipation was sufficient to prevent unbounded increase in the accumulated stress.

Numerical Results

A parameter space investigation was performed in which κ and γ were varied and the model was simulated for each pair of parameter values. A total of 399 simulations was performed, employing 21 values of κ uniformly spaced in the range $0 \leq \kappa \leq 1$ and 19 values of γ uniformly spaced in the range $0 \leq \gamma \leq 0.95$. Each simulation consisted of 5×10^5 loading timesteps which is typically an order of magnitude

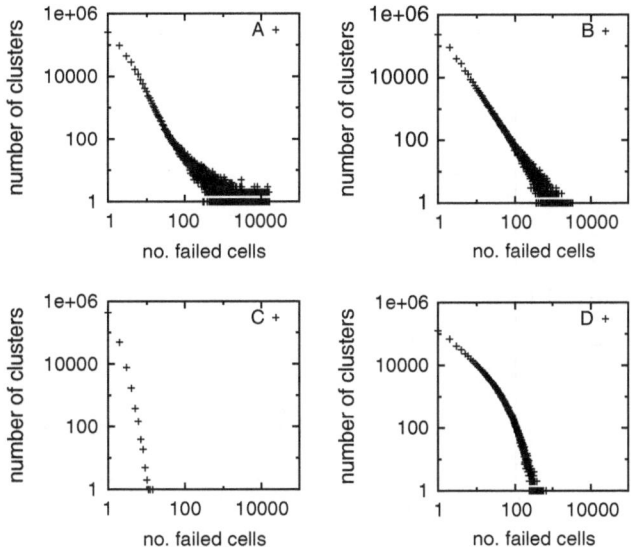

Figure 1

Examples of the four classes of interval event size distributions found in different sub-regions of the parameter space. Letters correspond to the labelled regions of Figure 2.

greater than the number of timesteps required for the model to self-organise into a statistically steady state. All simulations employed the same strength distribution, with a fractal dimension of $D = 2.3$.

Event Size Scaling

Data recorded during each simulation were subjected to a suite of analysis and the results from different simulations were compared. Analysis of interval event size distributions showed that four different classes of event scaling were displayed for models in different subregions of the parameter space. Figure 1 illustrates these four classes and Figure 2 shows the portion of parameter space corresponding to each class. For three of the classes power-law scaling of small to moderate events was observed, however the large event tail differed.

In a corner of parameter space corresponding to small values of the stress transfer ratio ($\kappa \to 0$) and small dissipation ($\gamma \to 1$), an over-abundance of large events relative to smaller events was observed. Hereafter, such a distribution is termed a *characteristic* event distribution. For nearly equal stress transfer to previously failed and unfailed neighbours ($\kappa \to 1$) there was an under-abundance of large events.

A power-law distribution for all event sizes was obtained only for simulations lying along a line through the parameter space. Figure 2 shows the approximate location of this line, to within the resolution of the parameter space exploration

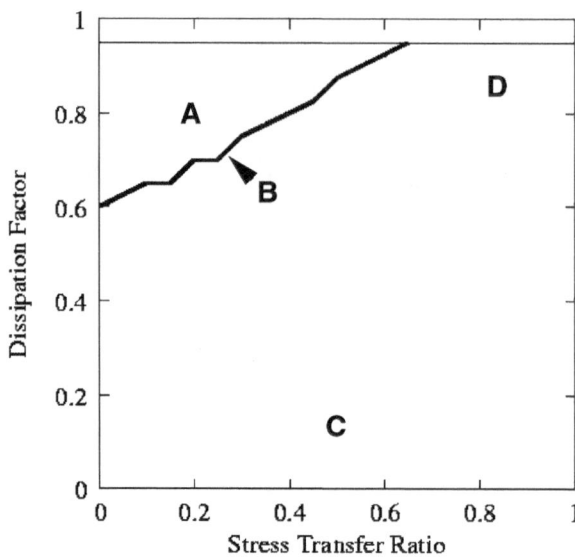

Figure 2
Parameter space plot indicating the relative locations of the four different classes of event size-scaling c.f.
Figure 1. The line sub-dividing the parameter space indicates the approximate location of the line of
power-law scaling simulations.

performed. There was a smooth transition from a characteristic distribution to one
with a roll-over tail as the line of power-law scaling simulations was crossed, i.e., the
numbers of larger events diminished gradually as one traversed parameter space from
$\kappa = 0$ to $\kappa = 1$.

The final class of event scaling, corresponding to large dissipation ($\gamma \rightarrow 0$),
involved a general roll-over in the event distribution for all event sizes. No power-law
scaling was evident. In the limit of total dissipation ($\gamma = 0$), the event distribution
consisted of a single datapoint corresponding to 10^5 individual cell failures, i.e., cells
failed independently as explained above.

Based upon this evidence alone, one suspects that the line of power-law scaling
simulations represents a boundary in parameter space separating models displaying
differing classes of dynamical behaviour. One way to verify this suspicion is to
examine other macroscopic properties of the models. If this line represents a change
in dynamical behaviour, then the line should also mark an observable change in
other macroscopic properties.

Evolution of Mean Stress

The evolution of the mean accumulated stress within the models was also
examined. Figure 3 depicts the mean stress evolution for the simulations whose event
distributions are shown in Figure 2. Models with a characteristic event distribution

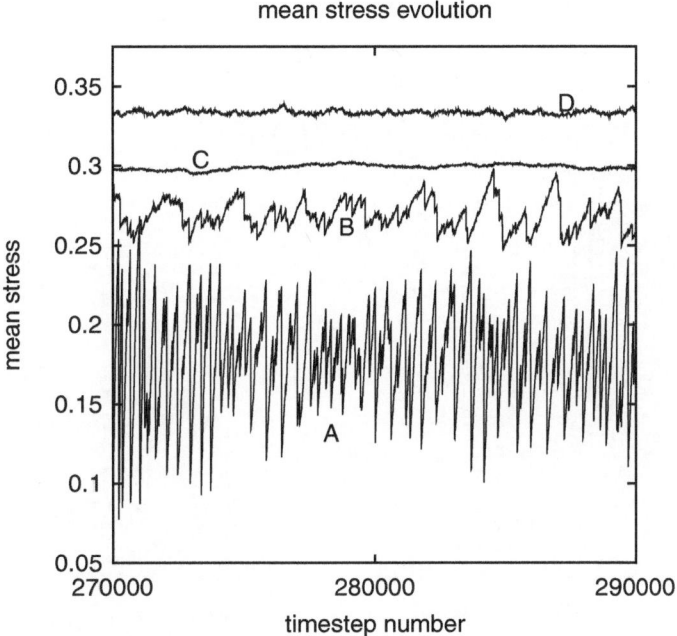

Figure 3

Examples of mean stress evolution for the same simulations depicted in Figure 1. Simulations with a characteristic event distribution displayed quasi-periodic cycles of stress accumulation culminating in large events. Simulations with an under-abundance of large events displayed a near-constant mean stress with no clear evidence for periodicity.

displayed large amplitude, saw-tooth fluctuations in mean stress. Closer examination indicates that mean stress evolution is consistent with quasi-periodic cycles of stress accumulation culminating in sudden stress drops due to the occurrence of large events. Simulations with a power-law event distribution displayed more irregular mean stress fluctuations with a smaller amplitude and a higher time-averaged mean stress. In contrast, models with an under-abundance of large events or highly damped models displayed very small amplitude mean stress fluctuations with no evidence for periodicity.

To summarise the results from all the simulations performed, the time-averaged mean stress, and the maximum and minimum values of the mean stress were plotted as a surface in the 2-D parameter space (see Fig. 4). Large mean stress fluctuations and a low time-averaged mean stress only occur in the corner of parameter space consisting of models with a characteristic event distribution. Elsewhere there is a near-constant high mean stress. The roll-over in time-averaged mean stress corresponds to the line of power-law scaling models, strengthening the hypothesis that this line represents a boundary between two different classes of dynamical behaviour.

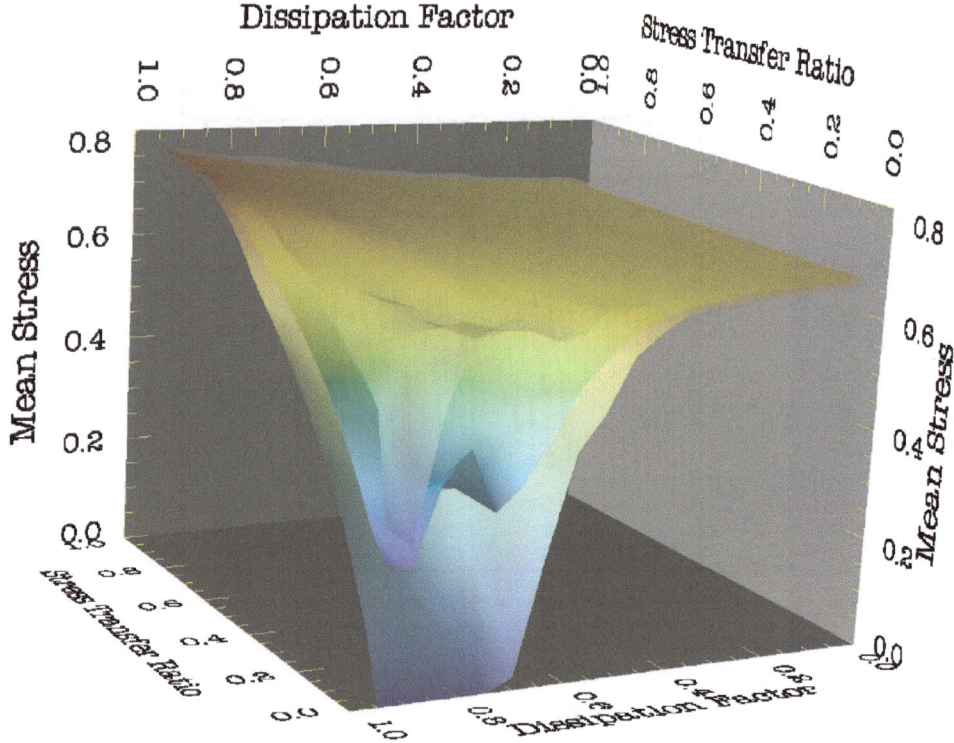

Figure 4
Parameter space plot of the time-averaged mean stress and maximum and minimum mean stress for each simulation. The roll-over in time-averaged mean stress corresponds with the line of power-law scaling simulations delineated in Figure 2.

Accelerating Precursory Activity

One of the motivations for performing this parameter space investigation was to explore the conditions under which accelerating precursory activity is a systematic fore-runner for large events in at least some types of earthquake analogue automata. In a previous study of two similar automata (JAUMÉ et al., 2000), accelerating precursory activity was identified in a crack-like automaton (with $\kappa = 0$ and $\gamma = 0.75$) whereas a constant rate of activity was observed for an automaton with $\kappa \sim 0.8$ and $\gamma = 1$. Given these results it was suspected that the models with large mean stress fluctuations and a characteristic event distribution, may display accelerating precursory activity. This suspicion was verified by determining the probability that a given large event is preceded by accelerating precursory activity, as opposed to a constant rate of activity. The method for computing this probability is as follows.

The 100 largest events occurring during the later half of each simulation were selected for analysis. For each large event, the cumulative stress release prior to that event was fitted to a power-law time-to-failure relation of the form:

$$\sum_{t=(t_f-t_w)}^{t_f} \sigma_R(t) = A - B(t_f - t)^m \quad , \tag{3}$$

where $\sigma_R(t)$ is the total stress released from failed cells at timestep t, and the critical time t_f was defined as the time of the large event. The time-window was varied in the range $500 \leq t_w \leq 3000$, ensuring that at least 500 preceding events were included in the sequence to be fitted. The power-law exponent was also varied in the range $0.05 \leq m \leq 1.00$.

For each value of t_w and m, the fit parameters A and B were computed via least-squares regression. The same sequence of cumulative stress release was also fitted to a linear relation corresponding to a constant rate of activity. The goodness-of-fit ($\mathscr{C}$) for any given choice of t_w and m was computed as the ratio of the r.m.s. error of the power-law fit to the r.m.s. error of the linear fit (as proposed by BUFE and VARNES, 1993). The fit with the lowest value of $\mathscr{C}$ was selected as the best power-law fit for the cumulative stress release prior to the large event under analysis.

Having computed the best power-law fit for each of the 100 large events in each simulation, the number of reasonable power-law fits was counted. The criteria for considering a given fit reasonable were that $\mathscr{C} < 0.7$ and $m < 0.8$. A fit satisfying these criteria is judged to be evidence for accelerating precursory activity prior to the large event, as opposed to a constant rate of activity. The number of reasonable power-law fits for each simulation was defined as the probability that a large event is preceded by accelerating precursory activity.

Figure 5 is a parameter space plot of this probability. A non-zero probability of accelerating precursory activity is obtained only in the corner of parameter space corresponding to simulations with large mean stress fluctuations and a characteristic event distribution. Elsewhere in the parameter space, a zero probability implies that activity occurs at a constant rate. If one were to employ Eq. 3 to attempt to forecast large events during a given simulation, this would only be practical for simulations with a non-zero fit probability.

Discussion of Stress Field Evolution

The results from this parameter space investigation support the claim that the line of power-law scaling simulations represents the boundary between two distinctly different classes of dynamical behaviour. The first class is characterised by quasi-periodic, characteristic events which are typically preceded by an acceleration in the rate of stress release. The mean stress of such models displays large, saw-tooth fluctuations consistent with quasi-periodic stress accumulation culminating in characteristic events. The second class of behaviour is characterised by a near-constant mean stress, an under-abundance of large events and no observable evidence for periodicity. Stress is released at a constant rate.

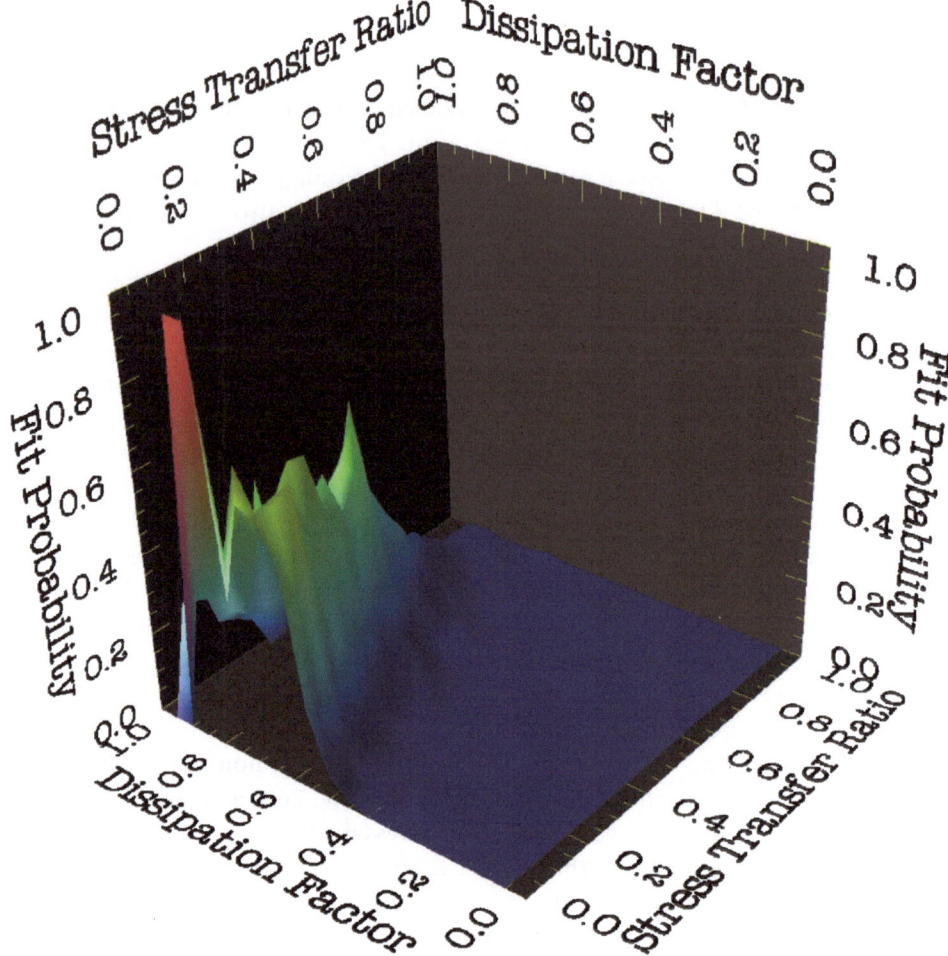

Figure 5
Parameter space plot of the probability that a given large event is preceded by accelerating precursory activity. A non-zero probability is only obtained for simulations with an over-abundance of larger events. Elsewhere, activity occurs at a constant rate.

This second class of behaviour is reminiscent of self-organised criticality (BAK and TANG, 1989) in the sense that the mean stress remains near-constant once the initial transient has subsided, i.e., once the model has self-organised into the critical state. Strictly speaking these models are not self-organised critical because they do not display a power-law distribution for all event sizes however this second class of models and SOC models is considerably more similar than the first class of characteristic models.

We interpret the first class of models as displaying a form of *recurrent criticality*. The models self-organise into a critical state during periods of stress accumulation,

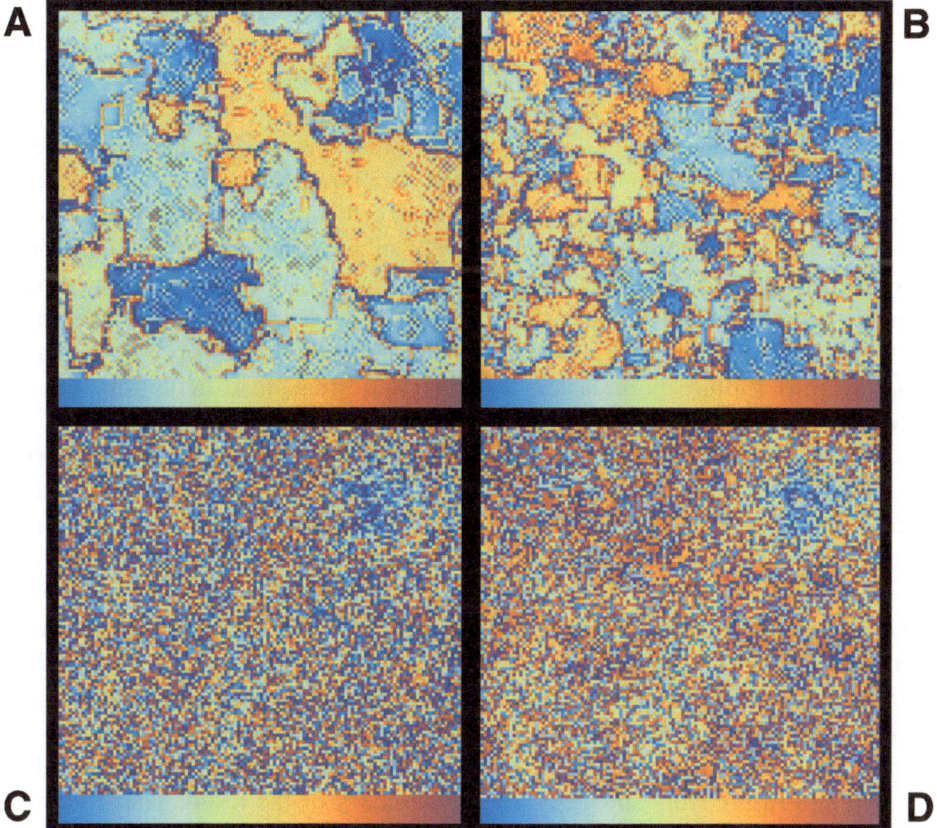

Figure 6

Indicative stress snapshots from each class of models. Notice that characteristic and GR-scaling simulations contain broad correlated regions which accumulate stress and fail together, unlike the other models in which small-scale stress heterogeneity dominates the stress field and the accumulation to failure in future events cannot be readily observed.

however the large events occurring when the critical state is attained, perturb the model away from the critical state. Subsequently the models again self-organise back to the critical state. Accelerating precursory activity has been identified as the signature of systems approaching a critical point (SORNETTE and SAMMIS, 1995) and hence, the occurrence of accelerating precursory activity in these models during stress accumulation, is evidence that the models are approaching a critical state. Since the models repeatedly approach and retreat from the critical state, the term recurrent criticality seems appropriate.

Intuitively one would expect that the differing macroscopic properties of the two classes should be a reflection of differing internal dynamics. One way to examine the internal dynamics is by observing the evolution of stress within the models. Figure 6 contains stress snapshots from the various portions of the parameter space. These snapshots are indicative of the stress field observed throughout the simulations. Recurrent criticality and power-law scaling models display quite different features to that of models outside the recurrent criticality regime.

Specifically, broad regions of recurrent criticality models are observed to accumulate stress and fail semi-independently. The boundaries of previous ruptures are delineated by crack-like features that are one cell wide and which have a higher stress level than surrounding cells immediately subsequent to rupture. This higher stress level is due to the high stress concentrations along rupture fronts which occur for smaller values of the stress transfer ratio ($\kappa \to 0$). The crack-like boundaries are not "frozen in" features; they evolve with time as subsequent ruptures propagate across previous rupture boundaries. By observing stress field evolution, one can predict which regions are most likely to fail in the near future.

Such predictions of future events are not possible for models outside the recurrent criticality regime. These models display a highly heterogeneous stress field with no evidence for broad regions separated by cracklike boundaries. The features of the stress field remain relatively stationary in time. Explicit examination of the stresses of each cell would be needed to identify which cells are likely to fail in a single event.

An analysis of stress correlation evolution prior to large events in these models was also performed. This analysis provides further insight into the evolution of stress within the different classes of models. In recurrent criticality models, systematic stress correlation evolution accompanies cycles of accelerating precursory activity. The results of this analysis and that of other earthquake analogue automata will be published in a future work, however a brief summary of the results is as follows.

During the interval preceding a large event, positive stress correlations for intermediate- to long-range progressively form, in agreement with the results of SORNETTE and SAMMIS (1995). Subsequent to a large event, the long-range correlations are abruptly destroyed and positive short-to-intermediate-range correlations are formed. These correlations are destroyed by smaller aftershock events, prior to an interval of monotonic increasing mean stress (i.e., stress accumulation). During the interval of stress accumulation the stress field is typically decorrelated at all scales. Once sufficient stress accumulates, long-range correlations again reform, prior to the next large event. This systematic evolution of stress field correlations is not observed for models outside the recurrent criticality regime. Such models display a decorrelated stress field at all times. There is no clear evidence for systematic evolution of stress correlations.

These results suggest that the degree of correlation of the stress field may be a factor determining the class of dynamical behaviour displayed by the models. One way to control the degree of stress correlations is to change the fractal correlation dimension (D) of the strength distribution. Since the strength distribution determines whether cells redistribute stress, altering the correlation dimension of the strength distribution indirectly alters the degree of correlations within the stress field.

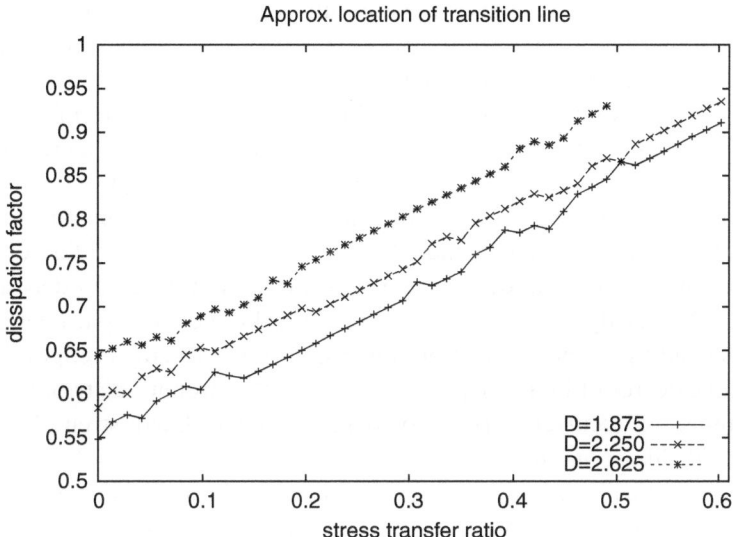

Figure 7

Approximate location of GR-scaling simulations for three different values of the fractal correlation dimension of the strength distribution. The size of the recurrent criticality regime reduces in size as the strength distribution becomes increasingly decorrelated.

Figure 7 is a plot of the line of power-law scaling simulations for suites of simulations employing three different values of the correlation dimension (D) of the strength distribution. As the correlation dimension is increased (the strength field becomes more decorrelated), the size of the recurrent criticality regime diminishes. In other words, recurrent criticality is observed for fewer simulations as the strength distribution becomes increasingly decorrelated. This is in agreement with the observation that a highly correlated stress field corresponds with recurrent criticality while a decorrelated stress field is associated with SOC-like dynamics.

Conclusions

A parameter space study of a class of statistical fractal automata is performed in which parameters are varied that govern the amount of dissipation of stress and the transfer of stress within the model region. Results from the parameter space investigation confirm the existence of a class of dynamics in which large events are systematically preceded by periods of accelerating precursory activity. This class of dynamics is characterised by power-law scaling of smaller events with an over-abundance of larger events (a characteristic event distribution). Mean stress evolution consists of quasi-periodic cycles of stress accumulation, culminating in a large event that abruptly decreases the amount of stress accumulated in the models. Systematic formation of long-range stress correlations, accompanied by accelerating precursory activity, precede large events.

This class of dynamics has been termed recurrent criticality to reflect that models displaying this class of dynamics repeatedly self-organise towards a critical state only to be perturbed away from that state by a large event. Recurrent criticality is distinct from self-organised criticality in that the mean stress does not remain near-constant with small fluctuations. A line of models displaying power-law scaling of events was found to delineate the boundary between recurrent criticality models and models displaying SOC-like dynamics. The location of this boundary in parameter space was found to depend upon the correlation dimension of the fractal strength distribution. Increasing the degree of long-range correlations in the strength distribution promotes the occurrence of recurrent criticality while a decorrelated strength distribution promotes SOC-like dynamics.

Recurrent criticality provides an explanation for both the power-law scaling of earthquakes and the observed acceleration in seismic activity preceding large events from a variety of tectonic settings. Based upon studies of these earthquake-analogue models, one may make qualitative statements regarding the physical properties necessary for recurrent criticality to occur within the Earth's crust. Firstly it is necessary that large earthquakes are capable of releasing a significant fraction of the stress accumulated within the region of interest, so that the region is perturbed away from criticality by such events. Secondly, stress transfer during events must be capable of forming long-range stress correlations. Stress transfer which exclusively roughens (or decorrelates) the stress field at all scales will promote SOC-like dynamics.

The identification of recurrent criticality as a stable dynamical mode of simplified earthquake models is a positive result for earthquake forecasting research, lending weight to the argument that accelerating precursory activity is a systematic intermediate-term precursory of large or great earthquakes. Assuming this to be the case, power-law time-to-failure analysis may provide an effective means for intermediate-term earthquake forecasting, at least in regions for which properties such as dissipation of seismic energy and fault geometry place the Earth's crust within the recurrent criticality regime of state space. It would be beneficial to continue to improve the methodology of power-law time-to-failure analysis, using past seismicity as a test-base, and to remain vigilant of seismic activity rates in regions with a significant seismic hazard.

Acknowledgments

This research was funded by the Australian Research Council and the University of Queensland. Simulations were performed on the Australian Solid Earth Simulator, a 40 GFlop SGI Origin 3800 supercomputer.

REFERENCES

BAK, P. and TANG, C. (1989), *Earthquakes as a Self-organised Critical Phenomenon*, J. Geophys. Res. *94*, B11, 15,635–15,637.

BAK, P., TANG, C., and WIESENFELD, K. (1988), *Self-organised Criticality*, Phys. Rev. *A 38*, 1, 364–374.

BUFE, C. G. and VARNES, D. J. (1993), *Predictive Modelling of the Seismic Cycle of the Greater San Francisco Bay Region*, J. Geophys. Res. *98*, B6, 9871–9883.

BURRIDGE, R. and KNOPOFF, L. (1967), *Model and Theoretical Seismology*, Bull. Seismol. Soc. Am. *57*, 341.

GUTENBERG, B. and RICHTER, C. F. (1956), *Magnitude and Energy of Earthquakes*, Ann. Geofis. *9*, 1.

JAUMÉ, S. C. and SYKES, L. R. (1999), *Evolving Towards a Critical Point: A Review of Accelerating Seismic Moment/Energy Release Prior to Large and Great Earthquakes*, Pure Appl. Geophys. *155*, 279–305.

JAUMÉ, S. C., WEATHERLEY, D., and MORA, P. (2000), *Accelerating Seismic Energy Release and Evolution of Event Time and Size Statistics: Comparison of Models and Observations*, Pure Appl. Geophys. *157*, 2209–2226.

SAMMIS, C. G. and SMITH, S. W. (1998), *Seismic Cycles and the Evolution of Stress Correlation in Cellular Automaton Models of Finite Fault Networks*, Pure Appl. Geophys. *155*, 307–334.

SORNETTE, D. and SAMMIS, C. G. (1995), *Complex Critical Exponents from Renormalisation Group Theory of Earthquakes: Implications for Earthquake Predictions*, J. Phys. I France *5*, 607–619.

TURCOTTE, D. L., *Fractals in Geology and Geophysics* (Cambridge Univ. Press, 2nd ed., 1997).

WEATHERLEY, D., JAUMÉ, S. C., and MORA, P. (2000), *Evolution of Stress Deficit and Changing Rates of Seismicity in Cellular Automaton Models of Earthquake Faults*, Pure Appl. Geophys. *157*, 2183–2207.

(Received September 27, 2002, revised March 3, 2003, accepted March 30, 2003)

 To access this journal online:
http://www.birkhauser.ch

C. Earthquake Generation and Cycles

Pure appl. geophys. 161 (2004) 2023–2051
0033–4553/04/102023–29
DOI 10.1007/s00024-004-2547-9

© Birkhäuser Verlag, Basel, 2004

| Pure and Applied Geophysics

Dynamical System Analysis and Forecasting of Deformation Produced by an Earthquake Fault

MARIAN ANGHEL[1], YEHUDA BEN-ZION[2], and RAMIRO RICO-MARTINEZ[3]

Abstract—We present a method of constructing low-dimensional nonlinear models describing the main dynamical features of a discrete 2-D cellular fault zone, with many degrees of freedom, embedded in a 3-D elastic solid. A given fault system is characterized by a set of parameters that describe the dynamics, rheology, property disorder, and fault geometry. Depending on the location in the system parameter space, we show that the coarse dynamics of the fault can be confined to an attractor whose dimension is significantly smaller than the space in which the dynamics takes place. Our strategy of system reduction is to search for a few coherent structures that dominate the dynamics and to capture the interaction between these coherent structures. The identification of the basic interacting structures is obtained by applying the Proper Orthogonal Decomposition (POD) to the surface deformation fields that accompany strike-slip faulting accumulated over equal time intervals. We use a feed-forward artificial neural network (ANN) architecture for the identification of the system dynamics projected onto the subspace (model space) spanned by the most energetic coherent structures. The ANN is trained using a standard back-propagation algorithm to predict (map) the values of the observed model state at a future time, given the observed model state at the present time. This ANN provides an approximate, large-scale, dynamical model for the fault. The map can be evaluated once to provide a short-term predictions or iterated to obtain a prediction for the long-term fault dynamics.

Key words: Fault dynamics, surface deformation, earthquake prediction.

1. Introduction

A first principles approach to modeling and forecasting the dynamics of an earthquake fault is not feasible at present because the governing physical laws, geometric and structural fault properties, and controlling variables (fault stresses and slips) are not fully available. A practical alternative is to build "phenomenological" models that attempt to estimate the overall character of the system's dynamics. These models quantify basic deterministic or stochastic relationships involving only a few

[1]Computer and Computational Sciences Division, Los Alamos National Laboratory, Los Alamos, NM, U.S.A. E-mail: manghel@lanl.gov

[2]Department of Earth Sciences, University of Southern California, Los Angeles, CA, U.S.A. E-mail: benzion@usc.edu

[3]Department of Chemical Engineering, Instituto Tecnologico de Celaya, Celaya, Guanajuato, Mexico. E-mail: ramiro@losalamos.princeton.edu

irreducible degrees of freedom, which may be used for short-term prediction. We show that for earthquake forecasting this can be done using the spatio-temporal strain patterns embedded in the observable surface displacements. Such an approach is based on the observation that the large-scale dynamics of the system often evolves on a manifold (or invariant measure for the case of strange attractors) with a dimension that is significantly smaller than that of the system's phase space. In such cases, a few macroscopic observables can approximate very well the present state of the system, and predictive models based on the dynamics of a reduced number of macroscopic observables can then be constructed.

In order to identify meaningful low-dimensional structures, we apply the Proper Orthogonal Decomposition (POD)—also known as the Principal Component Analysis (PCA) or Karhunen-Loéve expansion—to an ensemble of surface deformation data generated by the system's dynamics. The POD provides the most efficient way of capturing the dominant components of a dynamical process with only finitely many, and often surprisingly few, "modes" (HOLMES *et al.*, 1996). Using synthetic calculations for a strike-slip fault system, we demonstrate that a reduced number of deformation modes can explain, on the average, the large-scale dynamics of elastic surface deformations. The dynamics of these modes live on a low-dimensional "reduced" attractor in the neighborhood of which the system spends most of its time. The state of the system in this reduced space—model state space—is represented by a set of modal coefficients that measures the projection of the ground surface deformation onto each dominant mode. Our goal is to extract the nonlinear modal dynamics in this model space by constructing a map of the observed model state at a future time, given the observed model state at the present time. We will present preliminary results in which an artificial neural network has been used with promising success to learn the dynamics of the reduced model from these modal time series. The reconstructed map has been iterated to obtain predictions for the long-term model dynamics starting from its current model state. A rough test of the reliability of the model forecast to approximate the future of the fault system is also discussed.

Our method may be compared to the linear pattern dynamics introduced by RUNDLE *et al.* (2000). Their technique is based upon a Karhunen-Loéve expansion of the spatio-temporal seismicity data and is used to estimate a linear stochastic model for the evolution of a probability density function for seismic activity. In contrast to that method, which provides a local linear approximation in a probability space, we propose a global nonlinear approximation that describes the effective large-scale dynamics in a low-dimensional phase space.

2. *Description of the Earthquake System*

We distinguish between the assumed physical system and the model which is an approximate representation of the system. The assumed system is itself a simplified representation of fault systems in the earth, but will be considered a valid description of reality if its behavior is typical for the dynamics of an isolated fault. Our goal is to build data driven models with useful predictive skills which approximate the large-scale dynamics of the system. If this is possible, we can conclude that the methods we propose can then be extrapolated to real isolated faults and, perhaps, even to complex fault systems.

Our assumed system corresponds to a discrete strike-slip fault of length $L = 70$ km and width $W = 17.5$ km embedded in a 3-D elastic continuum (BEN-ZION and RICE, 1993; BEN-ZION, 1996). The fault consists of a uniform grid of dynamical cells where slip is governed by static/kinetic friction processes, surrounded by regions with imposed constant slip rate of $V_{pl} = 35$ mm/year, representing the tectonic loading. We divide the 70 km $\times$ 17.5 km computational grid into 128 $\times$ 32 square cells, with length Δx and depth Δz having equal dimensions of approximately 550 m, (BEN-ZION, 1996) which corresponds to the dimension of small, effectively disconnected, slip patches in mature fault zones. The brittle deformation at any fault position and time is governed by quasi-static, 3-D elastic dislocation theory and spatially varying "macroscopic" constitutive parameters that describe the static and kinetic friction processes. The stress at any fault position (cell) increases with time, t (measured in years), due to the gradual tectonic loading and the time-dependent brittle deformation at other fault locations. It can be written as a boundary integral over the slip deficit $\phi_{kl} = V_{pl}t - u_{kl}$,

$$\tau_{ij}(t) = \sum_{k,l \in B} F_{ij,kl} \times (V_{pl}t - u_{kl}(t)) = \sum_{k,l \in B} F_{ij,kl}\phi_{kl}, \qquad (1)$$

where $F_{ij,kl}$ is the elastic stress transfer function based on the solution of CHINNERY (1963), u_{kl} is the right lateral slip at fault cell kl (measured in mm), and the summation is over the brittle area, $B = 70$ km $\times$ 17.5 km, of the fault plane. The cell indexes i, k and j, l along the horizontal and vertical directions, respectively, define the spatial cell location. We assume that the static brittle strength is uniform along the fault and is set to a value of $\tau_s = 100$ bars. If the stress τ_{ij} reaches the static strength τ_s, a brittle failure occurs at this location and the cell slips an amount Δu_{ij}, until the local stress drops to a prescribed arrest stress level, $\tau_{a,ij}$. The stress transferred from the failed cell can lead to subsequent brittle failures (i.e., rupture propagation) if the stress anywhere increased to the failure threshold. These failures may, in turn, induce or reinduce more brittle slip events. After an initial slip event the strength drops to a dynamic level, $\tau_{d,ij} < \tau_{s,ij}$, and reinitiation of brittle slip on an already failed cell occurs when $\tau_{ij} \geq \tau_{d,ij}$ there. The brittle slip associated with a stress drop can be described by

$$\Delta u_{ij} = \frac{\tau_{f,ij} - \tau_{a,ij}}{F_{ij,ij}}, \qquad (2)$$

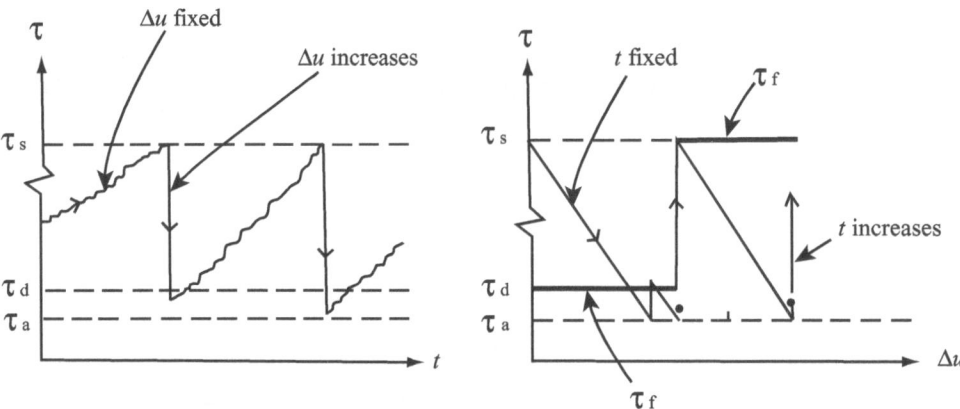

Figure 1
Schematic evolution of the stress and slip for an individual fault cell. The parameters that describe the static/kinetic friction law are: static strength, τ_s, dynamic strength, τ_d, and arrest stress, τ_a. The time dependent failure strength of the cell, τ_f, equals the static strength if the cell has not slipped yet and is adjusted to its dynamic value when the cell fails (from BEN-ZION and RICE, 1993).

where $F_{ij,ij}$ is the self stiffness of a cell and the failure stress $\tau_{f,ij}$ equals the static failure stress, τ_s, for a cell which has not slipped or is adjusted to its dynamic value, $\tau_{d,ij}$, every time a cell has failed. Figure 1 shows a schematic stress-time and stress-slip history for an individual cell. During an iteration loop, all failing cells are identified and their slips updated according to Eq. (2). At the end of each iteration the stress field is updated everywhere according to Eq. (1). The iterations end when there are no more brittle instabilities. This marks the end of an earthquake event whose strength is measured by its potency P defined (BEN-ZION, 2003) as the integral of slip over the rupture area,

$$P = \sum_{ij} \Delta u_{ij} \Delta A, \tag{3}$$

where $\Delta A = \Delta x \times \Delta z = 0.29 \, Km^2$ and Δu_{ij} is the total slip at cell (ij). During a seismic event the tectonic slip, $V_{pl}t$, remains constant. At the end of each event the failure strength everywhere on the fault recovers back to τ_s. In order to initiate the next seismic event, we advance in Eq. (1) the tectonic slip (or equivalently, advance the time t) until the fault cell closest to failure reaches its static failure strength. This triggers a new seismic event and restarts the failure iteration loops.

An important component of the fault rheology is the distribution of arrest and dynamic stresses along the fault. In the simulations used in this paper we chose random stresses,

$$\tau_{a,ij} = <\tau_a> + A\xi, \tag{4}$$

where $< \tau_a >$ is the fault-averaged arrest stress, ξ is a random number uniformly distributed in the range $[-0.5, 0.5]$, and A is the noise amplitude. The static strength, dynamic strength, and arrest stress are related to each other, at each point on the fault as

$$D = \frac{\tau_s - \tau_{a,ij}}{\tau_s - \tau_{d,ij}}, \tag{5}$$

where D is a dynamic overshoot coefficient. Its inverse, $1/D$, proportional to $\tau_s - \tau_{d,ij}$, is a measure of the dynamic weakening that characterizes static/kinetic friction models. The arrest and dynamic stress distributions do not evolve in time (quenched heterogeneities) and the fault dynamics is deterministic. We also assume no observational errors. We record the fault evolution after all transients have died out and the dynamics have reached a statistical steady state (all moments of the stress and displacement fields become time-independent). A thorough motivation of the fault system, based on a wealth of observations and numerical results, is provided in a series of papers by BEN-ZION and RICE (1993, 1995) and BEN-ZION (1996) while a detailed description of the dynamics can be found in ANGHEL 2004, BEN-ZION and RICE (1993) and BEN-ZION et al. (2003).

3. Qualitative Dynamics of Slips, Stresses, and Seismicity

We start by analyzing the evolution of seismicity and the dynamics of the first- and second-order moments of the stress and slip fields. Here we only investigate the dynamics of two fault realizations that differ largely in the value of the dynamic overshoot coefficient. System A with $D = 1.5$ has a large dynamic weakening, while system B with $D = \infty$ has no dynamic weakening. Both systems have random arrest stress distributions, as in (4), with $\langle \tau_a \rangle = 80$ bars and $A = 20$ bars amplitude. A detailed investigation of the dynamics, covering the system space between these two limits, is presented elsewhere (ANGHEL, 2004). In general, each system observable contains a different amount of information for understanding the underlying dynamics. As discussed below, for prediction purposes we are only interested in the dynamics at the large length and time scales. As the small-scale dynamics overwhelmingly dominates the statistics, it is not obvious a priori which observable provides information about the large-scale dynamics. In Figure 2 we present the time evolution of the potency released over a 30-year time interval. The fault motion in system B ($D = \infty$) seems to be dominated by erratic, random appearing evolution. There are no detectable time patterns. When significant dynamic weakening is present, as in system A ($D = 1.5$), a clear signature of a quasi-regular evolution emerges. This can be identified in the presence of very large events (these are large-scale events in which a large fraction of the fault slips), followed by intervals of seismic quiescence. There is a quasi-regular build up of correlations (BEN-ZION et al.,

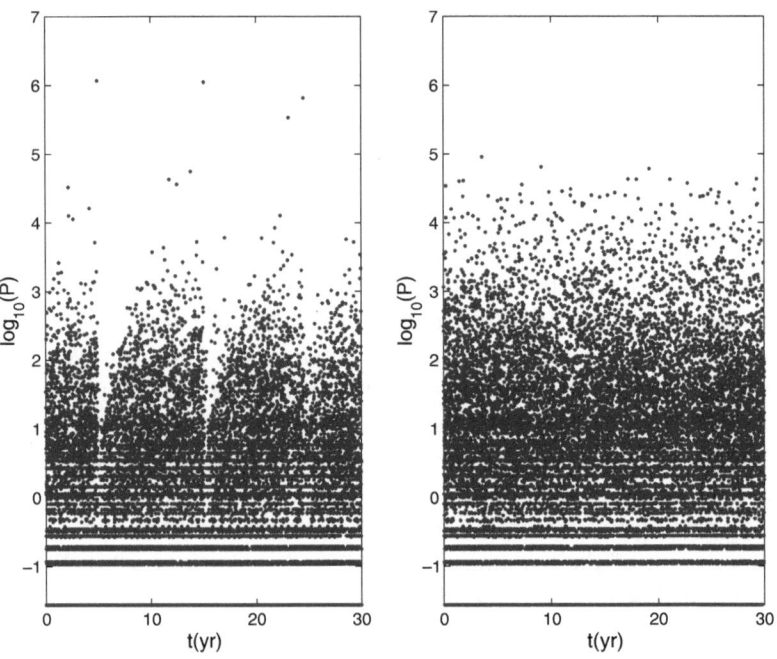

Figure 2

A 30-year evolution of the potency released by two fault systems. Results for system A, with small dynamic overshoot $D = 1.5$ are on the left and for system B, with large dynamic overshoot $D = \infty$ (and no dynamic weakening) are on the right. System A exhibits quasi-regular seismic cycles associated with the large events. There is no detectable structure in the evolution of seismicity in system **B**.

2003) which ultimately generates very large events, much larger in size than in system **B**. Although the evolution of seismicity is dominated by small events, there are significant evolution patterns reflecting the dynamics of the large-scale events.

These can be better seen in Figure 3 which shows the evolution of the average slip deficit,

$$\phi(t) = \frac{1}{M} \sum_{i,j} (\phi_{ij}(t) - \langle \phi_{ij} \rangle), \tag{6}$$

and slip deficit variance,

$$\sigma(t) = \frac{1}{M} \sum_{i,j} (\phi_{ij}(t) - \phi(t))^2. \tag{7}$$

Due to the inhomogeneity of the slip deficit, the slip fluctuations are measured from a local time average, $< \phi_{ij} >$, defined as

$$\langle \phi_{ij} \rangle = \frac{1}{T} \sum_{t=1}^{T} \phi_{ij}(t), \tag{8}$$

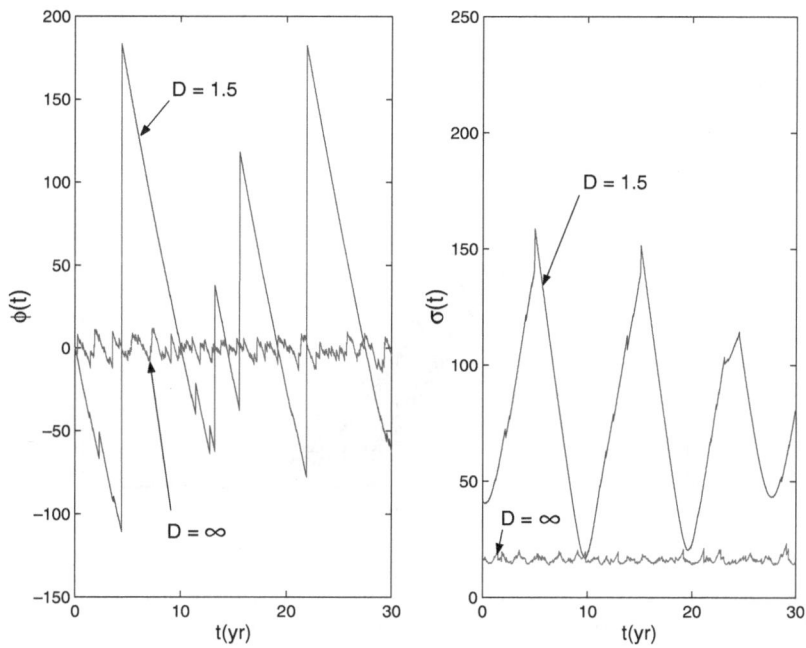

Figure 3
Time evolution of the average slip deficit (left) and its variance (right) for the $D = 1.5$ and $D = \infty$ fault systems.

where the length of the averaging time used is $T = 100$ years. For the $D = 1.5$ system we observe large seismic cycles and smooth long-term behavior between two consecutive large events. At the scale of the plot in Figure 3, the erratic and unpredictable small-scale dynamics is unobservable. Magnifying the plot will reveal small-scale jitters superposed over the smooth large-scale motion. This suggests a clear separation of time scales between the slow, large-scale motions and the fast, small-scale motions. In contrast, the $D = \infty$ system has small oscillations and noisy behavior; note the large difference in the scale of the variance evolution for these two systems. For the $D = \infty$ system there are no large-scale features and the dominant signature of small-scale events renders the system higher dimensional. Additional illustrations of the dramatic change in the dynamics for cases with and without dynamic weakening are given in BEN-ZION et al., 2003.

The emergence of a clear separation of length scales is also revealed in the frequency-size statistics shown in Figure 4. The $D = \infty$ system has a power-law distribution with an exponential cutoff, characteristic of systems close to a critical point. This confirms the analysis of FISHER et al. (1997) who showed that the dynamic weakening $\tau_s - \tau_{d,ij}$ is a tuning parameter and found that the system has

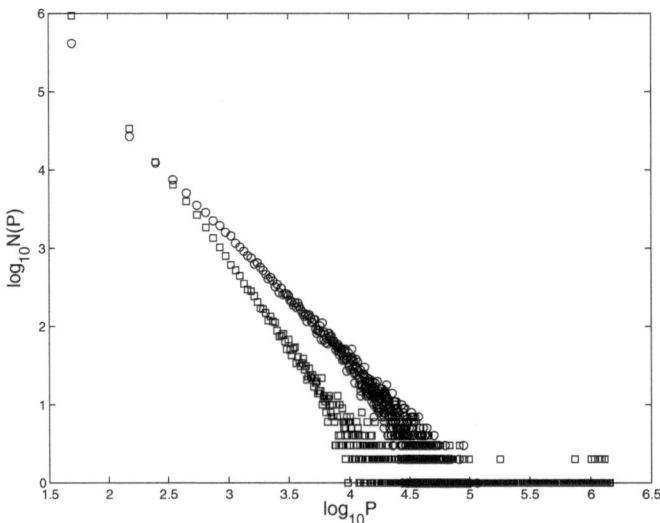

Figure 4
Log-log frequency-size histogram plots for the $D = 1.5$ system (squares) and the $D = \infty$ system (circles). The earthquake size is measured by its potency and the histograms use equal bins of size 100 in potency units ($0.29\text{mm} \times \text{Km}^2$).

an underlying critical point of second-order phase transition at zero dynamic weakening ($D = \infty$) and full conservation of stress transfer during failure events. The frequency-size statistics for the $D = 1.5$ system shows the existence of two separate earthquake populations. At small scales we observe a power-law distribution of event sizes. The other events cover a broad range of scales and their frequency is significantly higher than an extrapolation based on the power-law statistics of the small events. This behavior is reminiscent of the characteristic earthquake model (WESNOUSKY, 1994; BEN-ZION, 2003), which has a distinct probability density peak at the large size end of the frequency-size distribution. Moreover, a careful analysis (ANGHEL, 2004) has shown that most stress dissipation is produced at the large size end of the potency spectrum and is due to the large-scale motions. Averaged over small time intervals, the small events that dominate the frequency-size histogram produce no stress dissipation on average and only transfer stresses internally. This suggests a possible description of the dynamics in terms of large-scale dissipative motions that balance the stress increment due to an effective loading rate which absorbs most of the small-scale events into a renormalized plate velocity. This coarse-grained dynamics is inherently of lower dimension than the original dynamics. The question is whether the dimension collapse is significant to render data-driven model reconstruction algorithms practical.

4. Effective Dimension of the Large-scale Motions

For the class of systems discussed in this paper the knowledge of the slip deficit, $\phi_{ij} = V_{pl}t - u_{ij}$, where $i = 1, \ldots, N_x$ and $j = 1, \ldots, N_z$, fully specifies the state of the fault; the number of microscopic degrees of freedom is $M = N_x \times N_z$ and the state of the system is completely represented by a vector in a finite, but high-dimensional vector space $\mathbf{R}^M$, where $M = 4096$. We conjecture that a finite dynamic weakening favors the creation of spatial and temporal correlations that collapse the asymptotic, long time dynamics of the fault onto an attractor of a smaller effective dimension m; $m \ll M$. This is compatible with the recent analysis of Ben-Zion *et al.* (2003). The physics of the dimensional collapse consists the clear separation of time and length scales. In such cases we can in principle isolate the large-scale dynamics which, evolving slowly, will drive the rapidly relaxing small-scale dynamics. However, are real faults characterized by a large dynamic weakening? Full elastodynamic simulations of rupture propagation in a 3-D elastic solid (MADAR-IAGA,1976) indicate that $D \simeq 1.25$, corresponding to large dynamic weakening. Moreover, we have shown elsewhere (ANGHEL, 2004) that this separation of scales holds for a broad range of D values and is, therefore, generic. Consequently, restricting our study to system A, we shall address for the rest of this paper the following issues: (1) Whether we can build a predictive model for the large-scale motions and (2) What effect the neglected small-scale motions have on the predictability of the larger scales.

In order to model the large-scale motions we must find first if their dynamics is deterministic in character. In principle, the coarse-grained dynamics of any deterministic system is stochastic. When a separation of scales exists, we can assume that only the statistical properties of the small-scale motions influence the larger scales, through coefficients of turbulent viscosity, and that at any moment these statistical properties are determined by the larger scale motions (LORENZ, 1969). A model describing only the large-scale dynamics is assumed to be effectively deterministic. However, we must be careful, because, as pointed out by LORENZ (1969) there remain uncertainties in these statistics, and their progression to the very large-scales will ultimately produce large-scale errors and limit the forecasting skill of our model. We will therefore assume, as a working hypothesis, that a model describing only the larger scales motions is effectively deterministic. In order to test our assumption and find an upper bound on the dimension of the model, we start with a phase space analysis of the large-scale motions as is reflected in the scalar dynamics of the slip deficit variance.

The key to unraveling the dynamics from a scalar time series is to reconstruct a vector space, formally equivalent to the original attractor of the dynamics. For a deterministic system, takens embedding theorem (KANTZ and SCHREIBER, 1997) guarantees, that such a reconstruction is possible from a time-delay reconstruction of only one scalar observable. In our case, the embedding is performed using Cartesian coordinates comprised of the observed slip deficit variance and its time-delayed copies,

$$\mathbf{X}_n = [\sigma(n), \sigma(n - d), \ldots, \sigma(n - (m - 1)d)], \qquad (9)$$

where $\sigma(n) = \sigma(n\Delta t)$ is the slip deficit variance measured at equal sampling times $\Delta t = 0.1$ years, d is a time lag (an integer multiple of the common lag Δt), and m is an embedding dimension. The optimal time lag, $d = 3$, has been chosen at the first minimum of the time-delayed mutual information as suggested in FRASER and SWINEY(1986) and motivated in detail in KANTZ and SCHREIBER (1997). The problem is to determine the optimal embedding dimension m, for which the reconstruction by delay vectors provides an acceptable unfolding of the attractor so that the orbits composing the attractor are no longer crossing each other in the reconstructed phase space (ABARBANE,1996). The practical solution is to look for false neighbors in the embedding phase space at a given value of m (ABARBANEL, 1996; KANTZ and SCHREIBER, 1997). To understand this concept, consider the situation that an m dimensional delay reconstruction is an embedding, but an $(m - 1)$ dimensional delay reconstruction is not. If the embedding dimension is too small to unfold the attractor, a small $\mathbf{R}^{m-1}$ neighborhood will contain points that belong to different parts of the original attractor. Therefore, at a later time, the images of these points under the system's dynamics will split onto different groups, depending on from which part of the attractor the points originate. This lack of a unique location of all the images in $(m - 1)$ dimensions is reflected in finding false neighbors, meaning that determinism is violated. When increasing m, starting with small values, one can detect the minimal embedding dimension, m_e, by finding no more false neighbors. Then, we can invoke the result of Sauer *et al.* (1991) who showed that the attractor formed by $\mathbf{X}_n$ in the embedding space is equivalent to the attractor in the unknown space in which the system is living if m is larger than twice the box counting dimension, d_0, of the attractor. Often, an $m = d_0$ embedding dimension is enough for unfolding the attractor. Therefore, the minimal embedding dimension sets the following bounds, $(m_e - 1)/2 \leq d_0 \leq m_e$, for the dimension of the attractor in the true phase space. Figure 5 shows that the percentage of false neighbors for the $D = 1.5$ system falls under 1% when the phase space embedding of the slip variance data reaches dimension seven. This behavior suggests a topological dimension in the true phase space between three and six. We note that the percentage of false neighbors is not completely reduced to zero. The explanation for this behavior is discussed in Section 6. To appreciate this result, we show for comparison in Figure 5 the behavior of the system in the limit of no dynamic weakening. Even for an embedding dimension as high as ten, the percentage of false neighbors does not drop below 4%, and is an order of magnitude larger than the percentage for the $D = 1.5$ system at the same embedding dimension. This confirms our previous analysis and proves that in the limit of low dynamic weakening the system has high dimensional dynamics.

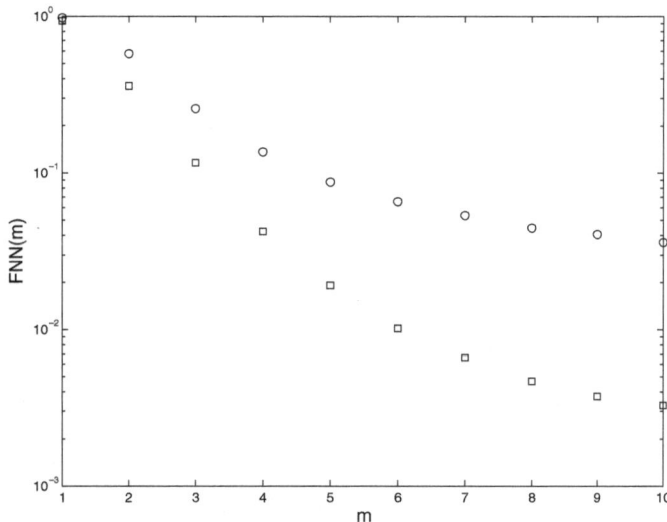

Figure 5

The fraction of false nearest neighbors (FNN) as a function of the embedding dimension $m = 1, \ldots, 10$. Compared to model B ($D = \infty$, circles), there is a dramatic decrease in the number of FNN for model A ($D = 1.5$, squares), as the dimension of the embedding phase space is increased. This suggests that a low-dimensional deterministic dynamics is a good approximation for the large-scale dynamics of the fault.

The foregoing analysis indicates convincingly the deterministic structure of the large-scale motions: for stochastic or very high dimensional dynamics the number of false neighbors will never effectively drop to zero. The low-dimension result is consistent with the embedding theorem (SAUER et al., 1991) which asserts that it is not the dimension of the underlying true phase space that is important for the minimal dimension of the embedding space, but only the fractal dimension of the support of the invariant measure generated by the dynamics in the true phase space (KANTZ and SCHREIBER, 1997). These results justify our earlier remarks that due to some collective behavior only a few dominant degrees of freedom remain and our conjecture that their dynamics is effectively deterministic.

Before addressing the problem of identifying these collective degrees of freedom, we would like to know if their dynamics is chaotic and, if so, what are the implications for their predictability. To answer this question we compute the maximal Lyapunov exponent. This exponent measures the exponential divergences of nearby trajectories and is an average of these local divergences over the entire data. A positive maximal Lyapunov exponent is a signature of chaos. Its computation is based on the algorithm described by Kantz and SCHREIBER (1997) and uses the implementation provided in the publicly available software package TISEAN (HEGGER et al., 1999) (all nonlinear time series algorithms described in this paper use the same excellent implementation). For a point $\mathbf{X}_n$ of the time series in the embedding space, the algorithm determines all neighbors $\mathbf{X}_{n'}$ within a neighborhood

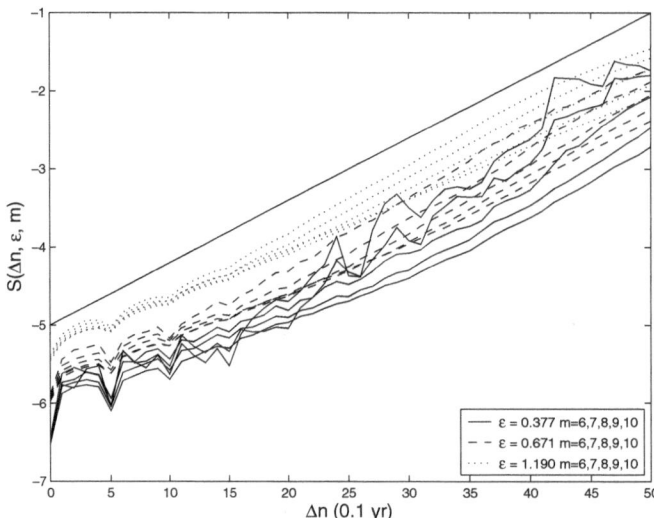

Figure 6
Estimates of the maximal Lyapunov exponent from the slip deficit variance time series data, $\sigma(t)$, for the $D = 1.5$ model. The logarithm of the stretching factor is computed for three different neighborhood sizes, $\epsilon = 0.377, 0.671, 1.19$, and $m = 6, 7, 8, 9, 10$ embedding dimensions. The robust linear behavior of $S(\Delta n)$ reflects the underlying determinism of the data and its slope is an estimate of the maximal Lyapunov exponent, $\lambda_{max} = 0.8 \text{years}^{-1}$ (the straight line has slope 0.08 and the unit of time is 0.1 years).

U_n of radius ϵ. Then, for each neighbor, we consider the distance $\delta(0) = |\mathbf{X}_n - \mathbf{X}_{n'}|$ and we read its evolution $\delta(\Delta n) = |\mathbf{X}_{n+\Delta n} - \mathbf{X}_{n'+\Delta n}|$ from the time series. If for all neighbors $\delta(\Delta n) \simeq \delta(0)e^{\lambda \Delta n}$, then λ is the local divergence rate. The local divergence rate varies along the attractor and the true maximal Lyapunov exponent is an average over many reference points n. Therefore, in practice we compute the average over the distances of all neighbors to the reference part of the trajectory as a function of the relative time Δn. The logarithm of the average distance at time Δn measures the effective expansion rate over the time span Δn:

$$S(\Delta n, \epsilon, m) = \left\langle \ln \left(\frac{1}{|U_n|} \sum_{\mathbf{X}_{n'} \in U_n} |\mathbf{X}_{n+\Delta n} - \mathbf{X}_{n'+\Delta n}| \right) \right\rangle_n, \qquad (10)$$

where the outer bracket denotes the averaging over the reference point $\mathbf{X}_n$, $|U_n|$ is the number of neighbors in U_n, and the argument of the logarithm is the average local expansion rate. If $S(\Delta n, \epsilon, m)$ vs. Δn exhibits a robust linear increase with identical slope for a range of ϵ values and for all m larger than some minimum embedding value, then its slope is an estimate of the maximal Lyapunov exponent λ_{max} per time step (KANTZ and SCHREIBER, 1997). Figure 6 shows three bundles of curves corresponding to three neighborhood sizes, $\epsilon = 0.377, 0.671$, and 1.19, as can be seen for $\Delta n = 0$. Each bundle shows the behavior of the expansion rate for $m = 6, 7, 8, 9$

and 10 and proves that the result is robust to changes in ϵ and does not depend on the embedding dimension when m is large enough. Our estimate for the maximal Lyapunov exponent is $\lambda_{max} = 0.8$ years^{-1}. We can therefore conclude that the large-scale motions have a positive maximal Lyapunov exponent and exhibit sensitive dependence on the initial conditions. Does this dependence set a fundamental limitation to long-term forecasting? Two initially close trajectories will diverge exponentially in the phase space with a rate given by the largest Lyapunov exponent λ_{max}. Therefore, for any finite uncertainty δ in the initial conditions, we can forecast the future state of the system only up to a maximum time,

$$T_p \simeq \frac{1}{\lambda_{max}} \ln\left(\frac{\Delta}{\delta}\right), \tag{11}$$

where Δ is the accepted tolerance level. For the earthquake system, an approximate estimate for the predictability time is $T_p = 1/\lambda_{max} \simeq 1.25$ years. This result is disappointing and, due to expected model reconstruction errors, which might overshadow the uncertainty in the initial conditions, raises questions regarding the value of any forecasting endeavor. However this estimate is naive and holds only for infinitesimal perturbations and in nonintermittent systems. Generally, the predictability time is scale-dependent (BOFFETTA *et al.*, 2002) and can be much longer than the rough estimation $T_p \simeq 1/\lambda_{max}$. As we will shortly see, at the large length scale of interest for forecasting, a better estimation of the predictability time is $T_p \simeq 10$ years.

5. Proper Orthogonal Decomposition

The dynamics of the first-and second-order moments of the slip and stress fault fields provide an adequate description of the large-scale motions. Unfortunately, their measurement poses very difficult problems: the stress field is not directly observable while the slip field requires the solution of an ill-posed inverse problem. Therefore, we propose an alternative analysis of the large-scale motions based on the spatio-temporal strain patterns embedded in the surface deformation fields that can be accurately measured by InSAR and GPS observations (BURGMANN *et al.*, 2002). Assuming uniform slip discontinuities over each rectangular cell, the components of the surface displacement vector U_α are found to be (OKADA, 1985),

$$U_\alpha(\mathbf{x}) = \sum_{i,j} \Gamma^\alpha(\mathbf{x}; \mathbf{x}_{ij}) \times (u_{ij} - V_{pl}t), \tag{12}$$

where $\Gamma^\alpha(\mathbf{x}; \mathbf{x}_{ij})$ is the surface displacement in the direction $\alpha = x, y, z$ at the observation point $\mathbf{x} = (x, y, 0)$ due to uniform unit slip on the fault cell (ij) located at $\mathbf{x}_{ij} = (x_{ij}, 0, z_{ij})$. The space-time signal is obtained by simultaneous measurements of the surface displacements on a 64×32 uniform rectangular grid which covers a

100 *km* × 50 *km* surface area centered around the fault. For each surface deformation we compute an ensemble of N snapshots, $U_\alpha(\mathbf{x}, n) = U_\alpha(\mathbf{x}, n\Delta t)$, $n = 1, \ldots, N$, and $\alpha = x, y, z$, every $\Delta t = 0.1$ years. The analysis of each deformation direction proceeds identically, therefore we will henceforth drop the deformation index α. Since we are only interested in decomposing the dynamics of fluctuations, it is convenient to separate the flow $U(\mathbf{x}, n)$ into a time-independent mean and a fluctuating part, i.e.

$$U(\mathbf{x}, n) = \langle U(\mathbf{x}) \rangle + u(\mathbf{x}, n), \tag{13}$$

where $\langle u(\mathbf{x}, n) \rangle = 0$ and $\langle \cdot \rangle$ indicates the ensemble average.

The identification of the active degrees of freedom in the surface deformation fields uses the Principal Orthogonal Decomposition (POD) (HOLMES *et al.*, 1996). The POD seeks an optimal representation (in the least-square sense) for the members of the ensemble $\{u(\mathbf{x}, n)\}_{n=1}^N$. It searches for generalized directions $\phi(\mathbf{x})$ in the configuration phase space, such that most of the ensemble fluctuations will be directed along $\phi(\mathbf{x})$. Mathematically, this is achieved by maximizing the average projection of the $\{u(\mathbf{x}, n)\}$ ensemble onto $\phi(\mathbf{x})$, i.e.,

$$\max_{\phi \in L^2, ||\phi||=1} \langle (u, \phi)^2 \rangle, \tag{14}$$

$(\cdot, \cdot)$ is the L^2 inner-product in the configuration space and the L^2 norm constraint, $||\phi|| = 1$, is required for the maximum to be defined. The solution to this variational problem is given by the eigenfunctions $\{\phi_i\}_1^M$ of the following integral equation (HOLMES *et al.*, 1996)

$$\int K(\mathbf{x}, \mathbf{y}) \phi_i(\mathbf{y}) \, d\mathbf{y} = \lambda_i \phi_i(\mathbf{x}) \quad , \quad \lambda_1 \geq \lambda_2 \geq \ldots \geq \lambda_M \geq 0, \tag{15}$$

whose kernel $K(\mathbf{x}, \mathbf{y})$ is the two point correlation matrix of the ensemble, $K(\mathbf{x}, \mathbf{y}) = \langle u(\mathbf{x}) u(\mathbf{y}) \rangle$. There are at most M eigenfunctions corresponding to the total number of degrees of freedom. The eigenvalues λ_i measure the mean square fluctuations of the ensemble in the directions defined by their corresponding eigenfunctions: $\lambda_i = \langle (u, \phi_i)^2 \rangle$. We can also think of λ_i as the average "energy" of the ensemble fluctuations projected onto the ϕ_i axis. Therefore, ranked in decreasing order of their eigenvalues, the eigenfunctions (sometimes called empirical eigenfunctions, coherent structures, or dominant modes) will identify the dominant directions in the configuration space along which most of the fluctuations take place. The first four modes describing the predominant motions of the u_x deformation field for system A with $D = 1.5$ are shown in Figure 7. The coherent, collective nature of the fluctuations represented by these modes is eloquently expressed by their spatial structure. The spatial coherence sharply decreases for the higher order modes (not shown). The modes probe the system at different scales, from the largest scales, captured by the most dominant modes, to the smallest scales, described by the higher order modes. This observation is hardly surprising, since for a translationally

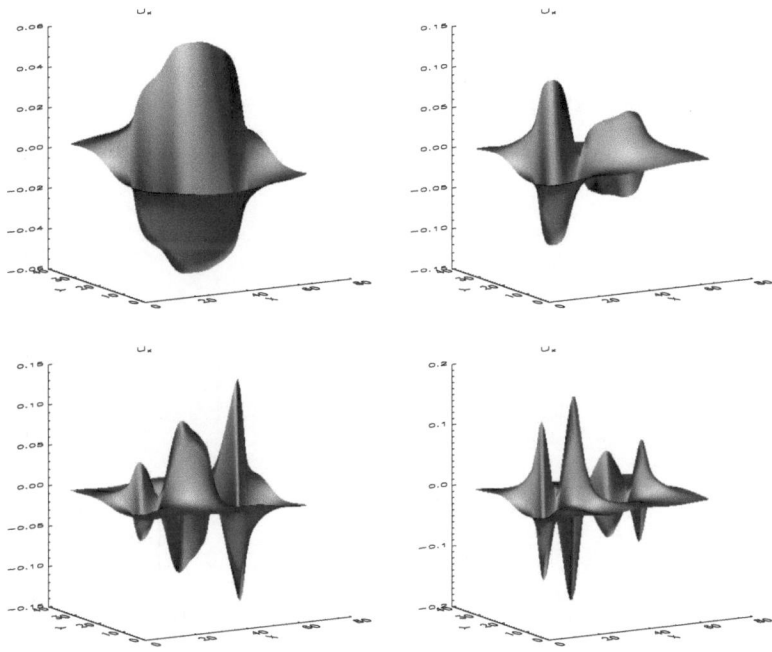

Figure 7

The first four POD spatial modes of the u_x surface deformation field for fault system A with dynamic overshoot $D = 1.5$. These modes define, in decreasing order of their eigenvalues, generalized deformation directions which support most of the variance produced by the dynamics of surface fluctuations. The coherent, collective nature of the modes is reflected in their overall shape.

invariant system the POD modes are simply the Fourier modes (HOLMES *et al.*, 1996).The basis $\{\phi_i\}_{i=1}^{M}$ is optimal in the sense that any reduced representation of the form

$$u(\mathbf{x}, n) \simeq \sum_{i=1}^{m} A_i(n)\phi_i(\mathbf{x}), m < M, \qquad (16)$$

describes typical members of the ensemble better than *any* linear representation of the same dimension m in any other basis: the leading m POD modes contain the greatest possible "energy" on average (HOLMES *et al.*, 1996). Therefore, due to the coherence and optimality of the POD basis, Eq. (16) provides the most efficient Euclidean (linear) embedding for the large-scale motions of the system. Moreover, the modal coefficients are uncorrelated on average, i.e.,

$$\langle A_i A_j \rangle = \lambda_i \delta_{ij}, \qquad (17)$$

which reflects the fact that orthogonality in the embedding space, $(\phi_i, \phi_j) = \delta_{ij}$, is related to the statistical properties of the time series (HOLMES *et al.*, 1996).

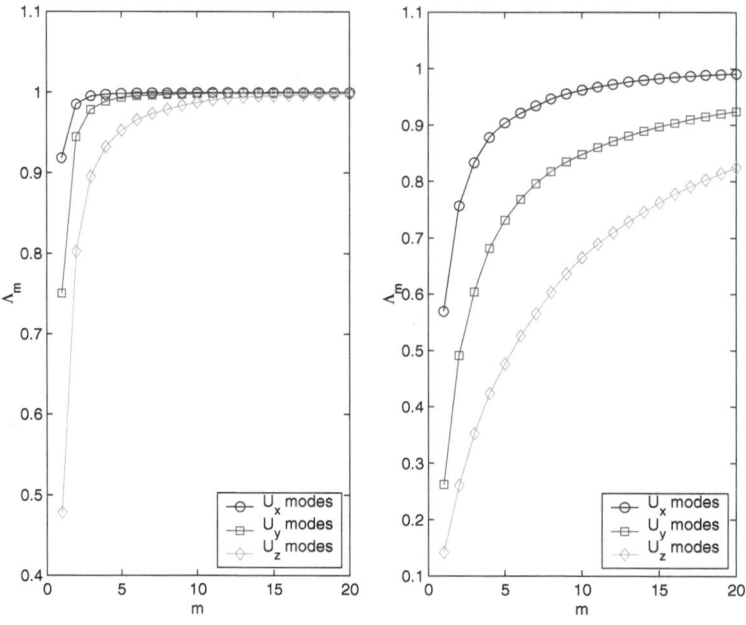

Figure 8

Cumulative normalized eigenvalue spectrum for each of the surface deformation modes for system A on the left and B on the right—only the first 20 POD modes are shown. In order to explain $f = 95\%$ of the variance in the u_x, u_y, u_z data sets for system A we need only 2, 3, and 5 modes, respectively. In contrast, for system B we need 9, 27, and 42 modes. Generally, the surface deformations along the strike direction, u_x, have the most efficient POD description.

The choice of the embedding dimension m is based on the computation of the cumulative normalized eigenvalue spectrum, Λ_m, defined as:

$$\Lambda_m = \frac{\sum_{j=1}^{m} \lambda_j}{\sum_{j=1}^{M} \lambda_j}. \tag{18}$$

The cumulative spectrum can help us define an effective POD embedding dimension by finding the minimum number of modes needed to capture some specific fraction $f < 1$ of the total variance of the data:

$$d_{\text{POD}} = \arg \min_m \{\Lambda_m : \Lambda_m > f\}. \tag{19}$$

Figure 8 shows the cumulative normalized spectrum for each of the surface deformation modes. For system A ($D = 1.5$), in order to explain $f = 95\%$ of the variance in the u_x, u_y, u_z data sets we need only 2, 3, and 5 modes respectively. This is consistent with our earlier estimates of the embedding dimension. In contrast, for system B ($D = \infty$), we need 9, 27, and 42 modes, respectively.

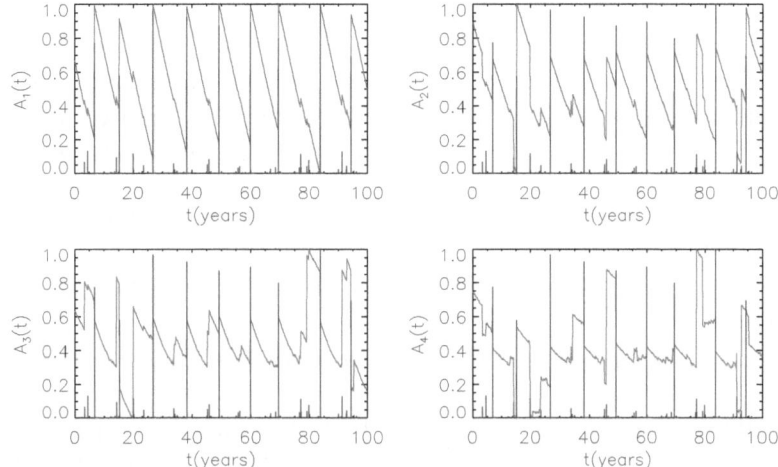

Figure 9

Modal time series describing the evolution of the first four u_x spatial modes shown in Figure 7. This time dependence is determined by projecting the u_x surface deformation onto each POD mode every 0.1 years. The vertical lines describe the time evolution of the cumulative potency released over each 0.1 year interval. Note the correlations between the time intervals of high potency release and the discontinuities present in the modal time series.

It is important to realize that the POD modes carry no dynamic information. Reshuffling the ensemble of configurations—and, therefore, destroying the dynamic content hidden in their time ordering—produces the same POD modes. Therefore, in order to identify the dynamics of each POD mode, we calculate the projections of the surface deformation fields onto the dominant deformation modes, i.e.,

$$A_i(n) = (u(\mathbf{x}, n), \phi_i(\mathbf{x})). \tag{20}$$

This way we generate a small number of modal time series $A_i(n), i = 1, \ldots, m$, that encode the evolution, interaction and dynamics of the spatial modes (HOLMES *et al.*, 1996). They encapsulate the projection of the system's dynamics onto the m-dimensional *model* space defined by Eq. (16).

In Figure 9 we represent the modal time series corresponding to the first four u_x spatial modes for fault system A. The vertical lines describe the time evolution of the binned potency released (the total potency released over the time interval Δt). For the first time series, $A_1(n)$, the location in time of the amplitude jumps coincides with the time of large events (within the temporal resolution defined by Δt). Moreover, we can show that the size of the amplitude jumps is proportional to the size of the event, i.e., $\Delta A_1(n) = A_1(n+1) - A_1(n) \simeq P_{\max}(n)$, where $P_{\max}(n)$ is the largest event in the n-th time interval. Therefore, if we can model the evolution of this mode, the time and size of the amplitude jumps will give us useful information about the time and size of the large earthquake events. Moreover, we notice that the higher order modes evolve on faster time scales, as is also noticeable from the faster decay of the temporal autocorrelations shown in Figure 11.

The picture that emerges from this decomposition evolves gradually from the slow, coherent large-scale motions, whose dynamics is captured by the most dominant POD modes, to the fast, incoherent small-scale motions described by the higher order modes. In Section 7 we describe how a neural net can be used to process these time series in order to extract a low-dimensional nonlinear dynamic model with short-term predictive capabilities.

6. *Geometry of the Attractor in the POD Basis*

The POD modes provide an optimal basis for the phase space embedding expressed in Eq. (16). The phase space vectors in this reconstruction, $\mathbf{X}_n = (A_1(n), \ldots, A_m(n))$, approximate the location of the system on the attractor at discrete time moments. Compared to the time-delay embedding, the POD basis unfolds the attractor with increased resolution as the embedding dimension increases. This property allows us to probe the structure of the attractor at different length scales by computing its correlation dimension. This measure was introduced by GASSBERGER and PROCACIA (1983) to quantify the self-similarity of geometrical objects. It is based on the definition of the correlation sum for a collection of points $\mathbf{X}_n$ in some vector space $\mathbf{R}^m$. This is the fraction of all possible pairs of points which are closer than a given ϵ in a particular norm. The basic formula is

$$C(m, \epsilon) = \frac{2}{R(N - 2W - 1)} \sum_{i=1}^{R} \sum_{j=1, |i-j|>W}^{N} \Theta(\epsilon - \| \mathbf{X}_i - \mathbf{X}_j \|_\infty), \qquad (21)$$

where we use the maximum norm, $\| \mathbf{X} \| = \max |x_i|$, Θ is the Heaviside step function, N is the total number of data points, and R is the total number of reference points. The sum counts the pairs $(\mathbf{X}_i, \mathbf{X}_j)$ whose distance is smaller than ϵ. Note that data points within a time window $2W$ of any reference point j are not included in the correlation sum. An appropriate choice of $W = 100$ reduces the spoiling effect of autocorrelations from the correlation sums as suggested by THEILER (1990). In the limit of an infinite amount of data and small ϵ, the attractors of deterministic systems show power-law scaling, i.e., $C(\epsilon) \simeq \epsilon^{d_2}$, and we can define the correlation dimension d_2 by

$$d_2 = \lim_{\epsilon \to 0} \lim_{N \to \infty} d_2(m, \epsilon). \qquad (22)$$

For each embedding dimension m, $d_2(m, \epsilon)$ measures the local slopes of the correlation sum at different length scales ϵ,

$$d_2(m, \epsilon) = \frac{\partial \ln C(m, \epsilon)}{\partial \ln \epsilon}. \qquad (23)$$

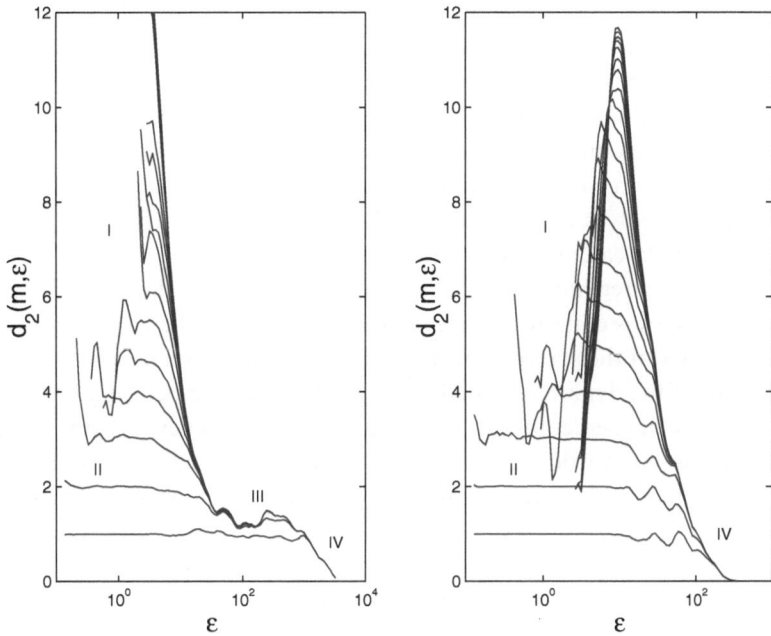

Figure 10

Local slopes $d_2(m, \epsilon)$ of the correlation sum for the multivariate time series describing the evolution of systems A (on the left) and B (on the right) for embedding dimensions $m = 1, ..., 20$. For system A ($D = 1.5$), we detect the emergence of a scaling region around $\epsilon \approx 100$, suggesting a correlation dimension $d_2 \simeq 1.3$ and self-similar geometry of the attractor at the large length scales. This behavior breaks down when the attractor is probed at smaller length scales where the dynamics is high dimensional. Within the same range of embedding dimensions, system B ($D = \infty$) shows no signature of a large-scale dimension collapse.

In Figure 10 for system A on the left and B on the right, we show the ϵ dependence of $d_2(m, \epsilon)$ for embedding dimensions $m = 1, \ldots, 20$. This plot allows the identification of a scaling range and estimation of the correlation dimension if such range occurs. This is reflected in the presence of a plateau in the ϵ dependence of $d_2(m, \epsilon)$ which changes minimally with embedding dimension m when $m > d_0$: the correlation sum probes the attractor at different length scale and tests for scale-invariance.

From the point of view of the physics involved, we distinguish four different types of behavior for $d_2(m, \epsilon)$ at different regions of length scale (KANTZ and SCHREIBER, 1997). For small ϵ and large m (region I in Figure 10) the lack of data points is the dominant feature and the values of $d_2(m, \epsilon)$ are subject to large statistical fluctuations. If ϵ is of the order of the size of the entire attractor (region IV), no scale invariance can be expected. In between, we can distinguish two regions. In region II, at small length scales but good statistics due to the small embedding dimension, the reconstructed points reflect the low amplitude and high frequency dynamics generated by the small events. At these length scales the dynamics is high dimensional and the reconstructed points fill the entire phase space available, therefore we expect $d_2(m, \epsilon) \simeq m$. Up to the embedding dimension $m = 4$ this estimate is recovered in the limit $\epsilon \to 0$. For increased embedding dimension there are large statistical fluctuations due to the lack of neighbors.

For system A, in region III located at larger length scales between regions II and IV, the dynamics evolves on a low-dimensional self-similar attractor whose presence is detected by the plateau in the correlation dimension at $d_2(m, \epsilon) \simeq 1.3$. Note that for system A the scaling regime, $\epsilon \to 0$, is not reachable. The breakdown of scaling at the smaller length scales is dynamic in nature and is not due to the lack of good statistics. The closer we look at the system, the more degrees of freedom become visible and the dimensionality of the dynamics is higher than our largest embedding dimension. This also explains why the percentage of false nearest neighbors never drops completely to zero with increased embedding dimension. On the same figure, within the range of embedding dimensions that were numerically accessible, system B shows no signature of a large-scale dimension collapse, in dramatic contrast with the behavior of system A.

The length dependent dimensions lead us to conjecture that we observe different subsystems on different length scales. From the self-similar and low-dimensional structure at the large length scales, the structure of system A attractor crosses over to a high-dimensional, stochastic-like structure when probed at very small length scales. Our strategy for building a model with good forecasting skills should be obvious now. The underlying idea is to approximate the large-scale motions of system A through a low-dimensional embedding into the subspace spanned by the most dominant POD modes and to extract a finite-dimensional model in the form of a set of ODEs of comparable dimension. Does a good nonlinear model for the large-scale motions exists? The answer depends on how rapidly the small-scale motions of the *true* system progress to reach the larger length scales. If this time is short, an effective model for the large-scale motions does not exist. If the growth time is sufficiently long, the large-scale motions *slave* the expected values of the small-scale motions and their dynamics is therefore well aproximated by a *deterministic* model. To address this question we now proceed to the model reconstruction task.

7. *Model Reconstruction and Short-term Earthquake Forecasting*

Using the POD decomposition we have identified a low-dimensional linear space in which system A evolves most of the time and we have reduced its dynamics to a small set of time series, $A_i(n)$, $i = 1, \ldots, m$, describing the evolution of the system in this reduced linear space. We now face the problem of determining the underlying dynamical process from the information available in these time series. They are assumed to be governed by a nonlinear set of ODEs and our modeling approach relies on the ability to identify an approximate m-dimensional model,

$$A_i(n+1) = F_i[A_1(n), A_2(n), \ldots, A_m(n)] \quad i = 1, \ldots, m, \tag{24}$$

that describes an explicit Euler approximation to the evolution and interaction of the spatial modes. To identify this nonlinear mapping we employ an artificial neural network (ANN). Generally, the neural network approach is used as a "black-box" tool in order to develop a dynamic model based only on observations of the system's input-output behavior (RICO-MARTINEZ et al., 1992, 1995). In the learning process the network adjusts its internal parameters to minimize the squared error between the network output and the desired outputs. A typical learning method is the error back-propagation algorithm which is a first-order gradient descent method (HAYKIN 1999).

All reconstruction results described here refer to the dynamics of the u_x surface deformation modes for model A with $D = 1.5$. We found it difficult to identify a simple model having the structure defined by Eq. (24). In order to improve the model forecasting skill it is useful to enlarge the structure of the model to include information pertaining to the past history of the modes. While the best model structure is still a subject of investigation, an analysis of the time patterns present in the modal evolution provides a partial understanding of this result. As Figure 9 shows, one essential feature of the modal dynamics is the presence of two time scales: within each earthquake cycle there are intervals of slow and fast motions with detectable quasi-regular behavior. This is reflected in the two-point autocorrelation functions (Fig. 11) defined as

$$C_{ii}(\tau) = \frac{\langle A_i(n)A_i(n+\tau)\rangle - \langle A_i(n)\rangle^2}{\langle A_i(n)\rangle^2}, \tag{25}$$

where the $\langle \cdot \rangle$ denotes the time average. The resulting plots exhibit oscillatory behavior with a slow amplitude decay over a longer time scale indicating that the system has two correlation time scales.

We have observed that providing the ANN with information about these long-term correlations of the modal dynamics produces nonlinear models with increased forecasting skill. To include this information, we have modified the structure of the model in Eq. (24) and replaced the modal coefficients A_i with time-delayed vectors $\mathbf{X}_i$,

$$\mathbf{X}_i = \underbrace{A_i(n), A_i(n-1), \ldots, A_i(n-d)}_{\text{short-time memory}}, \underbrace{A_i(n-2d), \ldots, A_i(n-(K-1)d))}_{\text{long-time memory}}, \tag{26}$$

where d is the delay and $(K-1)$ is the number of the time-delay intervals. Due to the presence in the modal dynamics of two different time scales, we have included both short, $A_i(n), A_i(n-1), \ldots, A_i(n-d)$, as well as long time, $A_i(n-2d), \ldots, A_i(n-(K-1)d)$, memory information in the time-delayed input

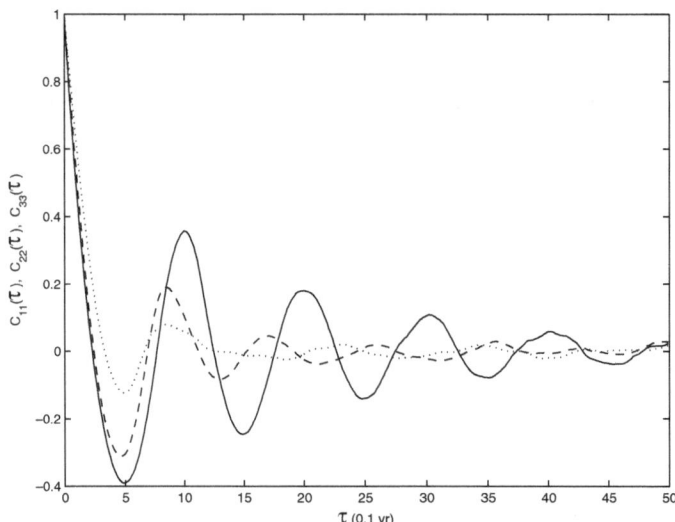

Figure 11

The temporal autocorrelations of the first three modes of the u_x deformation field in $D = 1.5$ model (mode 1-solid line, mode 2-dashed line, and mode 3-dotted line). The short-term oscillatory behavior and the slow long-term amplitude decay reveal the presence of two correlation time scales. The oscillations become faster and the long-term coherence decreases rapidly for the higher order modes.

vectors. The dimension of each time-delayed vector is $K + (d - 1)$. The ANN output then provides a prediction of the mode amplitude A_i at time $(n + 1)$,

$$A_i(n + 1) = F_i[\mathbf{X}_1(n), \mathbf{X}_2(n), \dots, \mathbf{X}_m(n)] \; i = 1, \dots, m, \tag{27}$$

based on input information describing the past mode histories.

The ANN is trained using a standard back-propagation algorithm (HAYKIN, 1999). When training succeeds, the ANN provides an approximate dynamical model (map) for the large-scale motions of the fault. The map can be evaluated once to provide short-term prediction or iterated to predict the long-term fault dynamics. Besides the parameters describing the structure of the ANN (nonlinear transfer functions, number of layers and neurons in each layer), the model structure itself has many parameters (m, d, K) that can be adjusted in order to improve the model performance. Our goal is to find a good model describing the evolution of the first POD mode, whose dynamic discontinuities trace accurately the time and size of the large events. Information regarding the higher order modes and their past histories is only necessary to uniquely determine the state and model trajectory in its phase space. One of the best models found to date describes the evolution of the first two u_x surface deformation modes: $m = 2$. For each spatial mode a time-delayed vector with parameters $K = 6$ and $d = 6$ was used. Because the dimensionality of the input space, $m(K + (d - 1)) = 22$, is higher than our estimated embedding dimension, we have decided to perform a POD analysis of the ensemble of input vectors. This time the POD decomposition performs the analysis of the dominant *temporal* patterns that are created by the modal dynamics. Similar to the spatial decomposition, we first compute the two point correlation matrix of the ensemble,

$$K_{ir,js} = \langle X_{ir} X_{js} \rangle, \tag{28}$$

where $i, j = 1, \ldots, m$ denote the input modes, $r, s = 1, \ldots, K + (d - 1)$ are the components of the time-delayed mode vectors, and $\langle \cdot \rangle$ is the average over the ensemble of input vectors. Due to the statistical independence of the modal coefficients, Eq. (17), the correlation matrix is to a good approximation block diagonal, i.e., $K_{ir,js} \simeq 0$ for $i \neq j$ (any modal cross-correlations are due to finite size effects). The eigenvalues and normalized eigenvectors of the correlation matrix,

$$\sum_{r,s} K_{ir,js} \Psi_{rs}^k = \mu_k \Psi_{rs}^k \quad , \mu_1 \geq \mu_2 \geq \cdots \geq \mu_{m(K+(d-1))} \geq 0, \tag{29}$$

will now provide an optimal representation of the dynamics of the spatial modes:

$$X_{ir}(n) = \sum_{k=1}^{m(K+(d-1))} B_k(n) \Psi_{ir}^k. \tag{30}$$

where

$$B_k(n) = (\mathbf{X}(n), \Psi^k) = \sum_{i=1}^{m} \sum_{r=1}^{K+(d-1)} X_{ir}(n) \Psi_{ir}^k. \tag{31}$$

We found that 99% of the variance of the input ensemble ($m = 2$, $K = 6$ and $d = 6$ model) can be represented by the first six temporal modes $\Psi^k, k = 1, \ldots, 6$. We have therefore chosen to approximate the input vectors presented to the ANN by their six-dimensional POD projection $\mathbf{B}(n) = (B_1(n), \ldots, B_6(n))$ onto the space spanned by the first six temporal eigenvectors. Denoting by $\mathbf{P_{X \to B}}$ this projection operator, the structure of the model has now become:

$$A_i(n+1) = F_i[\mathbf{P_{X \to B}}[\mathbf{X}_1(n), \mathbf{X}_2(n)]] = F_i[B_1(n), \ldots, B_6(n)] \; i = 1, 2. \tag{32}$$

A good forecasting performance was obtained for an ANN with two hidden layers of 10 neurons each. The input to the network is the six-dimensional projection $\mathbf{B}(n)$ and its two-dimensional output, $(A_1(n+1), A_2(n+1))$, is the model forecast for the first two mode amplitudes at the next time step. Because the large events have long recurrence times, the sequential selection of the training set will be dominated by input-output pairs that only express the quasi-linear behavior between two consecutive large events: the ANN will learn a linear model and will fail to forecast the intrinsically nonlinear large earthquake events. Therefore, we have designed the training set to include enough information describing the dynamics around the scarce large events. The modal time series were first rescaled to evolve in the interval $[0, 1]$.

We then classified as large events all events in which the first mode has a jump larger than 0.05 in the rescaled units. Next, we chose a time window $W = Kd$, and checked for the presence of a large event located at the center of this window, as it slides along the modal time series. If we found one we called this a nonlinear window, and all the input-output pairs describing the evolution of the system inside this time window were included in the training set. For each nonlinear window we have included training pairs from a linear window (a segment of the time series that did not have a large event). With this training set the ANN adequately learned to model the linear as well as the nonlinear features of the dynamics.

Due to non-convexity of the error surface in the network parameter space, we train many ANNs starting from different initializations of the neuron weights. To avoid overfitting, the training process stops when the error on the validation set starts to increase—the validation sample is a subset of the training set which is not actually used in training. At the end of the training cycle each ANN provides a model for the system dynamics. To find the "best" model we test the forecasting skill of each model realization, using time series segments not included in the training or validation set. Starting from an initial configuration describing for each input mode the current amplitude and its past $K + (d - 1)$ values, we iterate the ANN forward in time for a number of F steps. At each time step the output of the network was used to update and reconstruct the ANN input for the next time step. Assuming perfect initial conditions this procedure provides a trajectory whose forecasting accuracy is controlled only by the imperfections of our model. However, the reconstructed model is always an imperfect one and perhaps a perfect model describing the large-scale motions does not even exists. Moreover, the current state of the system in the model space is always obscured by observational uncertainty. To make things worse, our knowledge of the spatial modes themselves is limited by the finite size of the ensemble of surface deformations. For example, the accuracy of the two-point correlations cannot exceed $1/\sqrt{N}$, where N is the size of the ensemble: when we have only observations of finite duration, the "true" modes describing the large-scale motions will never be exactly known. How can we then evaluate the model forecasting skill? A detailed answer is beyond the scope of this paper and our goal here is only to show that despite all these difficulties, and many others not mentioned here, our imperfect models can provide stable and robust forecasts. Our remarks here follow SMITH (2000), to which we also refer for a clear discussion of uncertainty in initial conditions, model errors, ensemble verification, and predictability in general.

For nonlinear systems, uncertainty in the initial conditions severely limits the utility of single deterministic forecasts. Internal consistency requires that all nonlinear forecasts should be ensemble forecasts (SMITH, 2000). In this approach to forecasting, a collection of initial conditions, each consistent with the observational uncertainty, are integrated forward in time. When the system evolves on an attractor, the selection of the initial conditions is far from trivial. For a perfect model

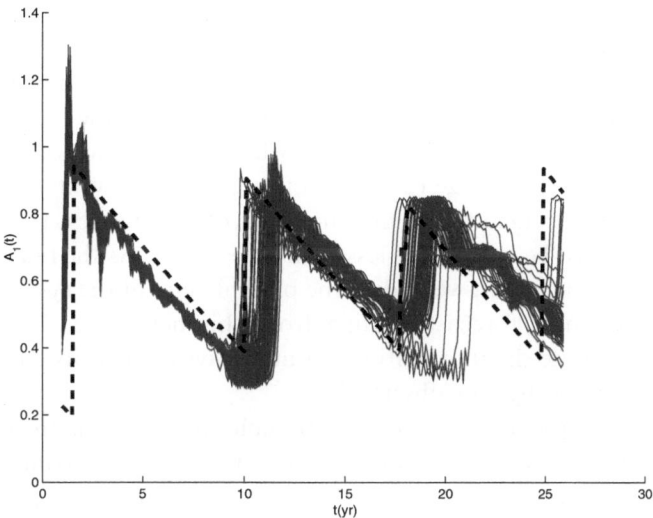

Figure 12

Model ensemble forecast for the evolution of the first u_x mode (coordinate A_1 in model phase space) as predicted by the iterated ANN (solid lines). Each trajectory of the ensemble starts from a different initial condition and is integrated forward in time for 250 steps. Each ensemble trajectory evolves into the model phase space, while the thick dashed line represents the true trajectory of the system projected onto the first coordinate of the model phase space (mode one). The ensemble is consistent with an arbitrary Gaussian, initial observational uncertainty, but its members do not live on the projection of the true system's attractor into the model phase space. Nevertheless, all members of the ensemble consistently forecast the incoming sequence of large events, proving the long-term forecasting reliability. We also notice a systematic bias in estimating the time to the next large event.

(a model that has the right dynamics and whose phase space is identical with the system's phase space) the members of the ensemble should be restricted to live on the system's invariant measure (attractor). Obviously, this is not *a priori* known. Unrestricted initial conditions, consistent only with the observational uncertainty, will extend into the full phase space and generate overdispersive ensembles. When this is the case, the predicted probabilities will not match the relative frequencies as demonstrated by GILMOUR (1998). Generally, all models, including ours, are imperfect and, therefore, a perfect ensemble and an accountable forecast method do not exist. Nevertheless, when the initial conditions are uncertain, ensemble forecasts are still required. A reasonable constraint would require the members of the ensemble to live on the projection of system's invariant measure (attractor) into the model phase space, although this choice is not necessarily optimal or even unique (SMITH, 2000).

Assuming no *a priori* knowledge of the true attractor projected into the model phase space, we present in Figure 12 an unconstrained ensemble of trajectories evolving under the dynamics of the ANN model. Each trajectory starts from a different initial state that contains small Gaussian perturbations from the exact initial conditions and is integrated forward in time for $F = 250$ steps. We compare the ensemble evolution (solid lines) with the true evolution of the most dominant surface mode (thick dashed line). The resulting ensembles of forecasts can be interpreted as a probabilistic prediction. The goal is to predict the time and the size of the jumps in the evolution of the first mode amplitude (which, as we have already discussed, corresponds to the time and size (potency) of large seismic events), and to estimate their forecast accuracy. Due to the time delay involved, the best time resolution of each trajectory cannot be in this case less than $\Delta t = 0.1$ yr. As the fault evolves to the next time step, 0.1 years later, we update the present state of the system and generate a new F step ensemble forecast starting from this new state. This procedure is intended to incorporate the information about the system and its current state as is continuously generated by new observations.

It is clear from Figure 12 that even though we lack a perfect model or an optimal ensemble, all members of the ensemble forecast the incoming sequence of large events. The forecast uncertainty of the ensemble grows very slowly showing long time reliability. The members of the ensemble spread out at a rate that depends on the local nonlinear structure of the model. This rate gives a local estimate of the stability of forecasts made in this region of the model's state space. It also controls the time scale (predictability time) on which the ensemble members scatter along significantly different trajectories. We observe regions of large predictability time that coexist with regions of relatively short predictability time (BOFFETTA *et al.*, 2002).

Significantly, the ANN has shown consistently the ability to improve its forecasting as the system approaches a large event. The predictability time is significantly larger than the microscopic expansion rate which is controlled by the maximal Lyapunov exponent. As hinted earlier, the predictability is scale-dependent and the large-scale motions are more predictable than the small-scale motions.

The predictability also depends on the location of the system on its true underlying attractor. But unlike the growth of initial uncertainty in model phase space, the local nonlinear structure of the system's attractor controls how rapidly the small-scale motions, not included in our model, progress to reach the large length scales. This expansion rate defines how rapidly the *truth*, dashed line in Figure 12, diverges from the best guess model trajectory. This is the trajectory starting from the true system state projected onto the model phase space. Of course, for the same model state there are infinitely many system states distinguishable only in their small length scale structure. Clearly, the accurate shadowing of the *truth* by the ensemble trajectories in Figure 12, shows that growth of the small-scale uncertainties does not severely limit the model forecasting skill. Their small growth rate makes possible the deterministic modeling of the large-scale motions. It sets an upper bound for the model predictability time, which can be approached by improving the parameterization of the small-scale dynamics.

8. Conclusions

We describe a conceptual framework for modeling and forecasting the evolution of a large strike-slip earthquake fault. The approach relies on the detection of spatio-temporal strain patterns embedded in the *observable* surface displacements: no detailed knowledge of the fault geometry, dynamics, or rheology is required. Rather than directly modeling the fault dynamics, we propose instead to model the dynamics of observable surface deformations which are linearly related to the original dynamics of the fault system.

The essence and novelty of the method lies in the discovery that the large length and time scales dynamics have a strong *low-dimensional, deterministic* component and are therefore amenable to representation by a deterministic model. We have also found that the large-scale motions provide reliable forecasting information about the large seismic events. These two fundamental results set the stage for standard data processing and model reconstruction techniques. First, we identify the large-scale motions with generalized directions (spatial modes) along which the dynamics has its *largest* fluctuations. Finding these directions is the natural task of the proper orthogonal decomposition applied to the ensemble of surface deformations generated during the evolution of the system. The most dominant spatial modes define the model phase space and provide an optimal embedding for the large-scale dynamics of the system. Second, the model reconstruction consists in finding a nonlinear set of ODEs whose trajectory in model phase space approximates the system trajectory projected into the model phase space. This is a *learning* task that can be successfully accomplished by an artificial neural network.

The method relies on the existence of some separation between the small (fast) and the large (slow) length and time scales. The physics responsible for this separation stems from the dynamic weakening used to model static/kinetic friction. We argue that in the presence of significant dynamic weakening the large-scale dynamics defines the low-dimensional backbone of the system's attractor. The small-scale dynamics is practically indistinguishable from stochastic dynamics and evolves in a small neighborhood of the attractor backbone. Based on this geometric picture, we argue that the statistical properties of the small-scale motions are determined by the larger-scale motions upon which they are superposed. In turn, the large-scale motions depend on the statistical properties of the small scales. As pointed out by LORENZ (1969), there remain uncertainties in the small-scale statistics, and hence in their influence upon the larger-scale. The predictability time of the large scale motions is controlled by the rapidity with which the small scales uncertainties progress to reach the very large-scales. This sets an upper bound to the predictability time of any large-scale model. Due to the robustness shown by the model ensemble forecast, we conclude that the intrinsic predictability time of the large-scale dynamics is very long, perhaps of the order of 10–20 years. This is ultimately the reason behind the deterministic behavior of the large-scale motions.

There are many difficult problems currently under investigation. For example, we presently study how robust the large-scale behavior is to changes in the model parameters. Preliminary results confirm the controlling role of the dynamic weakening and show that the deterministic character of the large-scale motions is robust to major changes in this parameter. We are also studying the effects of correlated heterogeneities, smooth brittle to ductile transitions, and continuous transition from static to kinetic friction. We do expect that all these effects are averaging out the small-scale dynamics and strengthen the deterministic structure of the attractor. Probably the most difficult problems to address are the construction of good models from short data observations and the problem of spatio-temporal chaos. In the last case, dimensions and Lyapunov exponents become intensive quantities and we seek to understand how the method scales with the size of the fault system. The current paper is concerned primarily with developing a methodology and the obtained results are preliminary. Continuing studies along the directions of this work may have a significant impact on the earthquake predictability problem.

Acknowledgments

We express our gratitude to Yannis Kevrekidis for useful discussions and insightful suggestions. This research was performed under the auspices of the U.S. Department of Energy at LANL (LA-UR-02-6309) under contract W-7405-ENG-36 and LDRD-DR-2001501 (MA) and the National Earthquake Hazard Reduction Program of the USGS under grant 02HQGR0047 (YBZ).

REFERENCES

ABARBANEL, D. I. H., *Analysis of Observed Chaotic Data* (Springer-Verlag, NY. (1996)).

ANGHEL, M., *On the effective dimension and dynamic complexity of earthquake faults.* Chaos, Solitons, and Fractals *19*, 399–420.

BEN-ZION, Y. and RICE, J. (1993), *Earthquake Failure Sequences along a Cellular Fault Zone in a Three-dimensional Elastic Solid Containing Asperity and Nonasperity Regions.* J. Geophys. Res. *98*, 14109–14131.

BEN-ZION, Y. and RICE, J. R. (1995), *Slip Patterns and Earthquake Populations along Different Classes of Faults in Elastic Solids*, J. Geophys. Res. *100*, 12959–12983.

BEN-ZION, Y., (1996), *Stress, Slip, and Earthquakes in Models of Complex Single-fault Systems Incorporating Brittle and Creep Deformations*, J. Geophys. Res. *101*, 5677–5706.

BEN-ZION, Y., ENEVA, M., and LIU, Y., (2003), *Large Earthquake Cycles and Intermittent Criticality on Heterogeneous Faults due to Evolving Stress and Seismicity.* J. Geophys. Res. 108, B6, 2307, DOI: 10.1029/2002JB002121

BEN-ZION, Y., (2003), Appendix 2, *Key formulas in earthquake seismology.* In *International Handbook of Earthquake and Engineering Seismology*, Part B Academic Press. 1857–1875.

BOFFETTA, G., CENCINI, M., FALCIONI M., VULPIANI, A. (2002), *Predictability: A Way to Characterize Complexity*, Phys. Rep. *356*, 367–474.

BURGMANN, R., ROSEN, P.A., and FIELDING, E.J. (2000), *Synthetic Aperture Radar Interferometry to Measure Earth's Surface Topography and its Deformation*. Annu. Rev. Earth Planet. Sci. *28*, 169–209.

CHINNERY, M. (1963), *The Stress Changes that Accompany Strike-slip Faulting*. Bull. Seismol. Soc. Am. *53*, 921–932.

FISHER, D.S., DAHMEN, K., RAMANATHAN, S., and BEN-ZION, Y. (1997), *Statistics of Earthquakes in Simple Models of Heterogeneous Faults*, Phys. Rev. Let. *78*, 4885–4888.

FRASER, A. and SWINEY, H.L. (1986), *Independent Coordinates for Strange Attractors, from Mutual Information*, Phys. Rev. A *33*, 1134–1140.

GILMOUR, I. (1998), *Nonlinear model evolution: ı-shadowing, probabilistic prediction and weather forecasting*. D. Phil. Thesis, Oxford University.

GRASSBERGER, P. and PROCACIA, I. (1983), *Measuring the strangeness of strange attractors*, Physica D *9*, 189–208.

HAYKIN, S. *Neural Networks: A Comprehensive Foundation*,(Prentice Hall, NJ. (1999).

HEGGER, R., KANTZ, H., and SCHREIBER, T. (1999), *Practical Implementation of Nonlinear Time Series Methods: The TISEAN package*, CHAOS *9*, 413–435.

HOLMES, P., LUMLEY, J. L., and BERKOOZ, G. *Turbulence, Coherent Structures, Dynamical systems and Symmetry*, (Cambridge University Press, Cambridge. (1996)).

KANTZ, H. and SCHREIBER, T. *Non-linear time Series Analysis* (Cambridge University Press, Cambridge 1997).

LORENZ, E. N. (1969), *The Predictability of a Flow which Possesses Many Scales of Motion*, Tellus *21*, 289–307.

MADARIAGA, R. (1976), *Dynamics of an Expanding Circular Fault*, Bull. Seismol. Soc. Am. *66*, 639–666.

MANNEVILLE, M. *Dissipative Structures and Weak Turbulence*, Academic Press, CA (1990)).

OKADA, Y. (1985), *Surface Deformations due to Shear and Tensile Faults in a Half-space*, Bull. Seismol. Soc. Am. *75*, 1135–1154.

RICO-MARTINEZ, R., KRISCHER, K., KEVREKIDIS, I. G., KUBE M. C., and HUDSON, J. L. (1992), *Discrete-vs. Continuous-time Nonlinear Signal Processing of Cu Electrodissolution Data*, Chem. Eng. Comm. *118*, 25–48.

RICO-MARTINEZ, R., KEVREDIDIS, I. G., and KRISCHER, K (1995) *Nonlinear system identification using neural networks: dynamics and instabilities*. In: Bulsari, A. B. ed. *Neural Networks for Chemical Engineers*. Elsevier Science, pp. 409–442.

RUNDLE, J. B., KLEIN, W., TIAMPO, K. F., and GROSS, S. (2000), *Linear Pattern Dynamics in Nonlinear Threshold Systems*, Phys. Rev. E *61*, 2418–2431.

SAMMIS, C. G. and SORNETTE, D. (2002), *Positive Feedback, Memory, and the Predictability of Earthquakes*, PNAS *99*, 2501–2508.

SAUER, T., YORKE, J. A., and CASDAGLI, M. (1991), *Embedology*. J. Stat. Phys. *65*, 579–616.

SMITH , L. (2000), *Disentangling uncertainty and error: On the predictability of nonlinear systems*. In (Mees A., ed. *Nonlinear Dynamics and Statistics*. Birkhäuser, pp. 31–64.

THEILER, J. (1990), *Estimating fractal dimension*. J. Opt. Soc. Am. A *7*, 1055–1073.

WESNOUSKY, S. G. (1994), *The Gutenberg-Richter or Characteristic Earthquake Distribution, which is it?* Bull. Seismol. Soc. Am. *84*, 1940–1959.

(Received September 27, 2002, revised April 25, 2003, accepted May 5, 2003)

Pure appl. geophys. 161 (2004) 2053–2068
0033–4553/04/102053–16
DOI 10.1007/s00024-004-2548-8

▌Pure and Applied Geophysics

3-D Modelling of Plate Interfaces and Numerical Simulation of Long-term Crustal Deformation in and around Japan

CHIHIRO HASHIMOTO[1], KENJI FUKUI[2] and MITSUHIRO MATSU'URA[2]

Abstract—We developed a 3-D simulation model for long-term crustal deformation due to steady plate subduction in and around Japan by incorporating viscoelastic slip-response functions into a realistic 3-D plate interface model, constructed on the basis of the topography of ocean floors and hypocenter distributions of earthquakes. The lithosphere-asthenosphere system is modelled by an elastic surface layer overlying a Maxwellian viscoelastic half-space. Kinematic interaction at plate interfaces is rationally represented by the increase of tangential displacement discontinuity (fault slip) across the interfaces. With this model, giving the steady slip rates at plate interfaces calculated from NUVEL-1A, we simulated long-term crustal deformation due to steady plate subduction in and around Japan. The simulated crustal deformation pattern is characterized by steep uplift at island arcs, sharp subsidence at ocean trenches and gentle uplift at outer rises. The numerical results show the strong dependence of the deformation pattern on the 3-D geometry of plate interfaces.

Key words: Subduction zones, 3-D plate geometry, dislocation theory, viscoelasticity, crustal deformation.

1. Introduction

At subduction zones, oceanic plates descend beneath continental plates at constant rates on a long-term average. The occurrence of large interplate earthquakes can be regarded as its perturbation. Thus, we may decompose tectonic stress change and crustal deformation during one earthquake cycle into two parts; the secular change due to steady slip motion over the whole plate interface and the cyclic change due to stick-slip motion in a seismogenic region. The steady slip motion along a curved plate interface brings about secular change in tectonic stress and long-term crustal deformation, characterized by island-arc uplift and trench subsidence (MATSU'URA and SATO, 1989; SATO and MATSU'URA, 1993). The stick-slip motion

[1] Institute of Frontier Research for Earth Evolution, Japan Marine Science and Technology Center Showa-machi 3173-25, Kanazawa-ku, Yokohama 236-0001, Japan. E-mail: hashi@jamstec.go.jp
 Present address: Department of Earth and Planetary Science, The University of Tokyo, Hongo 7-3-1, Bunkyo-ku, Tokyo 113-0033, Japan. E-mail: hashi@eps.s.u-tokyo.ac.jp
[2] Department of Earth and Planetary Science, The University of Tokyo, Hongo 7-3-1, Bunkyo-ku, Tokyo 113-0033, Japan. E-mail: matsuura@eps.s.u-tokyo.ac.jp

Table 1

Structural parameters used in numerical simulations

	h [km]	ρ [kg/m^3]	λ [GPa]	μ [GPa]	η [Pa s]
Lithosphere	40	3000	40	40	∞
Asthenosphere	∞	3400	90	60	5×10^{18}

h: thickness, ρ: density, λ and μ: Lamé elastic constants, η: viscosity

brings about cyclic stress change and crustal deformation in and around the seismogenic region (Hashimoto and Matsu'ura, 2000, 2002). In realistic simulation of the complete earthquake generation cycle at subduction zones, we cannot discard the effects of the steady slip motion.

In the present study, first, we construct a realistic 3-D structure model in and around Japan; a rheological structure of the lithosphere-asthenosphere system and 3-D geometry of plate interfaces. Next, we mathematically describe a basic idea for 3-D modelling of long-term crustal deformation due to steady plate subduction. Finally, we show the results of numerical simulation with the 3-D plate subduction model, and discuss the internal force bringing about long-term crustal deformation in and around Japan.

2. Construction of a Realistic Structure Model in and around Japan

In the present study, the lithosphere-asthenosphere system is modelled by an elastic surface layer overlying a Maxwellian viscoelastic half-space. The constitutive equation of the elastic surface layer is given by

$$\sigma_{ij} = \lambda^{(1)} \varepsilon_{kk} \delta_{ij} + 2\mu^{(1)} \varepsilon_{ij} \tag{1}$$

and that of the underlying viscoelastic half-space by

$$\dot{\sigma}_{ij} + \frac{\mu^{(2)}}{\eta} \left(\sigma_{ij} - \frac{1}{3} \sigma_{kk} \delta_{ij} \right) = \lambda^{(2)} \dot{\varepsilon}_{kk} \delta_{ij} + 2\mu^{(2)} \dot{\varepsilon}_{ij} \tag{2}$$

Here, σ_{ij}, ε_{ij}, and δ_{ij} are the stress tensor, the strain tensor, and the unit diagonal tensor, respectively. The dot indicates differentiation with respect to time. $\lambda^{(i)}$ and $\mu^{(i)}$ ($i = 1, 2$) denote the Lamé elastic constants of each medium, and η is the viscosity of the underlying half-space. The standard values of these structural parameters, which are used for the following computation, are listed in Table 1.

In and around Japan, the Pacific (PA) plate is descending beneath the North American (NA) and the Philippine Sea (PH) plates, and the Philippine Sea plate is descending beneath the North American and the Eurasian (EU) plates, as shown in Figure 1. The boundary between the North American and the Eurasian plates is not clear. For this reason we discard this plate boundary in our modelling.

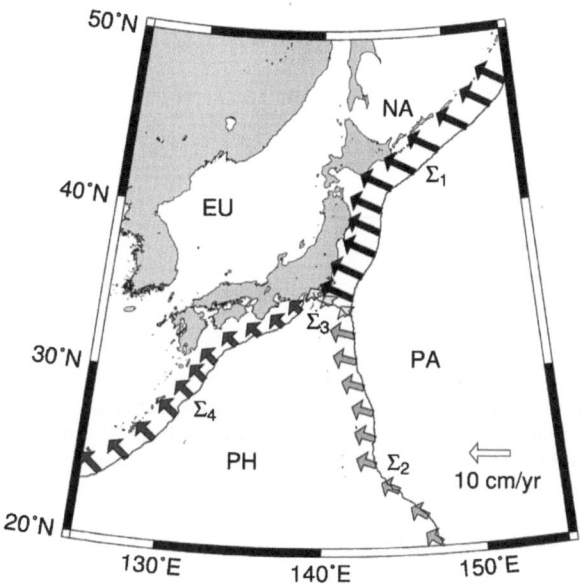

Figure 1

Plate boundaries and relative plate motion in and around Japan. The lithosphere is divided into four plates; the Pacific (PA), the North American (NA), the Philippine Sea (PH), and the Eurasian (EU) plates. These four plates are interacting with each other at four plate interfaces; Σ_1 (PA-NA), Σ_2 (PA-PH), Σ_3 (PH-NA) and Σ_4 (PH-EU). The relative plate motion vectors (thick arrows) are calculated from the relative rotation vectors in Table 2.

Now we take the model region extending from 125 °E to 155 °E in longitude and from 20 °N to 50 °N in latitude, and introduce the four plate interfaces, Σ_1 (PA-NA), Σ_2 (PA-PH), Σ_3 (PH-NA) and Σ_4 (PH-EU). The 3-D geometry of these plate interfaces is determined from the topography of ocean floors and hypocenter distributions of earthquakes. The original data of the hypocenter location are usually given by latitude v, longitude ε and depth z. First, we transform these data in a spherical coordinate system into a Cartesian coordinate system by using the Lambert conformal conic projection with the central point (v_0, ε_0) of the model region and the standard parallels v_1 and v_2. By this transformation an arbitrary point (v, ε) on the earth's surface (Fig. 2a) is projected onto the surface of a flat earth (Fig. 2b) as

$$\begin{cases} x^L = \rho \sin \theta \\ y^L = \rho_0 - \rho \cos \theta \end{cases} \tag{3}$$

with

$$\theta = n(\varepsilon - \varepsilon_0), \quad \rho = R_E F \tan^{-n}(\pi/4 + v/2), \quad \rho_0 = R_E F \tan^{-n}(\pi/4 + v_0/2) \tag{4}$$

and

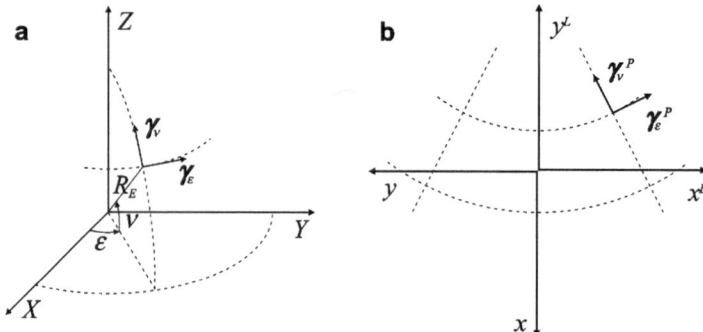

Figure 2

Relation between two different coordinate systems. (a) A spherical coordinate system. R_E denotes the radius of the earth. The tangential vectors γ_v and γ_ε represent the v-direction and the ε-direction at a point (v, ε) on the earth's surface, respectively. (b) A Cartesian coordinate system used to construct a realistic structure model. The direction vectors γ_v and γ_ε defined on the spherical earth are projected onto the flat earth as γ_v^P and γ_ε^P, respectively.

$$n = \frac{\ln(\cos v_1 / \cos v_2)}{\ln[\tan(\pi/4 + v_2/2) / \tan(\pi/4 + v_1/2)]}, \quad F = \frac{\cos v_1 \tan^n(\pi/4 + v_1/2)}{n}. \tag{5}$$

Here R_E denotes the earth's radius. In the present case, we take the central point to be (35 °N, 140 °E), and the standard parallels to be 30 °N and 40 °N. Next, for convenience, we perform the following coordinate transformation.

$$\begin{pmatrix} x \\ y \end{pmatrix} = \begin{pmatrix} -y^L \\ -x^L \end{pmatrix}. \tag{6}$$

Then the positive z-axis is taken to be in the direction of depth.

In the Cartesian coordinate system (x, y, z) we determine the 3-D geometry of plate interfaces by applying an inversion technique developed by YABUKI and MATSU'URA (1992) to observed data of hypocenter locations. The depth z to each plate interface is represented as a function of x and y by the superposition of basis functions;

$$z(x, y) = \sum_{k}^{K} \sum_{l}^{L} a_{kl} N_{4,k}(x) N_{4,l}(y). \tag{7}$$

Here, $N_{4,j}(s)$ is the B-spline function of order 4 (degree 3) with an equally spaced ($\Delta\zeta$) local support ($\zeta_j - 4\Delta\zeta \leq s < \zeta_j$). The B-spline function is defined by the following de Boor-Cox recurrence formula (DE BOOR, 1972; COX, 1972);

$$N_{r,j}(s) = \frac{(s - \zeta_{j-r}) N_{r-1,j-1}(s) - (\zeta_j - s) N_{r-1,j}(s)}{\zeta_j - \zeta_{j-r}} \tag{8}$$

with

$$N_{1,j}(s) = \begin{cases} 1/(\zeta_j - \zeta_{j-1}) & \zeta_{j-1} \leq s < \zeta_j \\ 0 & \text{otherwise} \end{cases}. \qquad (9)$$

Let $x_n = (x_n, y_n, z_n)$ be the hypocenter location of the n-th earthquake on the plate interface. Then, we obtain the following observation equations to determine the geometry of the plate interfaces;

$$z_n = z(x_n, y_n) + e_n \ (n = 1, \cdots, N). \qquad (10)$$

Here, e_n denotes the observation error of the n-th data. Substituting Eq. (7) into Eq. (10) and arranging the parameters a_{kl} in some order to make a M ($=K \times L$) dimensional column vector $a = [a_m]$, we can rewrite the observation equations as

$$z_n = \sum_{m=1}^{M} a_m M_m(x_n) + e_n \ (n = 1, \cdots, N) \qquad (11)$$

with

$$M_m(x_n) = N_{4,k}(x_n) N_{4,l}(y_n). \qquad (12)$$

Thus, we obtain the observation equation in vector form;

$$d = Ha + e. \qquad (13)$$

where H is a $N \times M$ dimensional coefficient matrix. For simplicity, we assume the errors e to be Gaussian, with zero mean and a variance-covariance matrix $\sigma^2 E$.

Then, from Eq. (13), we have a stochastic model which relates the data d with the model parameters a as

$$p(d|a; \sigma^2) = (2\pi\sigma^2)^{-N/2} \|E\|^{-1/2} \exp\left[\frac{-1}{2\sigma^2}(d - Ha)^T E^{-1}(d - Ha)\right]. \qquad (14)$$

Here, $\|E\|$ denotes the absolute value of the determinant of E. Given the data d, we may regard $p(d|a; \sigma^2)$ as a function of a and σ^2.

Now we introduce prior constraints on the roughness of a plate interface Σ in the form of a probability density function (pdf) with a hyperparameter ρ^2 as

$$p(a; \rho^2) = (2\pi\rho^2)^{-P/2} \|\Lambda_P\|^{-1/2} \exp\left(-\frac{1}{2\rho^2} a^T G^{-1} a\right). \qquad (15)$$

Here, the roughness of a plate interface Σ is defined by

$$r = \int_\Sigma \left\{ \left(\frac{\partial^2 z}{\partial x^2}\right)^2 + 2\left(\frac{\partial^2 z}{\partial x \partial y}\right)^2 + \left(\frac{\partial^2 z}{\partial y^2}\right)^2 \right\} d\Sigma \qquad (16)$$

and G is a $N \times M$ dimensional symmetric matrix, whose elements are directly calculated by substituting Eq. (7) into Eq. (16). Note that P is the rank of G, and $\|\Lambda_P\|$ is the absolute value of the product of non-zero eigenvalues of G. The hyperparameter ρ^2 controls the prior distribution of **a,** and hence the roughness of plate interface. Combining $p(\boldsymbol{d}|\boldsymbol{a}; \sigma^2)$ in Eq. (14) and $p(\boldsymbol{a}; \rho^2)$ in Eq. (15) by using Bayes' rule, we construct a highly flexible model with the hyperparameters σ^2 and ρ^2 called a Bayesian model;

$$p(\boldsymbol{a}; \sigma^2, \rho^2|\boldsymbol{d}) = cp(\boldsymbol{d}|\boldsymbol{a}; \sigma^2)p(\boldsymbol{a}; \rho^2). \tag{17}$$

The structure of the Bayesian model is controlled by the hyperparameters σ^2 and ρ^2. The determination of the best estimates of σ^2 and ρ^2 is done by minimizing a Bayesian information criterion (ABIC) proposed by AKAIKE (1980), which is defined by

$$ABIC = -2\mathrm{log}L(\sigma^2, \rho^2) + c \tag{18}$$

with

$$L(\sigma^2, \rho^2) = \int p(\boldsymbol{d}|\boldsymbol{a}; \sigma^2)p(\boldsymbol{a}; \rho^2)d\boldsymbol{a}. \tag{19}$$

Given the best estimates of the hyperparameters, we can easily obtain the best estimates of the model parameters **a** by using the following formula derived by JACKSON and MATSU'URA (1985);

$$\boldsymbol{a} = (\boldsymbol{H}^T \boldsymbol{E}^{-1} \boldsymbol{H} + \alpha^2 \boldsymbol{G})^{-1} \boldsymbol{H}^T \boldsymbol{E}^{-1} \boldsymbol{d} \tag{20}$$

with $\alpha^2 = \sigma^2/\rho^2$.

Figure 3 shows the 3-D geometry of plate interfaces determined from ISC hypocenter data by using the inversion technique described above. The main curved surface extending from north to south represents the upper boundary of the descending Pacific plate. The secondary curved surface branching from the main curved surface represents the upper boundary of the descending Philippine Sea plate. These curved surfaces are represented by the superposition of bicubic B-spline functions with equally spaced (8 km both in x and y) local supports. The plate interfaces have significant horizontal bends in the Tokachi-oki (TO) region, northeast Japan, in the Mikawa Bay (MB) and Bungo Channel (BC), southeast Japan. As demonstrated in the next section, the steady slip along these horizontal bends brings about large-scale crustal deformation around them.

3. Modelling the Crustal Deformation due to Steady Plate Subduction

Kinematic interaction at a plate interface is rationally represented by the increase of tangential displacement discontinuity across the interface (MATSU'URA and SATO, 1989). The displacement discontinuity (fault slip) is mathematically equivalent to the force system of a double-couple without moment (MARUYAMA, 1963; BURRIDGE and KNOPOFF, 1964), which has no net force and no net torque. Such a property must be satisfied for any force system acting on plate interfaces, since it is the internal force

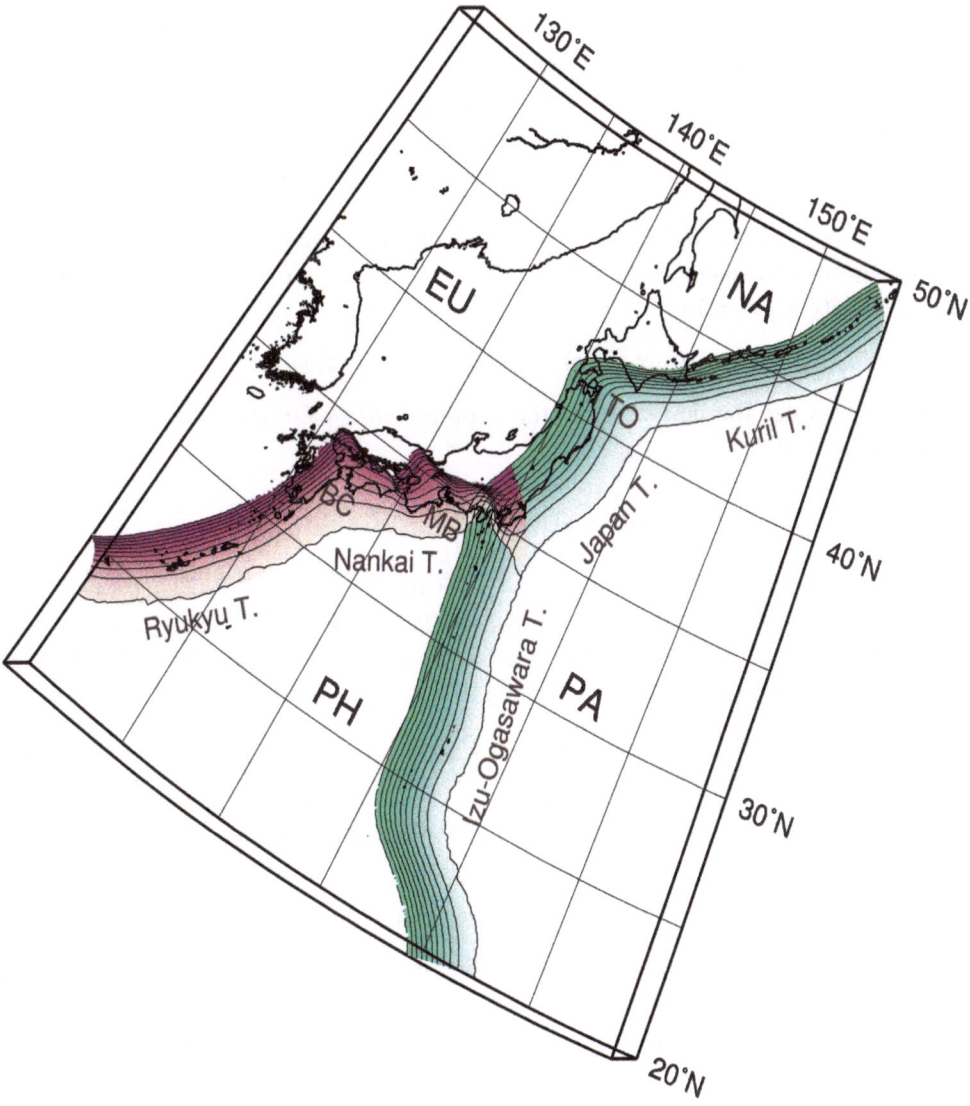

Figure 3
A 3-D structure model in and around Japan. The lithosphere-asthenosphere system is modelled by a 40 km-thick elastic surface layer overlying a Maxwellian viscoelastic half-space. The geometry of plate interfaces (upper boundaries of the descending Pacific and the Philippine Sea plates) is determined from ISC hypocenter data by using the inversion technique. Each plate interface is represented by the superposition of bicubic B-spline functions with equally spaced (8 km both in x and y) local supports. The intervals of the iso-depth contours are 10 km.

produced by a dynamic process within the earth. On the basis of this idea, we construct a 3-D simulation model for long-term crustal deformation due to steady plate subduction in and around Japan.

3.1 Viscoelastic Responses to Steady Plate Subduction

In general, the moment tensor density function corresponding to a unit step slip with a slip direction vector χ at a time τ and a point ξ on an interface with a normal vector n can be represented as

$$m_{pq}(\xi, \tau) = \mu[\chi_p(\xi)n_q(\xi) + \chi_q(\xi)n_p(\xi)]H(\tau), \tag{21}$$

where μ is the rigidity and $H(\tau)$ is the Heaviside step function. The i-component of viscoelastic displacement, $U_i(x,t; \xi, \tau)$, at a point x and time t due to the moment tensor density is given by

$$U_i(x,t; \xi, \tau) = F_{ipq}(x,t; \xi, \tau)m_{pq}(\xi, \tau). \tag{22}$$

Here, F_{ipq} denotes the derivative of the i-component of Green's tensor $G_{ip}(x,t; \xi, \tau)$ with respect to ξ_q;

$$F_{ipq}(x,t; \xi, \tau) = \frac{\partial}{\partial \xi_q} G_{ip}(x,t; \xi, \tau). \tag{23}$$

The concrete expressions of U_i ($i = x, y, z$) on the surface are given in MATSU'URA *et al.* (1981) and MATSU'URA and SATO (1989). Thus, we can calculate the displacement field u_i ($i = x, y, z$) due to arbitrary fault slip w along the plate interface by using the technique of hereditary integral as

$$u_i(x, t) = \int_{-\infty}^{t} \int_{\Sigma} \frac{\partial w(\xi, \tau)}{\partial \tau} U_i(x,t - \tau; \xi, 0)d\xi d\tau. \tag{24}$$

Now we decompose the total slip motion w into the steady slip at a plate convergence rate v_{pl} and its perturbation Δw as

$$w(\xi, \tau) = v_{pl}(\xi)\tau + \Delta w(\xi, \tau). \tag{25}$$

Substituting this expression into Eq. (24), we obtain

$$u_i(x,t) = \int_0^t \int_{\Sigma} v_{pl}(\xi)U_i(x,t - \tau; \xi, 0) \, d\xi d\tau + \int_0^t \int_{\Sigma} \frac{\partial \Delta w(\xi, \tau)}{\partial \tau} U_i(x,t - \tau; \xi, 0) \, d\xi d\tau. \tag{26}$$

Here, we supposed that the slip motion started at $t = 0$. The first and the second terms of Eq. (26) indicate the contributions from the steady slip motion and its perturbation, respectively. If we consider the long-term average of crustal deformation, we can neglect the second term. Furthermore, for sufficiently large t, we can regard U_i as a constant in time, and so the long-term displacement rates due to steady plate subduction are given by

$$\dot{u}_i(x) = \dot{u}_i^{\infty}(x) = \int_{\Sigma} v_{pl}(\xi)U_i(x,\infty; \xi, 0)d\xi. \tag{27}$$

Table 2

The relative rotation vectors in the Cartesian coordinate system, calculated from NUVEL-1A

	ω_X [radians/Myr]	ω_Y [radians/Myr]	ω_Z [radians/Myr]
PA-NA	−0.001768	0.008439	−0.009817
PA-PH	−0.0116	0.0120	−0.0003
PH-NA	0.009832	−0.003561	−0.009517
PH-EU	0.011071	−0.004765	−0.012823

In the present case, where the lithosphere-asthenosphere system is modelled by an elastic surface layer overlying a Maxwellian viscoelastic half-space, the slip response function U_i becomes zero at $t = \infty$, if the source is located in the viscoelastic half-space. Hence, in the case of steady subduction, we may consider only the effects of fault slip motion at the plate boundary cutting the elastic lithosphere.

3.2 Relative Plate Motion

Given the 3-D geometry of the plate interfaces and steady slip rates there, we can calculate the long-term crustal deformation due to steady plate subduction by using Eq. (27). In order to determine the steady slip rates v_{pl} on the plate interface, first, we calculate the relative plate motion at the plate boundaries. In the present study we use the relative plate motion calculated from NUVEL-1A (DeMets *et al.*, 1994), which provides the rotation vectors $\boldsymbol{\omega}$ for the relative plate motion in the Cartesian coordinate system (X, Y, Z) shown in Figure 2a. The X, Y and Z components of the relative rotation vectors for PA-NA, PA-PH, PH-NA and PH-EU are listed in Table 2.

The position vector of an arbitrary point with latitude v and longitude ε on the surface of the spherical earth is represented in the Cartesian coordinate system (X, Y, Z) as

$$X = (R_E \cos v \ \cos \varepsilon, R_E \cos v \ \sin \varepsilon, R_E \sin v)^T. \tag{28}$$

Then, given the relative rotation vector $\boldsymbol{\omega}$, the relative plate motion $\boldsymbol{v} = (v_X, v_Y, v_Z)^T$ at the point X is calculated by

$$\boldsymbol{v} = \boldsymbol{\omega} \times \boldsymbol{X}. \tag{29}$$

The longitudinal and latitudinal components (v_v, v_ε) of the relative plate motion vector $\boldsymbol{v}$ are written as

$$\begin{pmatrix} v_v \\ v_\varepsilon \end{pmatrix} = \begin{pmatrix} -v_X \sin v \cos \varepsilon - v_Y \ \sin v \sin \varepsilon + v_z \cos v \\ -v_X \sin \varepsilon - v_Y \cos \varepsilon \end{pmatrix}$$
$$= \begin{pmatrix} R_E(\omega_X \sin \varepsilon - \omega_Y \cos \varepsilon) \\ R_E[(\omega_X \cos \varepsilon - \omega_Y \sin \varepsilon) \sin v - \omega_Z \cos v] \end{pmatrix} \tag{30}$$

in the Cartesian coordinate system (X, Y, Z).

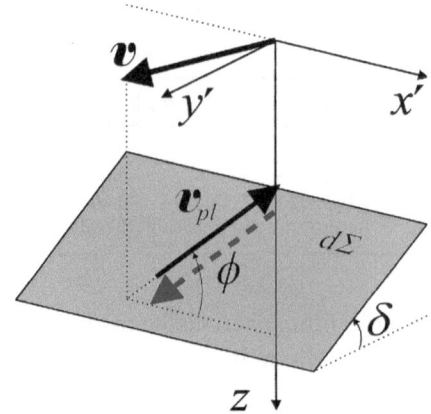

Figure 4

Determination of the steady slip vector v_{pl} on an element of plate interface Σ. The direction of the steady slip vector v_{pl} is given by the projected direction of the relative plate motion vector v onto Σ. The magnitude of the steady slip vector $|v_{pl}|$ is equal to $|v|$. The coordinate system (x', y') is defined by operating the same transformation matrix in Eq. (43) of the text to the coordinate system (x, y).

We project this velocity vector defined on the spherical earth onto the flat earth in the same way as described in Section 2. Denoting a projected point on the flat earth by $[x(v, \varepsilon), y(v, \varepsilon)]$, we can express the projected direction-vectors (γ_v^P and γ_ε^P) in Figure 2b) as

$$\gamma_v^P(v, \varepsilon) \frac{1}{\sqrt{(\partial x/\partial v)^2 + (\partial y/\partial v)^2}} \begin{pmatrix} \partial x/\partial v \\ \partial y/\partial v \end{pmatrix} \tag{31}$$

and

$$\gamma_\varepsilon^P(v, \varepsilon) \frac{1}{\sqrt{(\partial x/\partial \varepsilon)^2 + (\partial y/\partial \varepsilon)^2}} \begin{pmatrix} \partial x/\partial \varepsilon \\ \partial y/\partial \varepsilon \end{pmatrix} \tag{32}$$

Thus, we obtain the x- and y-components of the relative plate motion vector projected on the flat earth as

$$v = \begin{pmatrix} v_x \\ v_y \end{pmatrix} = v_v \gamma_v^P + v_\varepsilon \gamma_\varepsilon^P = \begin{pmatrix} \frac{v_v}{\sqrt{(\partial x/\partial v)^2 + (\partial y/\partial v)^2}} \frac{\partial x}{\partial v} + \frac{v_\varepsilon}{\sqrt{(\partial x/\partial \varepsilon)^2 + (\partial y/\partial \varepsilon)^2}} \frac{\partial x}{\partial \varepsilon} \\ \frac{v_v}{\sqrt{(\partial x/\partial v)^2 + (\partial y/\partial v)^2}} \frac{\partial y}{\partial v} + \frac{v_\varepsilon}{\sqrt{(\partial x/\partial \varepsilon)^2 + (\partial y/\partial \varepsilon)^2}} \frac{\partial y}{\partial \varepsilon} \end{pmatrix}. \tag{33}$$

In Figure 1 we show the relative plate motion vectors at plate boundaries, calculated from NUVEL-1A.

3.3 Steady Slip Rates on Plate Interfaces

Given the relative plate motion vectors on the flat earth's surface, we can determine the steady slip vectors (v_{pl}) on the plate interfaces in the following way. In the Cartesian coordinate system (x, y, z) defined in Section 2, the relative plate motion vector on the earth's surface is written as

$$v = (v_x, v_y, 0)^T. \tag{34}$$

To determine the direction of the steady slip vector v_{pl} on the plate interface Σ, first, we project the relative plate motion vector v onto Σ, as shown in Figure 4.

The geometry of the plate interface Σ is given by a function of $z = z(x, y)$, and so the unit normal vector is given by

$$n = \begin{pmatrix} n_x \\ n_y \\ n_z \end{pmatrix} = \frac{1}{\sqrt{(\partial z/\partial x)^2 + (\partial z/\partial y)^2 + 1}} \begin{pmatrix} \partial z/\partial x \\ \partial z/\partial y \\ -1 \end{pmatrix}. \tag{35}$$

In our model the plate interface is represented by the superposition of bicubic B-splines. Then, the spatial derivatives of z are expressed as

$$\frac{\partial z(x, y)}{\partial x} = \sum_k \sum_l a_{kl} \frac{\partial N_{4,k}(x)}{\partial x} N_{4,l}(y) \tag{36}$$

and

$$\frac{\partial z(x, y)}{\partial y} = \sum_k \sum_l a_{kl} N_{4,k}(x) \frac{\partial N_{4,l}(y)}{\partial y}. \tag{37}$$

The slip-direction vector χ must satisfy the orthogonality condition

$$n \cdot \chi = 0 \tag{38}$$

and the x- and y-components of χ should be proportional to those of the relative plate motion vector v

$$\chi = \frac{-1}{\sqrt{v_x^2 + v_y^2 + (v_x n_x + v_y n_y)^2 / n_z^2}} \begin{pmatrix} v_x \\ v_y \\ -(v_x n_x + v_y n_y)/n_z \end{pmatrix}. \tag{39}$$

As to the magnitude of the steady slip vector, we take $|v_{pl}|$ to be $|v|$ to satisfy the condition of mass conservation for the descending slab. Then, finally, the steady slip vector can be written as

$$v_{pl} = |v|\chi = \frac{-\sqrt{v_x^2 + v_y^2}}{\sqrt{v_x^2 + v_y^2 + (v_x n_x + v_y n_y)^2 / n_z^2}} \begin{pmatrix} v_x \\ v_y \\ -(v_x n_x + v_y n_y)/n_z \end{pmatrix}. \tag{40}$$

In the conventional expressions of crustal deformation due to faulting (e.g., MATSU'URA et al., 1981; MATSU'URA and SATO, 1997; HASHIMOTO and MATSU'URA, 2000), the fault slip vector is represented by using the dip angle δ, the slip direction ϕ, and the amount of slip. For convenience we show the expressions of these fault parameters in the present notation;

$$\begin{cases} \cos \delta = \dfrac{-n_z}{\sqrt{n_x^2 + n_y^2 + n_z^2}} \\[4mm] \sin \delta = \dfrac{\sqrt{n_x^2 + n_y^2}}{\sqrt{n_x^2 + n_y^2 + n_z^2}} \end{cases} \tag{41}$$

and

$$\begin{cases} \cos \phi = \dfrac{-v_x'}{\sqrt{v_x'^2 + (v_y'/\cos \delta)^2}} \\[4mm] \sin \phi = \dfrac{v_y'/\cos \delta}{\sqrt{v_x'^2 + (v_y'/\cos \delta)^2}} \end{cases} \tag{42}$$

with

$$\begin{pmatrix} v_x' \\ v_y' \end{pmatrix} = \frac{1}{\sqrt{n_x^2 + n_y^2}} \begin{pmatrix} n_y & -n_x \\ n_x & n_y \end{pmatrix} \begin{pmatrix} v_x \\ v_y \end{pmatrix} \tag{43}$$

4. Numerical Results and Physical Interpretation

In Section 2 we constructed a realistic 3-D structure model in and around Japan. On this structure model we developed a 3-D simulation model for long-term crustal deformation due to steady plate subduction in Section 3. Now we numerically compute the long-term surface displacement rates in and around Japan with the 3-D plate subduction model.

In Figure 5 we show the numerical results. Figure 5a shows the computed horizontal velocities of the Pacific and the Philippine Sea plates relative to the Eurasian plate. The 3-D plate subduction model realizes a uniform horizontal motion of each oceanic plate in the direction expected from NUVEL-1A. The horizontal deformation near the plate boundaries is not so significant. The convergence rate between the North American and the Pacific plates is about 8 cm/yr in northeast Japan, and that between the Eurasian and the Philippine Sea plates is about 4 cm/yr in southwest Japan. Figure 5b shows the computed vertical displacement rates in and around Japan. The pattern of vertical motion is characterized by steep uplift at island arcs, sharp subsidence at ocean trenches, and gentle uplift at outer-rises. This characteristic pattern accords with that of observed free-air gravity anomalies (SATO and MATSU'URA, 1992, 1993). The maximum subsidence rates at the Kuril trench, the Japan trench, and the Nankai trough are about 6 mm/yr, 2.5 mm/yr, and 1.5 mm/yr, respectively. The maximum uplift rates in Kuril, northeast Japan, and southwest Japan are about 2.5 mm/yr, 1.5 mm/yr, and 1.5 mm/yr, respectively. The pattern of the computed vertical

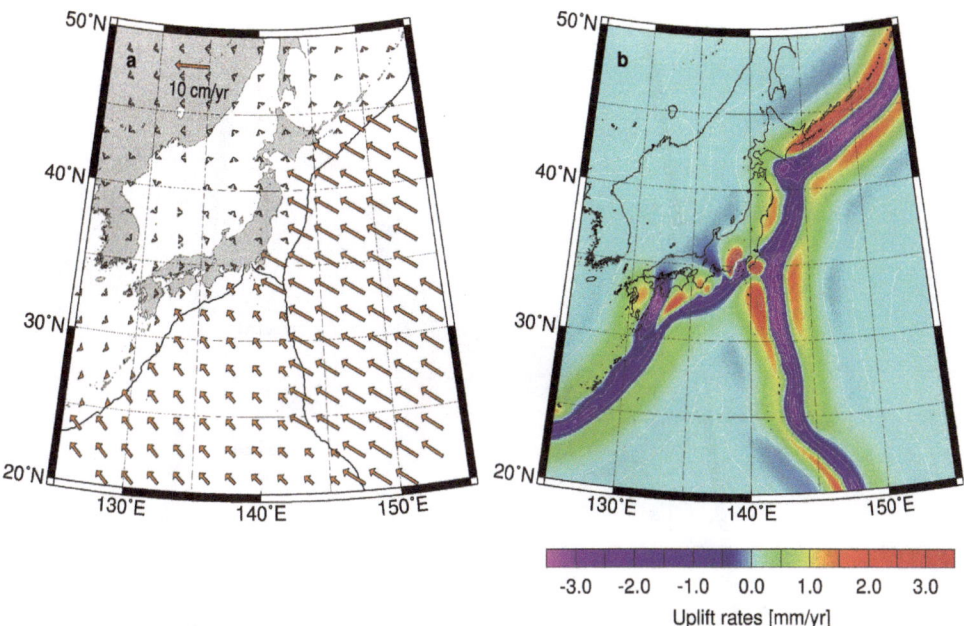

Figure 5

Simulated crustal motion. (a) Computed horizontal velocities of the Pacific and the Philippine Sea plates relative to the Eurasian plate. The convergence rate between the North American and the Pacific plates is about 8 cm/yr in northeast Japan, and that between the Eurasian and the Philippine Sea plates is about 4 cm/yr in southwest Japan. (b) Computed vertical displacement rates in and around Japan. The pattern of vertical motion is characterized by steep uplift at island arcs, sharp subsidence at ocean trenches, and gentle uplift at outer-rises. The contour intervals of the vertical displacement rates are 1 mm/yr.

displacement rates shows the significant effects of 3-D geometry of plate interfaces. For example, the large-scale horizontal bends of plate interfaces at the junction of the Kuril and the Japan trenches in north Japan and the junction of Nankai and the Ryukyu troughs in southwest Japan bring about significant subsidence in broad regions around there. The cause of such broad subsidence is in the horizontal extension of the overlying plates due to steady slip motion along the curved plate interfaces.

In general, the moment tensor density function m_{pq} corresponding to steady fault slip $w = v_{pl}\tau\chi$ at a point ξ on the plate interface with a unit normal vector n is represented as

$$m_{pq}(\xi, \tau) = \mu v_{pl}[\chi_p(\xi)n_q(\xi) + \chi_q(\xi)n_p(\xi)]\tau. \tag{44}$$

The derivatives of the moment tensor density m_{pq} with respect to the source coordinate ξ_q give an equivalent force

$$e_p(\xi, \tau) = -\frac{\partial}{\partial\xi_q}m_{pq}(\xi, \tau) = \mu v_{pl}\frac{\partial}{\partial\xi_q}[\chi_p(\xi)n_q(\xi) + \chi_q(\xi)n_p(\xi)]\tau. \tag{45}$$

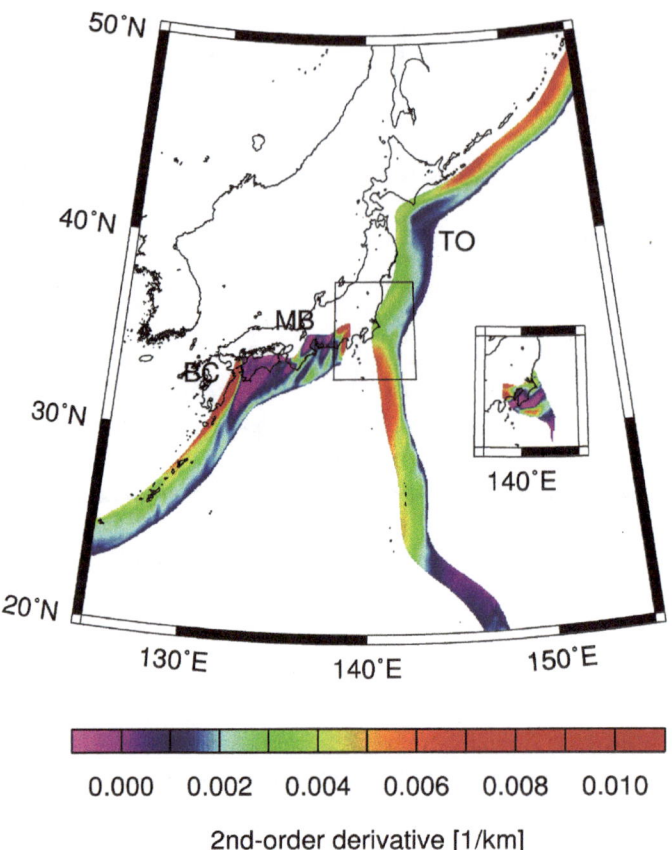

Figure 6

The pattern of calculated second-order derivatives of the depth function $z(x,y)(0 < z \leq 40$ km) of Σ along the direction of plate convergence. The inserted diagram shows the second-order derivatives of Σ_3 (PH-NA), where the Philippine Sea plate is descending beneath the North American plate and running on the descending Pacific plate.

This expression indicates that if the direction of the fault slip vector changes along a curved plate interface, it becomes the origin of internal force.

In the present case, the magnitude of the steady slip vector v_{pl} is nearly constant, but its direction χ changes along the curved plate interface Σ with a normal vector $\boldsymbol{n}$. To evaluate the relative strength of the internal force $\boldsymbol{e}$, we can use the second-order derivatives of the depth function $z(x,y)$ of Σ along the direction of plate convergence. Denoting the direction of plate convergence by

$$\boldsymbol{r} = \frac{\boldsymbol{v}}{|\boldsymbol{v}|} = \left(\frac{v_x}{\sqrt{v_x^2 + v_y^2}}, \frac{v_y}{\sqrt{v_x^2 + v_y^2}} \right)^T \tag{46}$$

the second-order derivatives of $z(x, y)$ along this direction are given by

$$\frac{d^2z}{dr^2} = \frac{\partial^2 z}{\partial x^2} \frac{v_x^2}{v_x^2 + v_y^2} + 2 \frac{\partial^2 z}{\partial x \partial y} \frac{v_x v_y}{v_x^2 + v_y^2} + \frac{\partial^2 z}{\partial y^2} \frac{v_y^2}{v_x^2 + v_y^2}. \qquad (47)$$

Figure 6 shows the pattern of the calculated second-order derivatives of the plate interfaces in and around Japan. We can find high-value belts at the depth range of 20–40 km of the plate interface along the Kuril-Japan-Izu-Ogasawara trench and the Ryukyu trough. These high-value belts are the origin of internal forces which bring about surface uplift and subsidence along the island arc-trench system. We can also find three low-value regions interrupting the high-value belts in Tokachi-oki (TO), Mikawa Bay (MB) and Bungo Channel (BC). These three regions correspond to the large-scale horizontal bends of the plate interfaces mentioned above. The characteristic patterns of crustal deformation associated with the vertical and the horizontal bends of plate interfaces indicate that the 3-D geometry of plate interfaces is essential to understand the kinematic interaction of adjacent plates in subduction zones.

References

AKAIKE, H. (1980), *Likelihood and the Bayes procedure*. In *Bayesian Statistics,* (eds Bernardo, J. M., DeGroot, M. H., Lindly, D. V. and Smith, A. F. M.) (Univercity Press, Valencia, Spain.1980) pp. 143–166.

BURRIDGE, R. and KNOPOFF, L. (1964), *Body Force Equivalents for Seismic Dislocations,* Bull. Seismol. Soc. Am. *54,* 1875–1888.

COX, M. G. (1972), *The Numerical Evaluation of B-splines,* J. Inst. Math. Appl. *10,* 134–149.

DE BOOR, C. (1972), *On Calculating with B-splines,* J. Approx. Theory *6,* 50–62.

DEMETS, C., GORDON, R. G., ARGUS, D. F., and STEIN, S. (1994), *Effect of Recent Revisions to the Geomagnetic Reversal Time Scale on Estimates of Current Plate Motions,* Geophys. Res. Lett. *21,* 2191–2194.

HASHIMOTO, C. and MATSU'URA, M. (2000), *3-D Physical Modelling of Stress Accumulation and Release Processes at Transcurrent Plate Boundaries,* Pure Appl. Geophys. *157,* 2125–2147.

HASHIMOTO, C. and MATSU'URA, M. (2002), *3-D Simulation of Earthquake Generation Cycles and Evolution of Fault Constitutive Properties,* Pure Appl. Geophys. *159,* 2175–2199.

JACKSON, D. D. and MATSU'URA, M. (1985), *A Bayesian Approach to Nonlinear Inversion,* J. Geophys. Res. *90,* 581–591.

MARUYAMA, T. (1963), *On the Force Equivalents of Dynamical Elastic Dislocations with Reference to the Earthquake Mechanism.* Bull. Earthq. Res. Inst. Tokyo Univ. *41,* 467–486.

MATSU'URA, M. and SATO, T. (1989), *A Dislocation Model for the Earthquake Cycle at Convergent Plate Boundaries,* Geophys. J. Int. *96,* 23–32.

MATSU'URA, M. and SATO, T. (1997), *Loading Mechanism and Scaling Relation of Large Interplate Earthquakes,* Tectonophy. *277,* 189–198.

MATSU'URA, M., TANIMOTO, T. and IWASAKI, T. (1981), *Quasi-static Displacements due to Faulting in a Layered Half-space with an Intervenient Viscoelastic layer,* J. Phys. Earth. *29,* 23–54.

SATO, T. and MATSU'URA, M. (1992), *Cyclic Crustal Deformation, Steady Uplift of Marine Terraces, and Evolution of the Island Arc-trench System in Southwest Japan,* Geophys. J. Int. *111,* 617–629.

SATO, T. and MATSU'URA, M. (1993), *A Kinematic Model for Evolution of Island Arc-trench Systems,* Geophys. J. Int. *114,* 512–530.

YABUKI, T. and MATSU'URA, M. (1992), *Geodetic Data Inversion Using a Bayesian Information Criterion for Spatial Distribution of Fault Slip,* Geophys. J. Int. *109*, 363–375.

(Received September 27, 2002, revised January 27, 2003, accepted February 10, 2003)

Pure appl. geophys. 161 (2004) 2069–2090
0033–4553/04/102069–22
DOI 10.1007/s00024-004-2549-7

❘ Pure and Applied Geophysics

GeoFEM Kinematic Earthquake Cycle Simulation in Southwest Japan

MAMORU HYODO[1,3] and KAZURO HIRAHARA[2]

Abstract—We construct a viscoelastic FEM model with 3-D configuration of the subducting Philippine Sea plate in Southwest Japan to simulate recent 300-year kinematic earthquake cycles along the Nankai-Suruga-Sagami trough, based on the kinematic earthquake cycle model. This 300-year simulation contains a series of three great interplate earthquakes. The inclusion of viscoelasticity produces characteristic velocity field during earthquake cycles regardless of the assumed constant plate coupling throughout the interseismic period. Just after the occurrence of interplate earthquakes, the viscoelastic relaxation creates the seaward motion in the inland region. In the middle period, the seaward motion gradually decreases, and the resultant velocity field is similar to the elastic one. Later, just before the next interplate earthquake, displacements due to the interplate coupling in the viscoelastic material are distributed more broadly in the forearc region than in the purely elastic one, since the viscoelastic relaxation due to the previous earthquake mostly disappears. The effects of such interplate earthquake cycles on five major inland faults in southwest Japan, where large intraplate earthquakes occurred during this period, are quantitatively evaluated using the Coulomb failure function (CFF). The calculated change in CFF successfully predicts the occurrence of the 1995 Kobe earthquake ($M\sim7$). The occurrence of other inland earthquakes, however, cannot be explained by the calculated changes in CFF, and especially the 1891 Nobi earthquake ($M\sim8$), the largest inland earthquake in Japan, which occurred at the time close to the local minimum of CFF. This implies that further improvements are necessary for our FEM modeling, such as the modeling of steady east-west compressive force and stress interactions between the inland faults.

Key words: 3-D viscoelastic structure, Southwest Japan, FEM, crustal deformation, Coulomb failure function, inland earthquake.

1. Introduction

Southwest Japan (SWJ) is located in the plate boundaries of the Pacific (PA), North American (NA), Philippine Sea (PH), and Amurian (AM) plates, and subject to the interaction due to the convergence of these four plates (Fig.1). Recently, the deployment of GEONET (GPS Earth Observation NETwork) by Geographical Survey Institute of Japan has revealed the contemporary deformation of the Japanese Islands. The characteristic deformation pattern of SWJ is shown in

[1] Graduate School of Sciences, Nagoya University, Chikusa, Nagoya 464-8602, Japan.
[2] Graduate School of Environmental Studies, Nagoya University, Chikusa, Nagoya 464-8602, Japan. E-mail: hirahara@eps.nagoya-u.ac.jp
[3] Now at The Earth Simulator Center, Japan Agency for Marine-Earth Science and Technology, 3173-25, Showa-machi, Kanazawa-ku, Yokohama, 236-0001, Japan. E-mail: hyodo@jamstec.go.jp

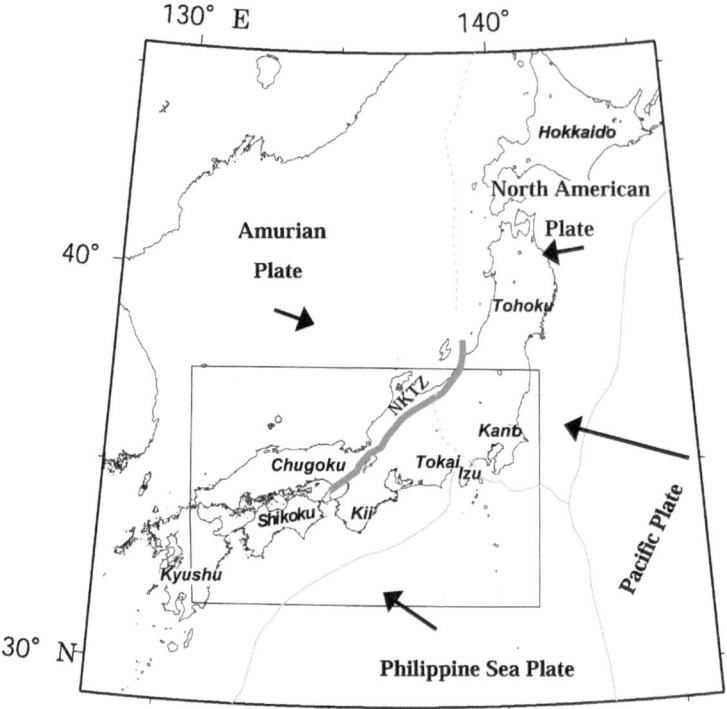

Figure 1

Plate boundaries in Japan and the investigated region in this study. Thin gray and dotted gray lines indicate the plate boundaries, and the thick gray line denotes a concentrated deformation zone, NKTZ, deduced by GEONET (SAGIYA et al., 2000). Black arrows indicate relative plate motions with respect to EU. The investigated region in this study, together with site names, is enclosed by a rectangle.

Figure 2. The displacement and strain rate field in SWJ, which are consistent with the Quaternary deformation, have the following two major characteristics. First, along the Nankai-Suruga trough, the Pacific coast of SWJ is displaced towards the northwest due to the subducting PH. Associated with the subducting PH, the occurrence of repeated interplate thrust earthquakes ($M\sim8$) has been documented since A. D. 684 with the recurrence interval of 90–209 years (Fig. 3). Recently, the next large event is anticipated in the Tokai district, because the last major events, the 1944 Tonankai and the 1946 Nankai earthquakes, ruptured the remainder of the subduction zone. Second, a remarkably high east-west strain rate belt runs NE-SW in the inland region of central Japan. Hereafter, we designate this high strain rate belt as NKTZ (Niigata Kobe Tectonic Zone) following SAGIYA et al. (2000). Within NKTZ, a number of Quaternary active faults concentrate and large intraplate earthquakes ($M\sim7$) with the recurrence interval of 1,000 years or more have occurred intensively. Most of these intraplate earthquakes have been caused by the tectonic east-west compressive force, which is consistent with the direction of strain rates observed by GPS.

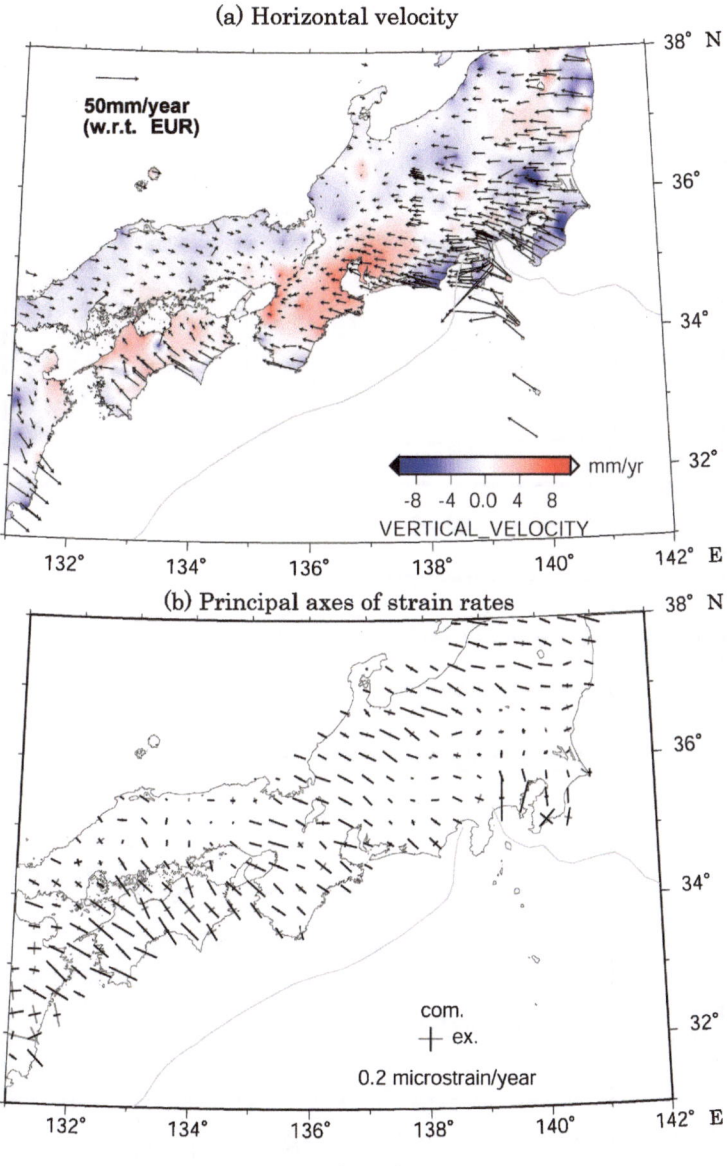

Figure 2

(a). Horizontal displacement vectors and vertical displacement rate of continuous GPS sites in central and SWJ. All vectors are relative to the Eurasian plate. The blue and red regions show the subsided and uplifted ones, respectively. (b) Principal axes of strain rates deduced from horizontal vectors in Figure 2(a). Solid and gray lines indicate compression and extension, respectively.

The historical earthquake data show the significant temporal correlation between the occurrence of the intraplate earthquakes and the great interplate ones along the Nankai trough. Namely, large intraplate earthquakes have occurred frequently in the

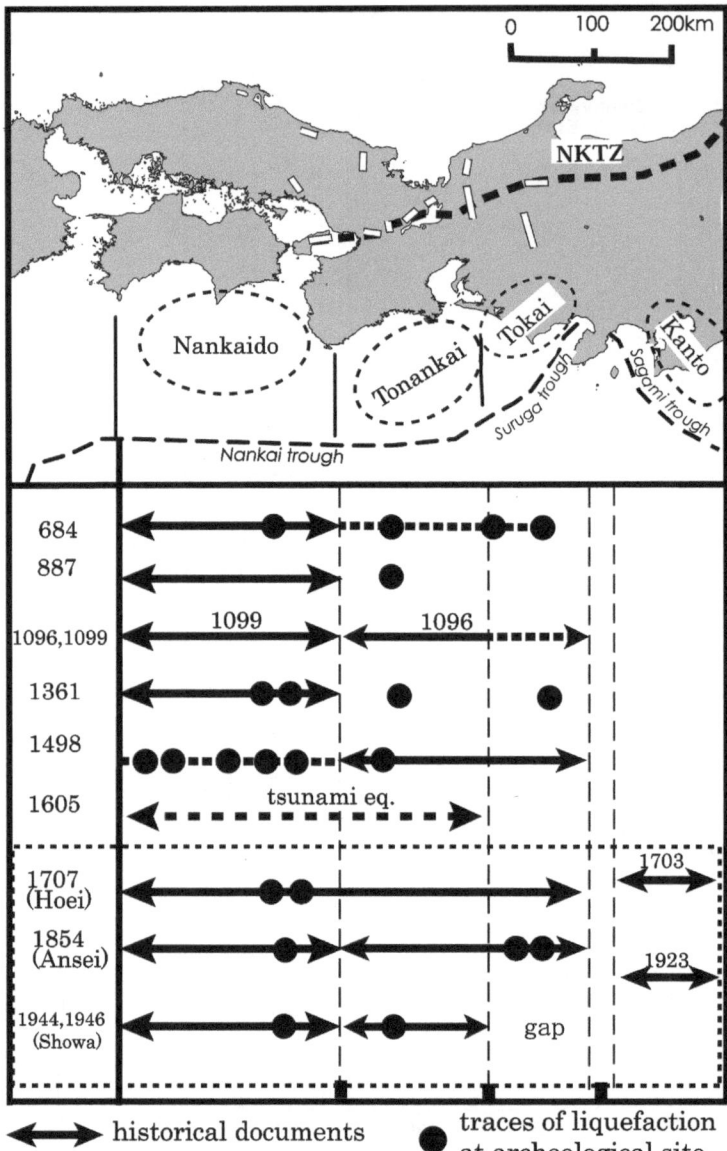

Figure 3
Fault segmentation and time sequence of great interplate earthquakes along the Nankai, Suruga and Sagami troughs identified from historical documents (horizontal arrows) and archeological studies (dots) modified from ISHIBASHI and SATAKE (1998). In the above map, white bars show major active faults in southwest Japan, and the thick dotted line indicates the location of the NKTZ (SAGIYA et al., 2000). The dotted circles represent the ruptured segments of Nankai, Tonankai and Tokai along the Nankai and Suruga troughs, and of Kanto along the Sagami trough.

period from several decades before to a decade after the occurrence of the great interplate earthquakes (HORI and OIKE, 1996).

As a mechanism explaining the temporal correlation between the intraplate earthquakes and interplate ones in SWJ, it has been suggested that the stress perturbation due to the earthquake cycles along the Nankai trough affect the steady tectonic east-west loading on inland faults (e.g., HORI and OIKE, 1999). However, this model is based on the purely elastic medium, and the strong structural heterogeneity and viscoelasticity, which are characteristic of subduction zones, are not taken into consideration. POLLITZ and SACKS (1997) examined the effect of the 1944 Tonankai and the 1946 Nankai earthquakes on the occurrence of the 1995 Kobe earthquake, the most recent large inland earthquake in SWJ, in a viscoelastic structure, and concluded the Kobe earthquake was triggered by the long-delayed viscoelastic response to the 1944 and the 1946 earthquakes. Their structural model, however, was a simple layered model with an elastic lithosphere and a viscoelastic asthenosphere, and the effect of reloading of the plate subduction and of previous interplate earthquakes along the Nankai trough was ignored.

In this paper we construct a realistic three-dimensional viscoelastic finite-element model with the complicated configuration of the subducting PH. We employ a 3-D viscoelastic FEM (Finite-Element Method) code, GeoFEM (IIZUKA et al., 2002) to re-evaluate the effect of the subducting PH and of great interplate earthquakes there during the past 300 years on the occurrence of large historical intraplate earthquakes in SWJ. This 300-year simulation contains a series of three great interplate earthquakes, the 1707 Hoei, the 1854 Ansei and the 1944 and 1946 Showa earthquakes along the Suruga-Nankai trough and two additional ones, the 1703 and the 1923 Kanto earthquakes along the Sagami trough, as understood in Figure 3.

2. Kinematic Earthquake Cycle Model Along the Nankai Trough

In the kinematic framework of earthquake cycle modeling, the interaction of two plates along the convergent plate boundary during the interseismic period is understood as the superposition of the steady-state subduction and the virtual normal slip on the seismogenic zone, which is generally called 'back-slip' (SAVAGE, 1983). According to MATSU'URA and SATO (1989), the deformation produced by the uniform steady slip contributes to the surface deformation. However, as in SUITO et al. (2002), we approximate the changes in crustal deformation in the interseismic period during a relatively short term including a few earthquake cycles only by backslip on the main thrust zone, because the steady slip contributes to the long-term deformation.

In applying the backslip model, what plate the southwest Japan arc overriding the subducting PH belongs to is a critical problem. Figure 2(a) shows the displacement velocities at the sites of the GEONET with respect to the Eurasian plate (EU).

Though SWJ has been assumed to be a part of EU (e.g., SENO et al., 1993), the sites in Chugoku district, the northwestern part of SWJ, have eastward velocities relative to the EU. This implies that the western part of Japan belongs to the AM (ZONENSHAIN and SAVOSTIN, 1981; WEI and SENO, 1998; HEKI et al., 1999), but not the EU. Nevertheless, MAZZOTTI et al. (2000) concluded that the forearc of SWJ is trapped between the converging PA, PH and AM, and happens to have the plate velocity close to that of EU based on an elastic dislocation modeling of the GEONET velocity field using the PH-EU velocity. Following their kinematic reference frame, we basically use the PH-EU velocity as a relative velocity between the subducting oceanic plate and the overriding one beneath SWJ.

Red arrows in Figure 4 delineate the subduction velocity vectors along the Nankai, Suruga and Sagami troughs assumed in this study. The PH-EU velocity vectors can be calculated by superposing two velocity vectors estimated from the relative rotation (Euler) vectors of PH-AM and AM-EU in Table 1. The predicted subduction velocities of PH are about 40–60 mm/yr along the Suruga-Nankai trough, which vectors are displayed by black arrows in Figure 4. However, beneath the Suruga trough, the subduction velocities indicated by red arrows are smaller, and their directions are different from those of EU-PH velocity vectors. This is mainly because the collision between the Izu peninsula and central Japan reduces the northward component of the subduction velocity of PH. In addition, a nascent plate boundary along the Zenisu Ridge now divides PH and Izu Micro-plate (IMP) (SAGIYA, 1999b; MAZZOTTI et al., 1999). Thus, the velocity of the subducting IMP from Suruga Bay to Sea of Enshu is effectively reduced compared with that of the PH. Along the Sagami trough, the subduction velocity best-fit for observed GPS velocities is 26 mm/yr toward N335° (HENRY et al., 2001). The westward component of this velocity vector is smaller than PH-EU relative vector, since the overriding plate in the Kanto district is changing from the plate with the velocity close to that of EU to the NA which has a westward velocity of about 10-mm/yr relative to the EU.

In Figure 4 the coupling zone on the plate interface is shown together with the iso-depth contours of the upper boundary of the PH. Following HEKI and MIYAZAKI (2001), the depth of full plate coupling, indicated by a dark shaded zone, is assumed to be 5–20 km, and the depths of transition, denoted by light shaded zones, are 0–5 and 20–30 km.

The slip deficit multiplied by the average recurrence interval of great interplate earthquakes along the Nankai trough is equal to the slip with the amount of 5–6 m, which is comparable to the observed slip of the 1944 Tonankai and the 1946 Nankai earthquakes. However, for the old interplate events before these, we can only know the occurrence time of events, and the precise estimate of slip amounts is difficult.

SHIMAZAKI and NAKATA (1980) suggested two models, which can estimate coseismic slip amounts from the occurrence time of earthquakes and constant plate velocity. One is termed as a 'slip-predictable model.' In this model the coseismic slip is determined by the slip deficit accumulated during the previous earthquake cycle.

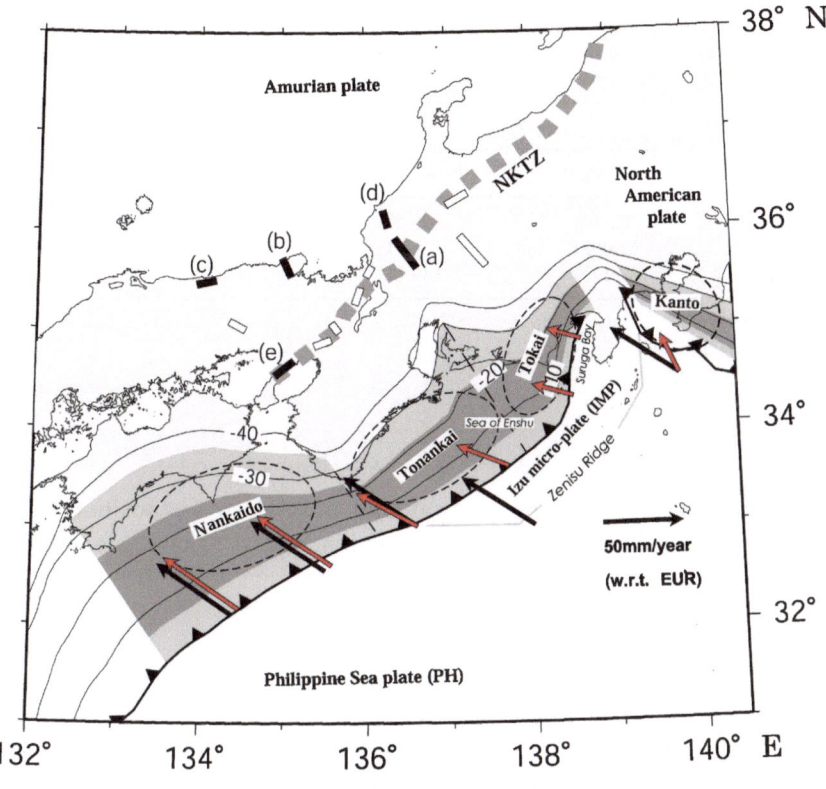

Figure 4

Subduction velocity along the Nankai, Suruga and Sagami troughs assumed in this study (red arrows). Black arrows are PH-EU relative velocities deduced from the rotation vector of PH-AM and AM-EU (see Table 1). The red arrows in the Tokai and Tonankai segments are predicted using IMP-CJP rotation vector of MAZZOTTI *et al.* (1999). In the Sagami trough, a red arrow shows the best-fit subduction velocity obtained by HENRY *et al.* (2001). Locked and transition zones in the Nankai, Suruga, and Sagami troughs are shown by dark and light shading. Solid lines represent the isodepth contours (km) of the top of the Philippine Sea plate. Thick lines represent the strikes and the locations of the faults (a)–(d) on which the failure stresses are calculated in this study.

The other model is described as a 'time-predictable model.' In this model the coseismic slip is equivalent to the slip deficit accumulated during the next earthquake cycle. They concluded that the earthquake recurrence along the Nankai trough is consistent with the 'time-predictable model' rather than the 'slip-predictable model' from the historical and geomorphological evidence. Therefore we use the 'time-predictable model' in this study to estimate the coseismic slip amounts.

During great interplate earthquake cycles along the Nankai trough, we estimate temporal changes in the failure stress on major inland faults. The failure of a fault depends not only on the shear stress but also the normal stress (e.g., KING *et al.*, 1994). Hence, in order to evaluate the likelihood of failure, we employ the changes in Coulomb failure function, which are defined as,

Table 1

Euler pole positions and rotation rates used to estimate the subduction veclocity of the oceanic plates beneath the SWJ.

Plate Pair	Pole		Angular Rate	
	Lat. [deg]	Long. [deg]	ω [deg/Ma]	σ_ω [deg/Ma]
AM-EU[*1]	22.30S	106.6E	−0.091	0.016
PH-AM[*2]	61.00N	168.8E	0.830	0.070
IMP-CJP[*3]	37.14N	139.6E	−4.520	0.800

[*1] Heki et al. [1999]
[*2] Miyazaki and Heki [2001]
[*3] Mazzotti et al. [1999]

$$\Delta CFF = \Delta\tau_s + \mu'\Delta\sigma_n, \tag{1}$$

where $\Delta\tau_s$ and $\Delta\sigma_n$ are changes in shear and normal stress (positive in tension), respectively. μ' is the effective coefficient of friction, and its value is generally assumed to be about 0.2–0.5.

3. Estimation of East-West Compressive Force

The GPS data show that the maximum compressive strain rate within NKTZ is -10^{-7}strain/yr, which is several times larger than that of the surrounding region, and the direction of the maximum compressive strain rate is about N100°E (Fig. 2(b)). Though there are several discussions concerning the origin of high strain rates within NKTZ, such as the collision of the AM and Northeast Japan arc (HEKI and MIYAZAKI, 2001), IIO *et al.* (2001) recently suggested that the lower crust beneath NKTZ is weakened by the high water content, and the elastic strength of the entire crust there is lower than that of the surrounding region. This implies that the convergence of the three plates, PA, AM and NA, may be accommodated as the concentrated deformation within NKTZ.

Thus, the origin of the east-west compressive tectonic force in inland areas has not yet been clearly understood, and hence the tectonic force cannot be modeled quantitatively by introducing three motions of plates, PA, NA and AM into our FEM model. Instead, we assign the steady east-west tectonic loading force in inland areas estimated from the observed GPS strain field in the following way. Assuming a simple compression, we can transform the GPS high strain rate within NKTZ into the stress accumulation rate. The shear and normal stress rates within the crust are expressed as,

$$\Delta\tau = \frac{1}{2}E \cdot \Delta\epsilon |\sin 2\theta| \tag{2}$$

$$\Delta\sigma = -\frac{1}{2}E \cdot \Delta\epsilon(1 + \cos 2\theta), \tag{3}$$

where E, $\Delta\varepsilon$ and θ are the Young's modulus (1.2×10^{11} Pa), maximum compressive strain rate (10^{-7} strain/yr), and the angle between the strike of an inland fault and minimum compressive axis of strain rate in NKTZ, respectively. We can estimate the accumulation rate of CFF due to the east-west compressive force, substituting eqs. (2) and (3) into eq. (1). Then, $\Delta\sigma$ is about several kPa/yr.

4. 3-D Viscoelastic FEM Model

Figure 5 displays our three-dimensional FEM model in SWJ. We use a 3-D FEM code, GeoFEM (IIZUKA *et al.*, 2002). There, we adopt iso-parametric hexahedral elements with eight nodes (ZIENKIEWICZ and CHEUNG, 1967). The total numbers of finite elements and nodes used here are 21,600 and 24,339, respectively. The size of the model is $1,400 \times 1,000 \times 200$ km. The average size of finite elements is about several ten kilometers. A fine discretization of meshes is necessary near the plate boundary because the dislocation is imposed on the finite element nodes on the plate interface, and relatively course meshes are used near the model boundaries. For constructing the subducting plate structure, we take account of the 3-D configuration of the upper surface of the subducting PH, which is estimated from the spatial distribution of subcrustal micro-earthquakes (ISHIDA, 1992; YAMAZAKI and OOIDA, 1985). The crust and the subducting PH with the thickness of 30 km are purely elastic, and the upper mantle is assumed to be a Maxwell viscoelastic body with a relaxation time of 4.96 years (SUITO and HIRAHARA, 1999). The material properties are tabulated in Table 2. We assume that the model surface is stress-free, and the other five outer boundaries can have slip components parallel to the planes. In this FEM model the earthquake cycles of PH are described by backslip model (see Section 2), and the dislocation is transformed to the equivalent nodal force by the split node technique (MELOSH and RAEFSKY, 1981).

As explained in Section 3, the steady east-west compressive force is not directly included in our FEM model. Accordingly, we do not include the westwards subducting PA along the Japan trench in our model. And the eastern part of our FEM model is coarser than the central and western part model as understood in Figure 5.

5. Results

In the viscoelastic simulation, the deformation field in any earthquake cycle is strongly dependent on the previous earthquakes or coupling changes. Hence, though we are interested in the deformation associated with the recent 300-year earthquake cycles along the Nankai, Suruga and Sagami troughs, we calculate the 700-years

(a) Horizontal view of mesh

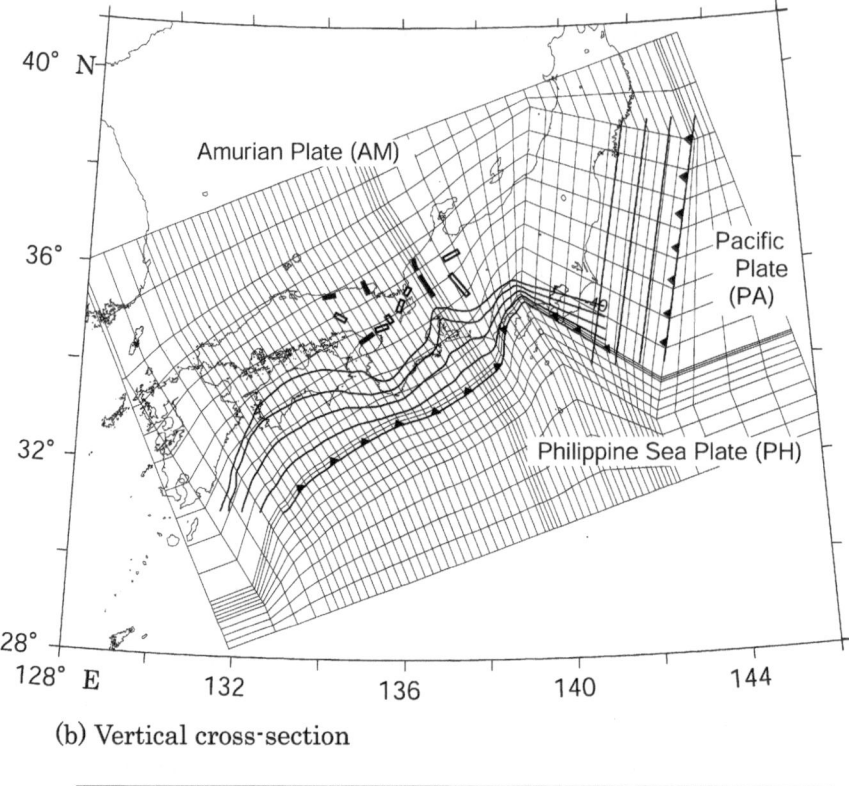

(b) Vertical cross-section

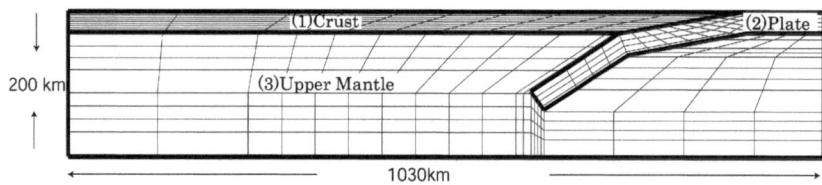

Figure 5

FEM meshes constructed in this study. (a) Horizontal view of the finite element mesh. The lines with black triangles indicate the locations of the Nankai-Suruga-Sagami troughs and the Japan trench. (b) Vertical cross section of FEM along the NW-SE direction. The crust (1) and plate (2) are elastic and the upper mantle (3) is a Maxwell solid, whose material properties are given in Table 2.

simulation, assuming the periodic occurrence of great interplate earthquakes along the Nankai-Suruga trough with the recurrence interval of 100 years, and earthquakes along the Sagami trough with the recurrence interval of 200 years during the first 400 years. In the subsequent 300 years, the great earthquake sequence follows the 'time predictable model' as stated in Section 2.

Table 2

Material properties of elastic crust, the PH slab and the viscoelastic upper mantle.

No.	Rigidity $[\times 10^{10}\text{Pa}]$	Poisson's ratio	Viscosity $[\times 10^{19}\text{Pa·s}]$	Relaxation time [yr]
(1) Crust	4.80	0.258	∞	∞
(2) PH	4.80	0.258	∞	∞
(3) Upper Mantle	6.39	0.283	1.0	4.96

5.1. Crustal Velocities during the Earthquake Cycle

For understanding the nature of the viscoelastic relaxation in the asthenosphere (WANG *et al.*, 2001), we examine, initially, the deformation velocity patterns in different stages during an earthquake cycle. As a reference, Figure 6 displays the calculated surface deformation velocities in the purely elastic material model. The red arrows represent the surface horizontal velocities due to the elastic coupling of the subducting plate interface, and red and blue regions denote uplift and subsidence, respectively. In the elastic model, the deformation velocity is limited near the coastal region, and its direction is similar to that of backslips indicated by red arrows in

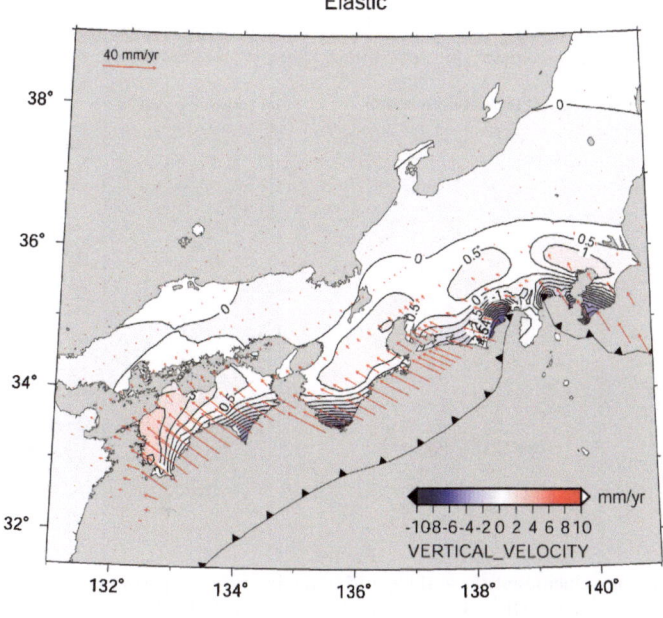

Figure 6

Calculated crustal deformation velocities due to the subduction of the Philippine Sea plate in a purely elastic medium. The red arrows show the horizontal rate vectors, which scale is indicated in the upper left corner. The blue and red regions show the subsidence and uplifted ones, respectively.

Figure 4. Clearly the deformation velocity and the direction are stationary throughout the postseismic and the interseismic periods in an earthquake cycle.

In contrast, the velocity field has significant time dependence in the viscoelastic simulation. Figures 7(a)–(d) show the surface velocity field at the different periods during the earthquake cycle, and Figures 8(a)–(c) indicate the corresponding velocity field in the vertical cross section along the line A-A' in Figures 7(a)–(c). As shown in Figure 7(a) and Figure.8(a), in the postseismic period six years after the occurrence of the 1854 Ansei event, which ruptured the entire Nankai-Suruga trough (Fig. 3),

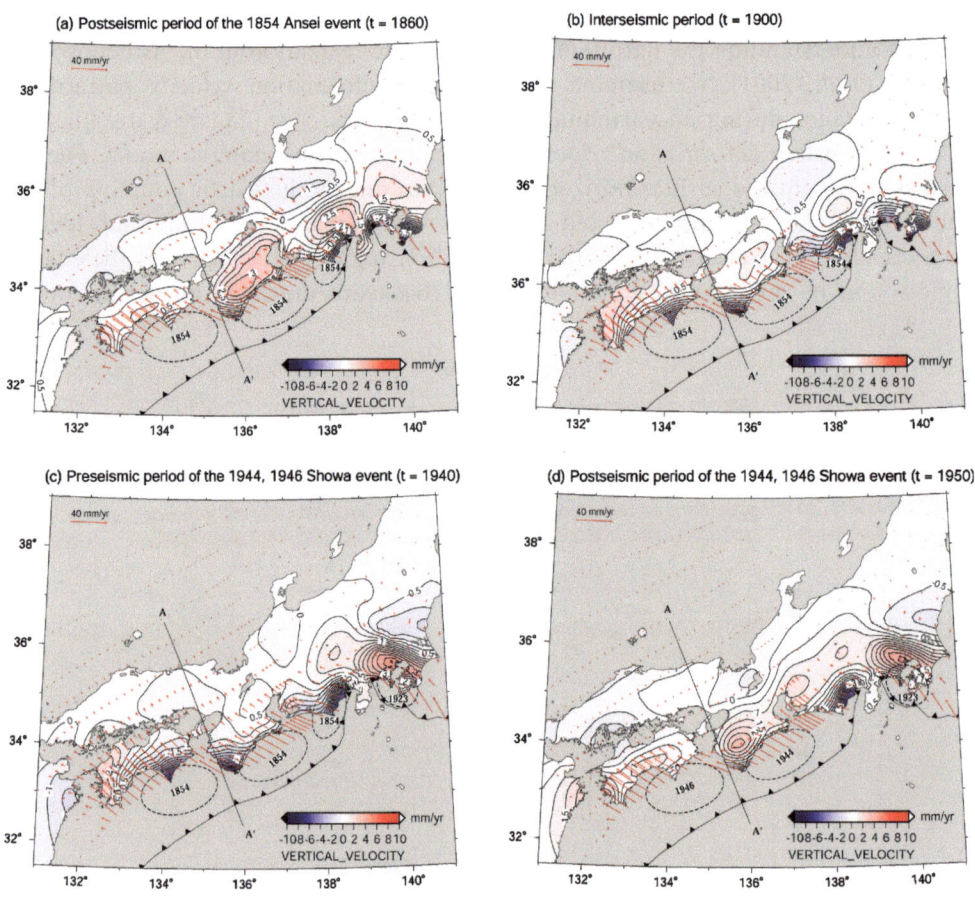

Figure 7

Calculated deformation velocities at the surface during the interseismic periods in the viscoelastic medium. (a) In 1860, the post-seismic period of the 1854 Ansei event. (b) in 1990, the interseismic period. (c) in 1940, the pre-seismic period of the 1944 and 1946 Showa events. (d) in 1950, the post-seismic period of the 1944 and 1946 Showa events. The dotted circles show the focal regions of great interplate earthquakes, which significantly influence the velocity field in each period. The red arrows, the blue and red regions have the same meanings as in Figure 6.

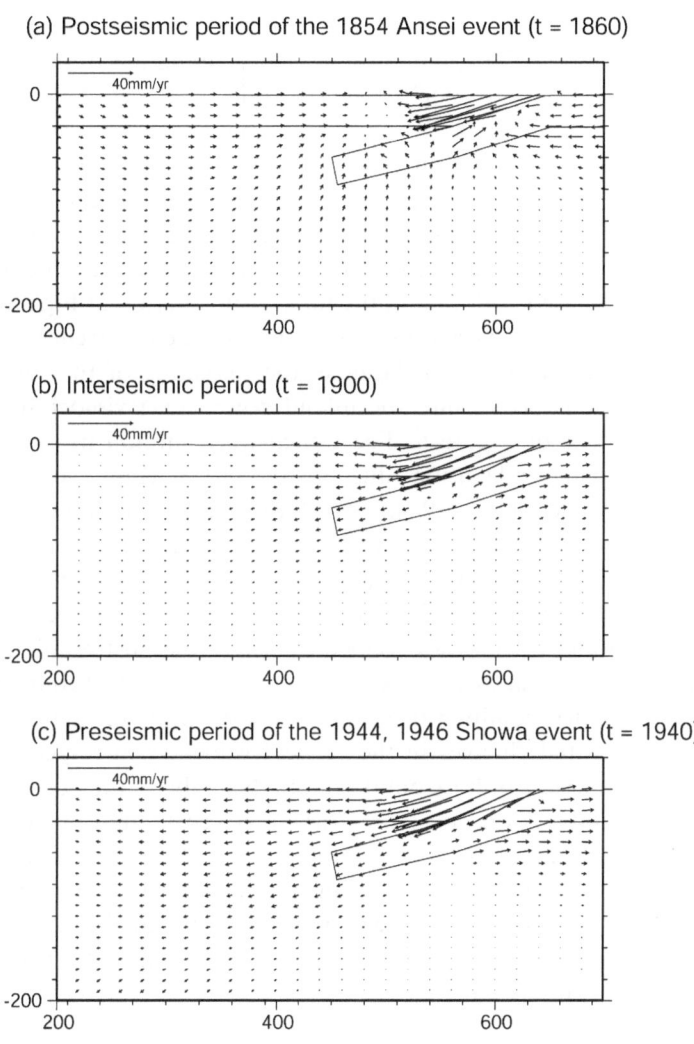

(a) Postseismic period of the 1854 Ansei event (t = 1860)

(b) Interseismic period (t = 1900)

(c) Preseismic period of the 1944, 1946 Showa event (t = 1940)

Figure 8

Vertical cross sections of calculated deformation velocity field along the line A-A' in Figures 7 (a)–(c). (a) In 1860, the post-seismic period of the 1854 Ansei event. (b) in 1990, the interseismic period. (c) in 1940, the pre-seismic period of the 1944 and 1946 Showa events. The black arrows indicate velocity vectors in the vertical cross section.

the inland region moves in the seaward direction although most of the coastal region moves landward as a result of simultaneous locking of the coupling zone. The seaward motion is a postseismic delayed response to the great interplate earthquake. This means the stress induced by the coseismic slip is redistributed into the elastic crust because of the viscous deformation of the upper mantle. The seaward motion of the inland region continues for several decades. In the subsequent interseismic period, 56 years after the occurrence of the Ansei event (Fig. 7(b) and Fig. 8(b)), the stress relaxation due to the coseismic slip of the previous earthquake gradually decreases, and the effect of the reloading of the subducting PH by backslip becomes dominant. Hence the resultant velocity field is similar to the elastic one in Figure 6. Moreover, later in Figure 7(c) and Figure 8(c), the asthenosphere is relaxed with respect to the coseismic deformation, and the backslip with a constant rate induces steady-state flow in the upper mantle. Resultingly, the asthenosphere loses the elasticity to resist the deformation in the overlying crust, and the horizontal displacement rate vectors extend in a broader region from the coastline further to the inland region, as compared with that in the interseismic period in Figure 7(b). In the later stage of the interseismic period more than half of the recurrence interval of the cycle after the occurrence of an earthquake, the reloading of the subducting PH, causes the deformation further into the inland region in a viscoelastic simulation rather in an elastic one of Figure 6. The trench-ward motion in the north Kanto exhibits the postseismic relaxation due to the 1923 Kanto earthquake. The postseismic velocity field of the 1944, 1946 Showa events in Figure 7(d) is quite different from that of the 1854 Ansei event in Figure 7(a), since the Tokai segment did not rupture during the Showa events. In the Tokai district, the northwestward velocities extend broader in the inland region similar to the preseismic period in Figure 7(c). In addition, trench-ward velocities due to the postseismic relaxation of the Showa events in the Chugoku and North Kii districts are masked by the extended northwestward velocities, resulting from the elastic coupling on the Tokai segment.

Corresponding spatial changes in the vertical velocities in respective stages are understood from the results of horizontal vectors. Namely, in the postseismic period, the opposite horizontal motions in Figure 7(a) create crustal uplift in a zone between the inland and the coastal region (Figure 8(a)). Later in the interseismic period, the uplift rate decreases with the decrease of the postseismic seaward motion in Figure 8(b). Finally, just before the next interplate earthquake, a broader region from the coast to the inland subsides because of the fluid-like behavior of the asthenosphere as shown in Figure 8(c).

5.2 Changes in Failure Stress on Inland Faults

Figure 9 shows temporal changes in failure stress (CFF) on inland faults during the recent 300-year earthquake cycles along the Nankai trough, in which the regional steady east-west tectonic loading is also considered in the inland region. CFF is evaluated at the depth of 13.5 km, which corresponds to the boundary between the brittle upper crust and the ductile lower crust, and the initial rupture of large intraplate earthquakes nucleates there. In each figure the solid line denotes the secular component due to the steady east-west compression, and the dotted line represents the change in CFF due to the superposition of the calculated earthquake cycle along the Nankai, Suruga and Sagami troughs and of the steady east-west

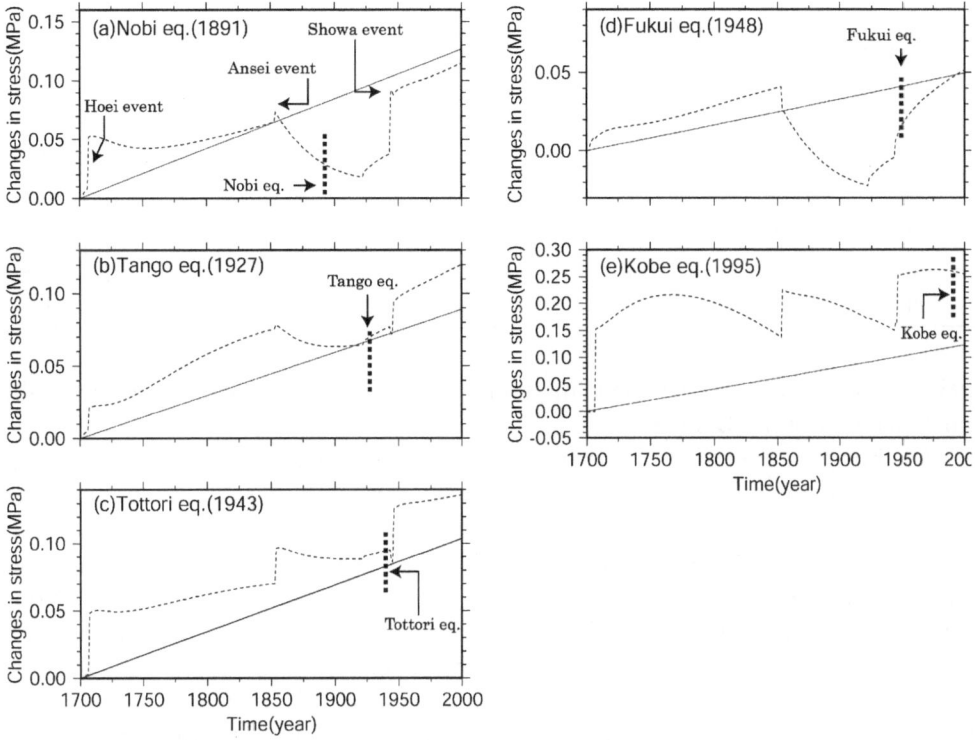

Figure 9

Changes in Coulomb failure function (CFF) on inland faults for (a) the 1891 Nobi, (b) the 1927 Tango, (c) the 1943 Tottori, (d) the 1948 Fukui and (e) the 1995 Kobe earthquakes. The solid lines indicate the changes in CFF due to the steady east-west loading force. The dotted line indicates the total CFF changes due to the east-west compressive force and the changing stress through the great earthquake cycles along the Suruga-Nankai trough.

compression. The effective frictional coefficient, μ', is assumed to be 0.3. As indicated by Iio *et al.* (2001), the lower crust beneath NKTZ, where many inland faults concentrate, is filled with water content. Therefore, the effective frictional coefficient there might be low. The step changes in CFF are due to the coseismic slips of great interplate earthquakes on the plate interface.

We examine the characteristics of changes in CFF on five recent major intraplate earthquake faults (a)–(e), of which strikes and locations are shown in Figure 4. The fault parameters of these faults are tabulated in Table 3.

5.2.1 NW-SE trending left-lateral faults; (a) Nobi, (b) Tango and (d) Fukui.

The 1891 Nobi earthquake is the largest intraplate earthquake (M~8.0) in Japan. It has a complicated rupture process, namely, the rupture started on the Neodani fault and propagated to the Umehara faults, the southern segment of the Neodani fault. Here, since we are interested in the triggering effect of the earthquake cycles of

Table 3

Fault parameters of inland faults considered in this study. The numbers of inland faults correspond to those in Fig.4.

No.	Name	Lat. [deg]	Long. [deg]	Strike [deg]	Dip [deg]	Rake [deg]	Year/ Month/Day
(a)	Nobi	35.92	136.37	145.0	90	0	1891/10/28
(b)	Tango	35.53	135.11	337.5	90	0	1927/03/07
(c)	Tottori	35.45	133.95	80.0	90	180	1943/09/10
(d)	Fukui	36.03	136.37	345.0	90	0	1948/06/28
(e)	Rokko-Awaji	34.53	134.91	53.0	90	180	1995/01/17

PH, we consider only the Neodani fault for the fault parameters of the Nobi earthquake given in Table 3.

As shown in Figure 9(a), for both the 1707 Hoei and the 1854 Ansei events that ruptured the entire Nankai-Suruga subduction zone, the failure stress (CFF) on the Neodani fault coseismically increases. However, the amount of coseismic increase and following postseismic changes in CFF are quite different. Thus the CFF rapidly decreased after the occurrence of the 1854 Ansei event, and the Nobi earthquake actually occurred in 1891 when the CFF was small, that is, in the time with the less likelihood of failure. In order to clearly understand the effect of earthquake cycles along the Nankai-Sagami trough on this fault, we show changes in stress due to earthquake cycles along the Nankai-Sagami trough on Neodani fault in Figure 10(a). In this figure the black, red and blue lines indicate CFF ($\mu' = 0.3$), shear and normal stresses, respectively. In the 1707 Hoei event, the ratio of coseismic slip on each segment along the Nankai-Suruga trough is equivalent to the ratio of the subduction velocity on each segment because of the assumption of the 'time predictable recurrence model.' Namely, the coseismic slip in the eastern Tokai segment is half of those in other western segments. Hence the shear and failure stress (CFF) on this fault largely increase, and the stress changes recover the initial level just before the Ansei event. In contrast, though the 1854 Ansei event ruptured the entire Nankai-Suruga trough, the coseismic slip amount on the Tokai segment predicted by the 'time-predictable model' is similar to those on the Nankai and Tonankai segments since the 1944 and 1946 Showa events left the Tokai segment unbroken. Thus the Ansei and the Showa earthquake cycles are no independent cycles, but a coupled one, and changes in stress will recover the initial level just before the next great interplate earthquakes which is assumed here to rupture the entire Suruga-Nankai subduction zone in 2046. This is the reason why the changes in stress after the occurrence of the 1854 Ansei event are complicated.

Likewise, the change in CFF can be explained on (b) Tango and (d) Fukui earthquake faults, which are located in space close to the Neodani fault and have the same left-lateral faulting. As in the case of the Nobi earthquake, the 1927 Tango and the 1948 Fukui earthquakes occurred when CFF were not in the local maximum in Figures 9(b) and 9(d).

5.2.1 NEE-SWW or NE-SW trending right-lateral faults; (c)Tottori and (e) Rokko-Awaji

There are NE-SW trending faults such as (c) Tottori and (e) Rokko-Awaji which are conjugate to those in the previous section. East-west compressive loading force causes right-lateral faulting for these faults. As shown in Figure 10(b), subduction of the plate and the rebound due to the interplate great earthquakes produce substantial changes in the normal stress and minor ones in the shear stress on these faults, because the convergence direction of PH is perpendicular to the strikes for these faults. Hence, the change in CFF is caused mainly by the normal stress. In addition, the seismic gap on the Tokai segment in the Showa events does not contribute to these faults since they are located far from the Tokai segment. Accordingly, the CFF was enlarged by the release of the normal stress associated with the Hoei, Ansei and Showa events along the Nankai trough. And for the Rokko-Awaji fault in Figure 9(e), the CFF for the 1707 Hoei event increased coseismically and during

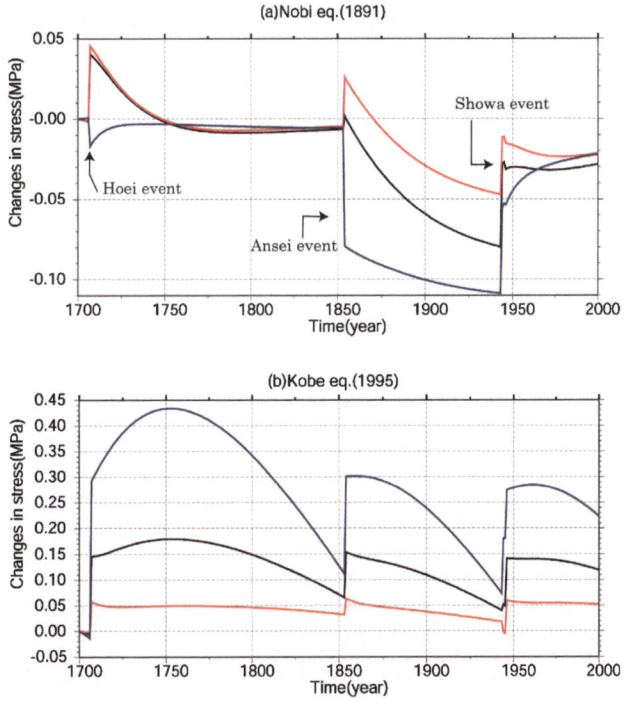

Figure 10

Changes in stress due to the earthquake cycles along the Suruga-Nankai trough on inland faults for (a) the 1891 Nobi and (b) the 1995 Kobe earthquakes. The black, red and blue lines indicate CFF ($\mu' = 0.3$), shear stress and normal stress due to the earthquake cycles along the Nankai-Suruga trough, respectively.

about 50 years after the event, and changed to decrease thereafter. This pattern of the change in CFF shows the characteristic feature of postseismic stress relaxation in a viscoelastic medium.

In Figure 9(e), the 1995 Kobe earthquake occurred near the peak of CFF, the time of the strongest likelihood of failure, which is the most effectively working case of failure function criterion in this study. The peak of CFF is produced by the combined effects of gradually decaying postseismic response and of the reloading of PH. The postseismic increase of tensional normal stress prevails over the increase of compressive normal stress due to the reloading of PH, which produces the gradually decaying increase of CFF in the postseismic period and a peak in 50 years after the occurrence of Showa events. As stated in the Introduction, POLLITZ and SACKS (1997) reached the same conclusion that the Kobe earthquake was triggered by the long-delayed viscoelastic response to the Showa events. They analyzed, however, only the effect of the Showa events in a simple layered viscoelastic medium. Our analysis includes the whole effect of recent three great interplate earthquake cycles along the Nankai trough in a realistic 3-D viscoelastic medium. Figure 9(e) shows the different CFF change in each great interplate earthquake, though they simply assumed the same pattern of CFF change is iterated as understood in Figure 4(b) in their paper. Figure 9(e) shows that the Kobe earthquake was likely to occur 60 years after the 1707 Hoei event, simultaneously with the occurrence of the 1854 Ansei event, and 50 years after the 1944 and 1946 Showa events. Actually, the Kobe earthquake occurred in 1995.

6. Discussion

In our FEM simulation only one event (e) occurred at the time of the local maximum of CFF. To the contrary, other inland events occurred at the time when CFF decreased, that is, at the time of less likelihood of the earthquake occurrence in view of CFF. Accordingly, our FEM simulation of earthquake cycle for great earthquakes along the Nankai trough could not well explain the occurrence of inland earthquakes. We discuss several factors in modeling which may produce unsatisfactory results and suggest how to improve the modeling in future studies.

Initially, we discuss the recurrence model to estimate the amount of earthquake slips and the assumption of the full interseismic coupling. In this study we show the results based on the 'time-predictable model'. We calculated the earthquake sequences based on the 'slip-predictable model', though the results are not shown here. There are differences in CFF itself between two models, although no large difference in the pattern of CFF, so that the results did not depend on the choice of the model. Further, the observations of crustal deformation have detected various (several days - several decades) fluctuations of interplate coupling associated with the afterslip following the 1946 Nankai earthquake (SAGIYA,

1999a) or slow thrust slip events such as the 2001 Tokai event (OZAWA *et al.*, 2001) and so on. The possible afterslip following the 1946 Nankai earthquake continued for several decades, and its total slip amount reaches 2–3 m at the deeper portion of coseismic rupture zone (SAGIYA, 1999a). In this study we ignored this afterslip and, instead, assumed the slip amount corresponding to the afterslip is coseismically released. Accordingly, the inclusion of the afterslip might alter the postseismic CFF pattern of the 1946 Nankai earthquake. To the contrary, the slow slip events (several days - one year) are equivalent to the interplate earthquakes with M_w 6–7, and much smaller than the great interplate ones ($M_w \sim 8.0$). Hence, the inclusion of the temporal changes in the interplate coupling caused by the slow slip events does not significantly contribute to the failure stress on inland faults.

Second, we used a value of 0.3 for the effective friction coefficient. The change of this value does not change considerably for the inland earthquakes (a), (b) and (d) in which the change in shear stress is substantial. As mentioned, IIO *et al.* (2001) pointed out that there possibly exists considerable water content in the lower crust in NKTZ, so that the lower value of 0.3 is considered to be plausible.

In this study we used the motion of the PH relative to the EU. However, for the Tokai segment along the Suruga trough, we adopted a half amount of convergent rate and the different convergence direction compared with those in other segments according to HEKI and MIYAZAKI (2001), which has been recently derived by considering the motion of IMP (SAGIYA, 1999b). Even if considering the continuous rate without IMP, the overall pattern of change in CFF and the result are not changed.

Finally, we discuss the problem on the steady east-west compressive force loading on the inland faults. In this study we do not address the origin of this east-west tectonic force, but assume the rate of the tectonic stress from the observed GPS strain rate field in NKTZ. Our 3-D FEM model includes only the subduction of PH, but not that of PA. In this respect, our model is not self-consistent. As explained in Section 3, the origin of this tectonic force has not been clarified todate. As a first step for clarifying the origin, we are now trying to model the high strain rate in NKTZ by considering the subducting PA and the lateral heterogeneity of the crust in a viscoelastic model suggested by IIO *et al.* (2001). Our preliminary result shows that the effect of the subducting PA is dominant from the Pacific coastal region to central inland Japan in a viscoelastic medium as in the case of Figure 7(c) and Figure 8(c) for the subducting PH. For the wider range of transferring strain against the pattern expected in the elastic medium as in Figure 6, it requires that the PA has been subducting without any great interplate earthquakes for a long time. In fact, no great interplate earthquakes have occurred for more than 300 years in the region where the PA subducts. The concentration of high strain rate in NKTZ requires the thinning of the elastic crust there as proposed by IIO *et al.* (2001). Thus the modeling of the high strain rate in NKTZ seems to require several complicated factors.

In the Chugoku district, the western SWJ, there is no large strain rate field as shown in Figure 2(b). In this study we tentatively assume the same strain rate for the faults of (b) and (c) as for those of (a), (d) and (e). This is another problem to be improved in future works.

Together with the modeling of east-west compressive force, the inclusion of historical large inland earthquakes in SWJ may significantly influence the stress state on each inland fault. FREED and LIN (2001) examined the effect of the 1992 Landers earthquake ($M_w = 7.3$) in Southern California on the occurrence of the 1999 Hector Mine earthquake ($M_w = 7.1$) using the viscoelastic FEM modeling. Their results show that postseismic lower-crustal or upper-mantle flow caused by the 1992 Landers earthquake can lead to postseismic failure stress increases reaching 0.1–0.2 MPa at the Hector Mine hypocenter that is located only 20 km from the Landers hypocentre. Therefore, this implies that stress transfer due to the occurrence of nearby inland earthquakes in SWJ may have a similar influence as that of the great interplate earthquakes along the Nankai trough, which is evaluated in this study (Figs. 9 and 10). In this study, however, since our main concern is the relation between the occurrence of the inland intraplate earthquakes and the great interplate earthquakes, we do not consider the interaction between the inland earthquakes.

7. Conclusion

The spatial distribution and temporal changes of the crustal deformation in SWJ have been simulated through viscoelastic FEM modeling based on the kinematic earthquake cycle model. Subsequently, the effect of earthquake cycles along the Nankai-Suruga-Sagami trough on the triggering of recent large earthquakes in the inland fault is quantitatively investigated, assuming steady east-west tectonic loading force which is estimated from the observed GPS strain rate field. The results are as follows:

(1) The inclusion of viscoelasticity leads to a variety of patterns of velocity field during earthquake cycles regardless of the assumption of the constant plate coupling throughout the interseismic period. Just after the occurrence, the viscoelastic relaxation of a great interplate earthquake creates the seaward motion of the inland region. In the middle period of an earthquake cycle the seaward motion is gradually decreased, and the resultant velocity field looks like the elastic one. Further later, just before the next interplate earthquake, displacements due to the interplate coupling in the viscoelastic material distribute broader in the forearc region than in the purely elastic one, since the viscoelastic relaxation due to the previous earthquake nearly disappears.

(2) The calculated changes in failure stress on many inland faults do not match the actual occurrence of inland earthquakes on these faults, while only the 1995 Kobe earthquake occurred at the time of the local maximum of the failure stress. This

implies that further improvements are necessary for our FEM modeling, such as the modeling of steady east-west compressive force or fault interactions between the inland faults.

Acknowledgment

The authors are grateful to anonymous reviewers for their careful reviews and valuable comments. This study is financially supported by 'Earth Simulator Project' of Ministry of Education, Culture, Sports, Science and Technology (MEXT), and 'Special Project for Earthquake Disaster Mitigation in Urban Areas' of MEXT.

REFERENCES

FREED, A. M. and LIN, J. (2001), *Delayed Triggering of the 1999 Hector Mine Earthquake by Viscoelastic Stress Transfer*, Nature *411*, 180–183.

HEKI, K., MIYAZAKI, S., TAKAHASHI, H., KASAHARA, M., KIMATA, F., MIURA, S., VASILENKO, N. F., IVASHCHENKO, A., and AN, G. (1999), *The Amurian Plate Motion and Current Plate Kinematics in Eastern Asia*, J. Geophys. Res. *104*, 29,147–29,155.

HEKI, K and MIYAZAKI, S. (2001), *Plate Convergence and Long-term Crustal Deformation in Central Japan*, Geophys. Res. Lett. *28*, 2313–2316.

HENRY, P., MAZZOTTI S., and LE PICHON, X. (2001), *Transient and Permanent Deformation of Central Japan Estimated by GPS, 1. Interseismic Loading and Subduction Kinematics*, Earth Planet. Sci. Lett. *184*, 443–453.

HORI, T. and OIKE, K. (1996), *A Statistical Model of Temporal Variation of Seismicity in the Inner Zone of Southwest Japan Related to the Great Interplate Earthquakes along the Nankai Trough*, J. Phys. Earth *44*, 349–356.

HORI, T. and OIKE, K. (1999), *A Physical Mechanism for Temporal Variation in Seismicity in the Inner Zone of Southwest Japan Related to the Great Interplate Earthquakes along the Nankai Trough*, Tectonophys. *308*, 83–98.

IIO, Y., SAGIYA, T., KOBAYASHI, Y., and SHIOZAKI, I. (2001), *Water-weakended Lower Crust and its Role in the Concentrated Deformation in the Japanese Islands*. In Proc. Internat. Symp. *Slip and Flow Processes in and below the Seismogenic Region*, 389–397.

IIZUKA, M., SEKITA, D., SUITO, H., HYODO, M., HIRAHARA, K., PLACE, D., MORA, P., HAZAMA, O., and OKUDA, H. (2002), *Parallel Simulation System for Earthquake Generation: Fault Analysis Modules and Parallel Coupling Analysis*, Concurrency Computat.: Pract. Exper. *14*, 499–519.

ISHIBASHI, K. and SATAKE, K. (1998), *Problems on Forecasting Great Earthquakes in the Subduction Zones around Japan* (in Japanese with English Abstract), Zisin *50*, 1–21.

ISHIDA, M. (1992), *Geometry and Relative Motion of the Philippine Sea Plate and Pacific Plate beneath the Kanto-Tokai District, Japan*, J. Geophys. Res. *97*, 489–513.

KING, G. C. P., STEIN, R. S., and LIN, J. (1994), *Static Stress Changes and the Triggering of Earthquakes*, Bull. Seismol. Soc. Am. *84*, 935–953.

MAZZOTTI, S., HENRY, P., LE PICHON, X., and SAGIYA, T. (1999), *Strain Partitioning in the Zone of Transition from Nankai Subduction to Izu-Bonin Collision (Central Japan): Implications for an Extensional Tear within the Subduction Slab*, Earth Planet. Sci. Lett. *172*, 1–10.

MAZZOTTI, S., LE PICHON, X., HENRY, P., and MIYAZAKI, S. (2000), *Full Interseismic Locking of the Nankai and Japan-West Kurile Subduction Zones: An Analysis of Uniform Elastic Strain Accumulation in Japan Constrained by Permanent GPS*, J. Geophys. Res. *105*, 13,159–13,177.

MAZZOTTI, S., HENRY, P., and LE PICHON, X. (2001), *Transient and Permanent Deformation of Central Japan Estimated by GPS, 2. Strain Partitioning and Arc-Arc Collision*, Earth Planet. Sci. Lett. *184,* 455–469.

MELOSH, H. J. and RAEFSKY, A. (1981), *A Simple and Efficient Method for Introducing Faults into Finite-Element Computation*, Bull. Seismol. Soc. Am. *71,* 1391–1400.

MATSU'URA, M. and SATO, T. (1989), *A Dislocation Model for the Earthquake Cycle at Convergent Plate Boundaries*, Geophys. J. Int. *96,* 23–32.

MIYAZAKI, S. and HEKI, K. (2001), *Crustal Velocity Field of Southwest Japan: Subduction and Arc-Arc Collision*, J. Geophys. Res. *106,* 4,305–4,326.

OZAWA, S., MURAKAMI, T., KAIZU, M., TADA, T., SAGIYA, T., YARAI, H., and NISHIMURA, T. (2001), *Anomalous Crustal Deformation in the Tokai Region in 2001.* 2001, PROGRAMME and ABSTRACTS, Seis. Soc. Japan, 2001 Fall meeting, C02 (in Japanese).

SAGIYA, T. (1999a), *Crustal Deformation Cycle and Plate Interaction in Shikoku District*, Japan Mantly Earth *24,* 26–33 (in Japanese).

SAGIYA, T. (1999b), *Interplate Coupling in the Tokai District, Central Japan, Deduced from Continuous GPS Data*, Geophys. Res. Lett. *15,* 2315–2318.

SAGIYA, T., MIYAZAKI, S., and TADA, T. (2000), *Continuous GPS and Present-day Crustal Deformation of Japan*, Pure Appl. Geophys. *157,* 2303–2322.

SAVAGE, J. C. (1983), *A Dislocation Model of Strain Accumulation and Release at a Subduction Zone*, J. Geophys. Res. *88,* 4984–4996.

SENO, T, STEIN, S., and GRIPP, A. E. (1993), *A Model For the Motion of the Philippine Sea Plate Consistent with Nuvel and Geological Data*, J. Geophys. Res. *98,* 17,941–17,948.

SHIMAZAKI, K. and NAKATA, T. (1980), *Time-predictable Recurrence Model for Large Earthquakes*, Geophys. Res. Lett. *7,* 279–282.

SUITO, H. and HIRAHARA, K. (1999), *Simulation of Postseismic Deformation caused by the 1896 Riku-u Earthquake, Northeast Japan: Re-evaluation of the Viscosity in the Upper Mantle*, Geophys. Res. Lett. *26,* 2561–2564.

SUITO, H., IIZUKA, M., and HIRAHARA, K. (2002), *3-D Viscoelastic Modeling of Crustal Deformation in Northeast Japan*, Pure Appl. Geophys. *159,* 2239–2260.

POLLITZ, F. F. and SACKS, S. (1997), *The Kobe, Japan, Earthquake: A Long-Delayed Aftershock of the Offshore 1944 Tonankai and 1946 Nankaido Earthquakes*, Bull. Seismol. Soc. Am. *87,* 1–10.

WANG, K., HE, J., DRAGERT, H., and JAMES, T. S. (2001), *Three-dimensional Viscoelastic Interseismic Deformation Model for the Cascadia Subduction Zone*, Earth Planet Space *53,* 295–306.

WEI, D. P. and SENO, T. *Determination of the Amurian Plate motion. In Mantle Dynamics and Plate Interactions in East Asia*, Geodyn. Ser. Vol. 27 (eds. M. F. J. Flower et al., AGU, Washington, D. C. 1998), 419 pp.

YAMAZAKI, F. and OOIDA, T. (1985), *Configuration of Subducted Philippine Sea Plate beneath the Chubu District, Central Japan* (in Japanese with English Abstract), Zisin *38,* 193–201.

ZONENSHAIN, L. P. and SAVOSTIN, L. A. (1981), *Geodynamics of the Baikal Rift Zone and Plate Tectonics of Asia*, Tectonophysics *76,* 1–45.

ZIENKIEWICZ, O. C. and CHEUNG, Y. K. (1967), The Finite Element Method in Structural and Continuum Mechanics, McGraw-Hill, New York.

(Received September 27, 2002, revised January 1, 2003, accepted February 10, 2003)

To access this journal online:
http://www.birkhauser.ch

Pure appl. geophys. 161 (2004) 2091–2102
0033–4553/04/102091–12
DOI 10.1007/s00024-004-2550-1

| Pure and Applied Geophysics

Finite Element Analysis of Fault Bend Influence on Stick-Slip Instability along an Intra-Plate Fault

H. L. XING[1], P. MORA[1], and A. MAKINOUCHI[2]

Abstract—Earthquakes have been recognized as resulting from stick-slip frictional instabilities along the faults between deformable rocks. A three-dimensional finite-element code for modeling the nonlinear frictional contact behaviors between deformable bodies with the node-to-point contact element strategy has been developed and applied here to investigate the fault geometry influence on the nucleation and development process of the stick-slip instability along an intra-plate fault through a typical fault bend model, which has a pre-cut fault that is artificially bent by an angle of 5.6° at the fault center. The numerical results demonstrate that the geometry of the fault significantly affects nucleation, termination and restart of the stick-slip instability along the intra-plate fault, and all these instability phenomena can be well simulated using the current finite-element algorithm.

Key words: Finite-element method, fault bends, stick-slip instability, frictional contact between deformable bodies, friction instability, earthquake.

1. Introduction

Earthquakes have long been recognized as resulting from stick-slip instabilities along the faults between deformable rocks, although the rupture process itself is generally complex due to the nonuniform distribution of stress and strength on faults. KING and NABELEK (1985) examined the source processes of the following eight events: the 1966 Parkfild earthquake ($Ms = 6.5$), the 1973 Luhuo earthquake ($Ms = 7.5$), the 1975 Lice earthquake ($Ms = 6.7$), the 1976 Tangshan earthquake ($Ms = 7.8$), the 1976 Caldiran earthquake ($Ms = 7.4$), the 1979 Coyote Lake earthquake ($Ms = 5.7$), the 1980 El Asnam earthquake and the 1984 Morgan Hill earthquake ($Ms = 6.1$), and concluded that the initiation and/or termination of earthquake rupture was usually controlled by geometrical irregularities of faults such as fault bends. KATO *et al.* (1999) experimentally investigated the effect of fault shape on the rupture process of stick-slip using a granite sample with a pre-cut fault that

---
[1] Earth Systems Science Computational Centre, The University of Queensland, St. Lucia, Brisbane, QLD 4072, Australia.
E-mail: xing@quakes.uq.edu.au
[2] The Institute of Physical and Chemical Research (RIKEN) , 2-1 Hirosawa, Wako, Saitama 351-0198, Japan.

was artificially bent, and drew the same conclusion. From the above practical observations and laboratory results, the geometry of the faults significantly affects the faulting process and must therefore be considered in earthquake research.

The finite-element method is now widely applied to numerical analysis of certain science and engineering problems. Several in-house and commercial codes were applied to simulate specific phenomena related with earthquakes (XING, 2002a, b and references thereafter). As for the numerical investigation of the effect of fault geometry on the earthquake process, only a few results have been reported. OGLESBY and DAY (2001) analyzed the effect of fault geometry on the dynamic propagation behavior of the Chi-Chi earthquake with the prescribed nucleation and the slip-weakening friction law, in which the fault was assumed to dip with a constant angle. To further investigate the occurrence of the earthquake and to predict it in the future, an arbitrarily shaped contact element strategy, named the node-to-point contact element strategy proposed with a static-explicit algorithm by the authors (e.g. XING and MAKINOUCHI 2000; XING *et al.* 1998a,b), was previously applied to model the friction contact behaviors between deformable rocks with stick and finite nonlinear frictional slip (XING and MAKINOUCHI, 2002a, b). This paper will focus on investigating the influence of a fault bend on the stick-slip instability along the intra-plate fault using the above algorithms and to help lay a foundation for further research on the intra-plate earthquake nucleation process and earthquake prediction.

2. Finite Element Formulation

The updated Lagrangian rate formulation is employed to describe the nonlinear contact problem. The rate type equilibrium equation and the boundary at the current configuration are equivalently expressed by a principle of virtual velocity of the form (e.g. XING and MAKINOUCHI, 2000)

$$\int_V (\dot{\tau}_{ij} - D_{ik}\sigma_{kj} + \sigma_{ik}L_{jk} - \sigma_{ik}D_{kj})\delta L_{ij}\, dV = \int_{S_F} \dot{F}_i \delta v_i\, dS + \int_{S_c^1} \dot{f}_i^1 \delta v_i\, dS + \int_{S_c^2} \dot{f}_i^2 \delta v_i\, dS,$$

(1)

where V and S denote, respectively the domains occupied by the total deformable body B and its boundary at time t; S_F is a part of the boundary of S on which the rate of traction $\dot{F}_i$ is prescribed; δv is the virtual velocity field which satisfies the condition $\delta v = 0$ on the velocity boundary; $\dot{\tau}_{ij}$ is the Jaumann rate of Cauchy stress; L is the velocity gradient tensor, $L = \partial v / \partial x$; D and W are the symmetric and antisymmetric parts of L, respectively; $\dot{f}^\alpha$ is the rate of contact stress on contact interface S_c^α of the body α and calculated as follows.

Friction is by nature a path-dependent dissipative phenomenon that requires the integration of the constitutive relation. In this study, a standard Coulomb friction model is applied analogous by to the flow plasticity rule which governs the slipping behavior. The basic formulations are summarized below (XING *et al.*2000; XING and MAKINOUCHI, 2002a,b) (a variable with tiled ($\sim$) above a variable denoting a relative component between slave and master bodies, and *l, m = 1,2; i, j, k = 1, 3* in this paper if without special notation).

Based on experimental observations, an increment decomposition is assumed

$$\Delta \tilde{u}_m = \Delta \tilde{u}_m^e + \Delta \tilde{u}_m^p, \tag{2}$$

where $\Delta \tilde{u}_m^e$ and $\Delta \tilde{u}_m^p$ represent the stick (reversible) and the slip (irreversible) part of $\Delta \tilde{u}_m$, respectively. In addition, the slip is governed by the yield condition

$$F = \sqrt{f_m f_m} - \bar{F}, \tag{3}$$

where $\bar{F}$, the critical frictional stress, $\bar{F} = \mu f_n$; $f_m(m = 1, 2)$ is the frictional stress component along the tangential direction m; μ is the friction coefficient, it may depend on the normal contact pressure f_n, the equivalent slip velocity $\dot{u}_{eq}^{sl}$ and the state variable φ, i.e. $\mu = \mu(f_n, \dot{u}_{eq}^{sl}, \varphi)$ (e.g., DIETERICH, 1979; RUINA, 1983; SCHOLZ, 1998).

If $F < 0$, contact is in the sticking state and treated as a linear elasticity, i.e.,

$$f_m = E_t \tilde{u}_m^e = E_t \Sigma \Delta \tilde{u}_m^e, \tag{4}$$

where E_t is a constant in the tangential direction.

When $F=0$, the friction changes its character from the stick to the slip. The frictional stress can be described as

$$f_m = \eta_m \bar{F} \text{ and } \eta_m = f_m^e / \sqrt{f_l^e f_l^e} \tag{5}$$

where $f_m^e = E_t(\tilde{u}_m - \tilde{u}_m^p|_0)$, and $\tilde{u}_m^p|_0$ is the value of $\tilde{u}_m^p$ at the beginning of this step.

The linearized form of Eq. (5) can be rewritten as

$$df_l = \frac{\bar{F} E_t}{\sqrt{f_m^e f_m^e}} (\delta_{lm} - \eta_l \eta_m) d\tilde{u}_m + \eta_l \mu \left(df_n + \frac{\partial \mu}{\partial f_n} df_n \right) + \eta_l f_n \left(\frac{\partial \mu}{\partial \dot{u}_{eq}^{sl}} d\dot{u}_{eq}^{sl} + \frac{\partial \mu}{\partial \varphi} d\varphi \right). \tag{6}$$

In addition, the penalty parameter method is chosen to satisfy the normal impenetrability condition when contact occurs, thus the normal contact stress can be calculated as

$$f_n = \boldsymbol{f} \cdot \boldsymbol{n} = E_n g_n \ (\neq 0 \text{ only for } g_n < 0). \tag{7}$$

Here E_n is the penalty parameter to penalize the penetration (gap) in the normal direction; g_n is the penetration (gap) in the normal direction.

In summary, from Eqs. (4), (6) and (7), the contact stress acting on a slave node can be described as (denote $\dot{f}_3 = \dot{f}_n$)

$$\dot{f}_i = G_{ij} \dot{\tilde{u}}_j + \dot{f}_{\varphi i}, \tag{8}$$

where G is the frictional contact matrix; $\dot{f}_{\varphi i}$ is from the contribution of the terms related with φ, when it is not a function of $\tilde{u}$; If $d\varphi$ is only the function of the unknown variable $d\tilde{u}$, $\dot{f}_{\varphi i} = 0$, i.e., all its contribution can be included in G at current state.

A node-to-point contact element strategy was proposed to handle the frictional contact problems between deformable bodies (XING and MAKINOUCHI, 1998, 2000, 2002a, b; XING *et al.* 1998) and is briefly introduced here. Assume a slave node s has contacted with point c on a surface element (master segment) E' and the surface element E' consists of γ nodes ($\gamma = 4$ in this paper if without special notation), thus the term related with contact in Eq. (1) can be described as ($\alpha = 1, (\gamma + 1)$, $\beta = 1, (\gamma + 1)$)

$$\dot{f}_i(\delta u_{si} - \delta u_{ci}) = \delta \dot{u}_{sci\beta} \lfloor \bar{K}_{fik} \rfloor_{\beta\alpha} \dot{u}_{sck\alpha} + R_\beta \dot{f}_{\varphi i}, \tag{9}$$

where

$$\lfloor \bar{K}_{fik} \rfloor_{\beta\alpha} = R_\beta e_i \cdot \left\{ G_{hk} R_\alpha e_h + \left(H_{jm} \hat{e}_j \left((\bar{C}_{ll} R_{\alpha,m} - \bar{C}_{ml} R_{\alpha,l}) e_k \right. \right. \right.$$
$$\left. \left. \left. \cdot \tilde{x} + R_\alpha (\bar{C}_{ll} \hat{e}_m - \bar{C}_{ml} \hat{e}_l) \cdot e_k \right) \right) \right\}$$
$$(h = 1, 3, \ l \neq m \text{ and no sum on } l), \tag{10}$$

here $\bar{C}_{ml} = C_{ml} - g_n n \cdot \hat{e}_{m,l}, C_{ml} = \hat{e}_m \cdot \hat{e}_l$, $\not{p} = \bar{C}_{11}\bar{C}_{22} - \bar{C}_{12}\bar{C}_{21}$, $\tilde{x} = x_s - x_c$, $E_{ijm} = \hat{e}_{i,m} \cdot \hat{e}_j$, $H_{jm} = \hat{f}_i E_{ijm}/\not{p}$, $R = [1 \ -N_1 \ -N_2 \ \cdots \ -N_\gamma]^T$, $N_p \ (p = 1, \gamma)$ is the shape function value of the point c on the surface element E', $\hat{e}_i$ and e_i are respectively the base vectors of the local natural and the local Cartesian coordinate systems on the master segment.

The time integration method is one of the key issues to formulate a nonlinear finite-element method. It is well known that the fully implicit method is often subjected to bad convergence problems, mostly due to changes of contact and friction states. In order to avoid this, we employ an explicit time integration procedure as follows. It is assumed that under a sufficiently small time increment all rates in Eq. (1) can be considered constant within the increment from t to $t + \Delta t$ as long as no drastic change of states (for example, elastic to plastic at an integration point, contact to discontact or discontact to contact on the contact interface, stick to slide or slide to stick in friction on the contact interface) takes place. The R-minimum method (e.g., YAMADA, 1968; XING and MAKINOUCHI, 2002c) is extended and used here to limit the step size in order to avoid such a drastic change in state within an incremental step.

Thus all the rate quantities used to derive Eq. (1) are simply replaced by incremental quantities as

$$\Delta u = v \Delta t, \quad \Delta \tau = \dot{\tau} \Delta t, \quad \Delta L = L \Delta t. \tag{11}$$

Finally, in combination with the above equations, Eq. (1) can be rewritten as

$$(\boldsymbol{K} + \boldsymbol{K}_f)\Delta \boldsymbol{u} = \Delta \boldsymbol{F} + \Delta \boldsymbol{F}_f, \qquad (12)$$

here $\boldsymbol{K}$ is the standard stiffness matrix corresponding to body B; $\Delta \boldsymbol{F}$ is the external force increment subjected to body B on S_Γ; $\Delta \boldsymbol{u}$ is the nodal displacement increment; $\boldsymbol{K}_f$ and $\Delta \boldsymbol{F}_f$ are the stiffness matrices and the force increments of all the node-to-point contact elements, for one node-to-point contact element E, they can be described as

$$\boldsymbol{K}^E_{f\,\beta\alpha} = -\int\limits_{S^E_c} \bar{\boldsymbol{K}}_{f\,\beta\alpha}\, dS, \quad \Delta \boldsymbol{F}^E_{f\,\beta} = \int\limits_{S^E_c} R_\beta \dot{f}_\varphi \Delta t\, dS. \qquad (13)$$

3. Numerical Investigation of Fault Bend Influence

3.1 The Fault Bend Model Analyzed

BRACE (1966) pointed out that earthquakes must be the results of stick–slip frictional instabilities along pre-existing fault or plate interfaces. The earthquake is the 'slip', and the 'stick' is the interseismic period of elastic strain accumulation. A frictional sliding instability between rock surfaces in the laboratory corresponds at least qualitatively to shallow depth earthquake instability along an existing fault (SCHOLZ, 1998). Thus, to investigate the fault geometry influence on the stick-slip instability along the intra-plate fault, a typical fault bend model ($300 \times 300 \times 50$ mm^3) is analyzed here which has a pre-cut fault that is artificially bent by an angle of $5.6°$ at the center E of the fault. The details of the geometry, the boundary conditions are shown in Figure 1. There exists no relative motion along the interface at both ends (as depicted using the thick black lines CD and FG in Fig. 1). This can be easily achieved using the 'stick' algorithm in the code. While, for the other part of the fault interface (i.e., segments DE and EF in Fig. 1), the widely applied rate- and state-dependent friction law (DIETERICH, 1979; RUINA, 1983) is utilized here with the following parameters: $\mu = 0.60 + (0.010 - 0.025)$ $\ln(V/0.001), d\varphi/dt = 0$. Thus the total fault consists of four fault segments: CD, DE, EF and FG in Figure 1. Here all the materials have the same properties: density $\rho = 2.60$ g/cm^3 Young's modulus $E = 44.8$ GPa and Poisson's ratio $\gamma = 0.12$. As for the loading conditions, two loading stages are applied here: firstly, the pressure along surfaces A and B are loaded to 10 MPa, then sustaining this pressure on surface A, while all the nodes on surface B are moved in the x direction at the velocity $Vx = -0.001$ mm/sec.

3.2 Influence of Fault Bend on Stick-slip Instability Behaviors

The corresponding calculated results of the relative velocity, the frictional force and the contact force at the different positions of the interface are analyzed, and

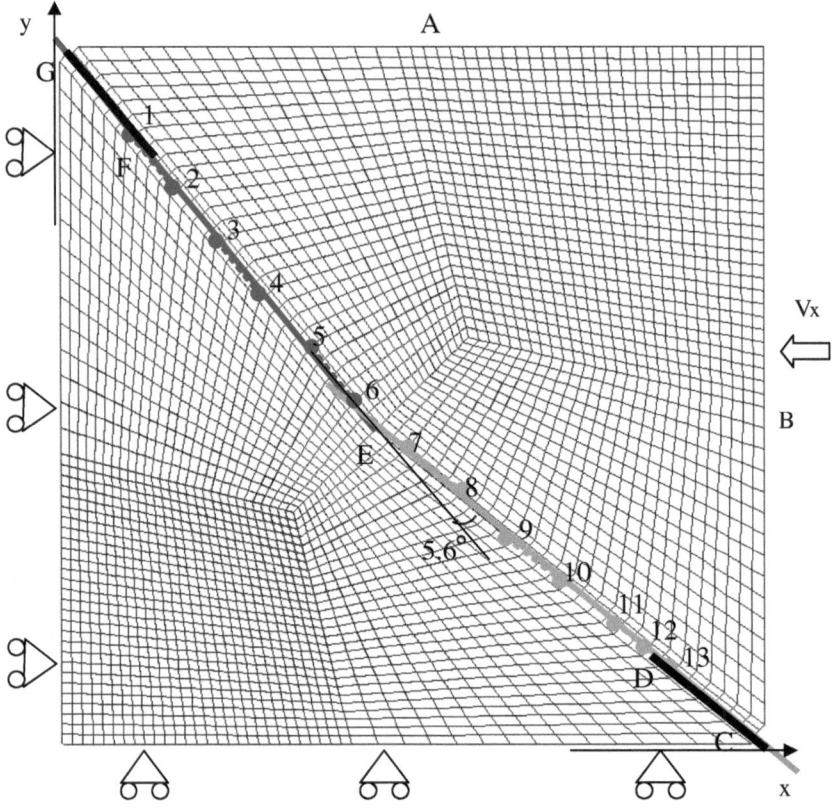

Figure 1

The mesh, the boundary conditions and the prescribed nodal positions used for the fault bend model in the x-y cross section. The fault geometry in the x-y cross section is depicted using the red line EG and the green line CE, while the no relative motion zone is designated using the thick black line. Thus the total fault consists of four segments as shown here: CD, DE, EF and FG. In addition, all the nodes at the surface marked by a triangular with two circles (⚥) are free along the direction of two circles but fixed at the other direction.

those along the central line of the body at the nodes denoted from 1 to 13 in Figure 1 are depicted in Figures 2-4, respectively.

Figure 2 shows variation of the relative velocity at the different positions during the second loading stage. There are two obvious changes that are denoted R1 and R2 in Figure 2(a) and described in more detail in Figures 2(b)-(d). This is mainly due to the nonuniform distribution of the normal contact force along the fault segments DE and EF, as shown in Figures 4(a) and (b). At the beginning of the loading stage all the nodes remain stick for a rather long time; this corresponds to the nucleation stage. During this stage, with similar initial value, the normal contact forces on the upper segment EF increase more quickly than those on the lower segment DE due to the effect of the fault bend, and the friction

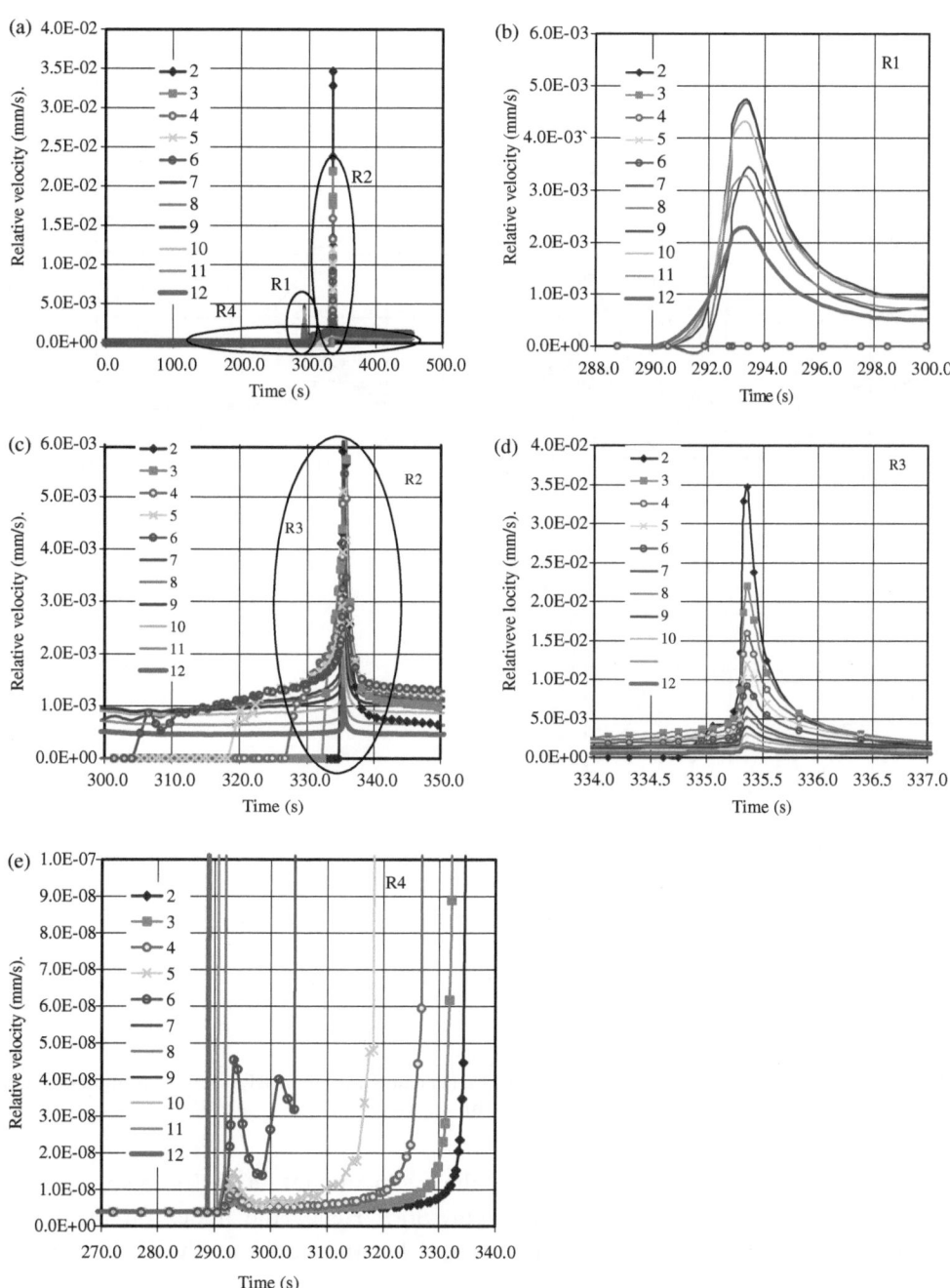

Figure 2
Relative velocity variation (a) during the total processes; (b) magnification of R1; (c) magnification of R2;
(d) magnification of R3 of (c); (e) magnification of R4.

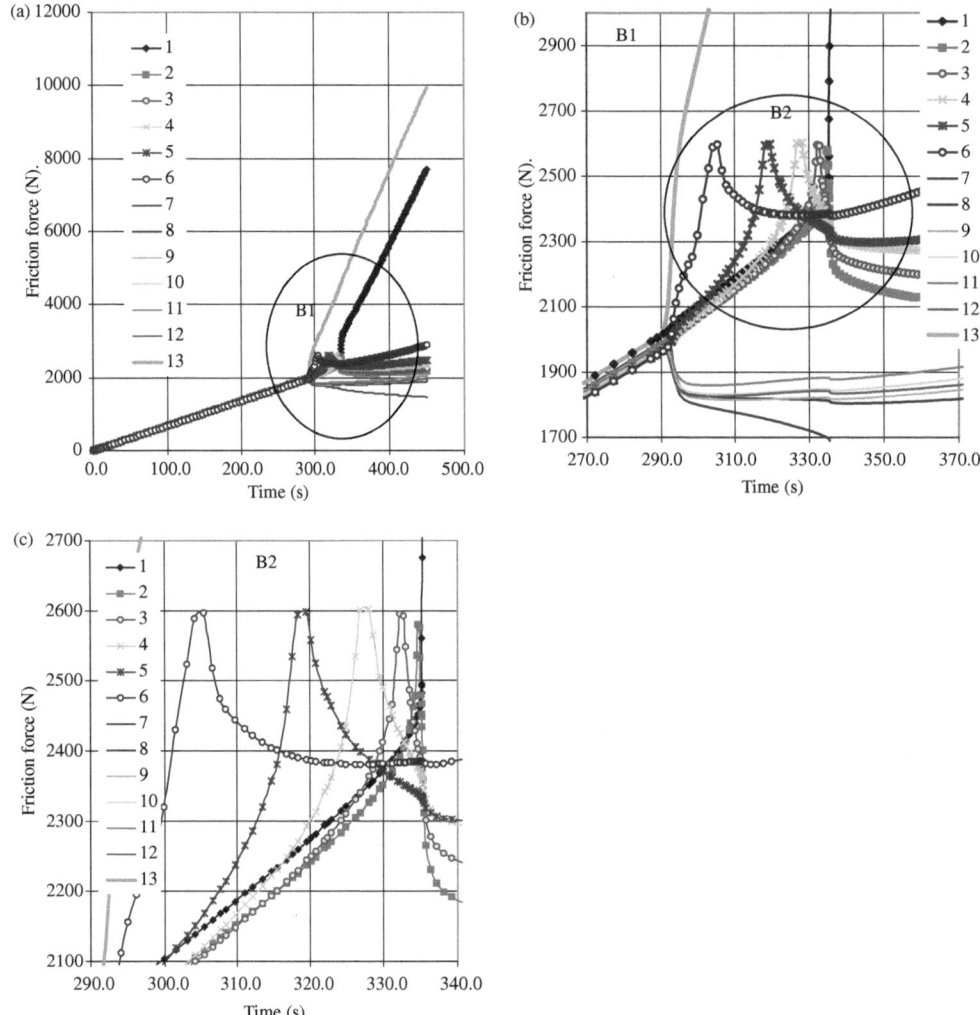

Figure 3
Friction force variation (a) during the total processes, (b) magnification of B1, (c) magnification of B2 of (b).

forces at the different positions along the segments DE and EF also rise but remain a similar value as shown in Figure 3(a). Hence the nodes on the lower segment DE reach the slip state earlier. And for the nodes at the fault segment

►

Figure 4
Normal contact force variation (a) during the total processes; (b) magnification of C1; (c) magnification of C2 of (b).

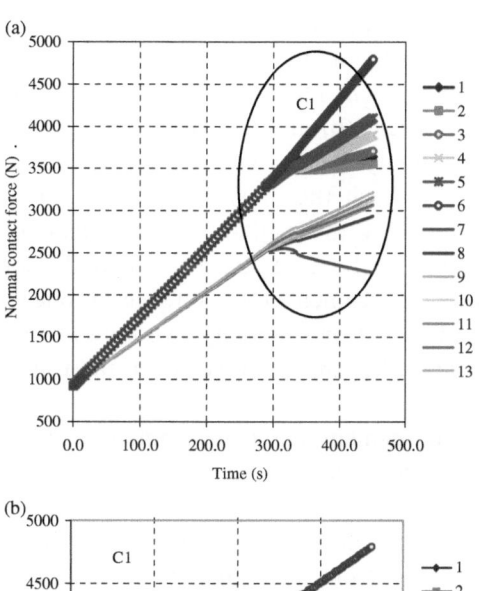

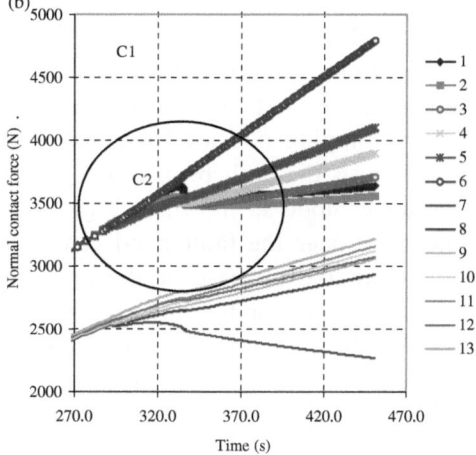

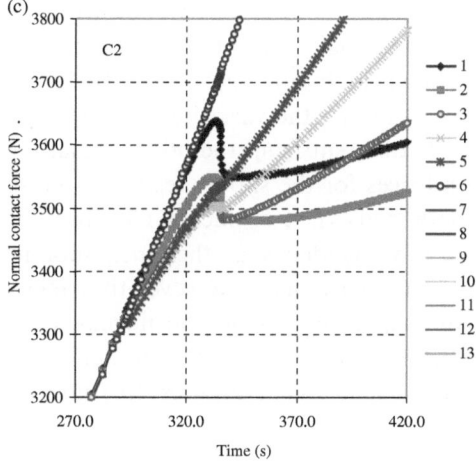

DE, the lower positions they are at, the earlier they enter the slip from the stick (i.e., one by one from the node at the position D to that at position E in Fig. 1). Finally, all the nodes on segment DE enter the slip but the other nodes are still in the stick as shown in Figures 2(b) and (e). In addition, once a node gets into the slip state, its relative velocity increases and reaches the maximum value, then decreases. Moreover, they reach the maximum value of the relative velocity at a comparable moment and obey the similar variation tendency at the R1 zone as shown in Figure 2(b). This is because the accumulated energy (friction force) of the corresponding node on segment DE is released once it enters the slip (as shown in Figs. 3(a) and (b)). For simplicity, we denote the second loading stage until now as stage A1, and after this as stage A2 in the following if without special notation.

Once all the nodes at segment DE (those nodes at positions from 7 to 12 in Fig.1) get into the slip state, it takes time for the stick nodes on segment EF to start entering the slip due to the rather large difference of the normal contact forces between segments DE and EF (as shown in Fig. 4(b)), i.e., the fault bend prevents these stick nodes from changing to the slip at the moment. This corresponds to the nucleation state of the second obvious velocity variation stage. Due to the influences of the slipping nodes of segment DE on segment EF, the relative velocity for the node near the fault bend center E (i.e., the slipping nodes) is larger than that for the upper node as shown in Figs. 2(e), correspondingly, the friction coefficient for the node near the fault bend center E is smaller than that for the upper node. Thus, with the increase of the prescribed displacement, the stick-slip instability is restarted again for the nodes on segment EF from the fault bend center E to the upper as shown in Figure 2(c). The closer to the center E the node is, the more time it takes to reach its maximum velocity value, and the smaller its maximum velocity value is (see Figures 2(c) and (d)). This mainly results from the influences of its neighbor nodes. For the node locating further from the center E, the more energy (i.e., the friction force) is released in a shorter time (see Figs. 3(b) and (c)), finally the nodes reach their maximum values at a corresponding time as shown in Figures 2(c) and (d).

For the slip stages of A1 or A2, the variations of the normal contact forces near the fault center E are quite different from the others, resulting in the corresponding variations of the friction forces for the slip nodes as shown in Figure 3(b).

For the nodes at both ends (i.e., segments CD and FG), the friction forces increase and remain the same tendency as their neighbor nodes on segments DE and EF when those are in the stick state, however they increase sharply once their neighbor nodes enter the slip state. As do the normal contact forces during the corresponding nucleation of stages A1 or A2, although afterwards they vary quite differently with each other as shown in Figures 4(b) and (c).

4. Discussion and Conclusions

The influence of a prescribed fault bend on the relative slip velocity, the transition of stick-slip state, the normal contact force and the friction force along the intra-plate fault between the deformable rocks are investigated. The numerical results show that: (1) There exist two stages of the slip process (i.e., the stick-slip instability) as shown in Figure 2. (2) The stick-slip instability initiates on the lower fault segment DE and stops near the fault bend center E, then restarts near the fault bend on the upper fault segment EF. (3) The first instability only occurs on the lower fault segment DE while the second propagates along both segments DE and EF, and the maximum value of the relative velocity at stage A2 is considerably larger than that at stage A1. All the above three phenomena are mainly due to the nonuniform distribution of the normal contact force as shown in Figure 4. During the total loading processes, the normal contact forces along the entire interface rise, however due to the effect of the fault bend, those on the upper segment EF increase more quickly than those on the lower segment DE during the second loading stage. Hence the nodes on the lower segment enter the slip earlier but with a smaller value of the relative velocity.

In summary, the fault bend significantly influences the nucleation, termination and restart of the stick-slip instability along the intra-plate fault, although the details may vary with the different boundary conditions and need further research. Thus the fault geometry should be carefully considered in the earthquake prediction. In addition, from the above numerical results, the current finite-element algorithm can favorably simulate the key phenomena induced by the fault bend and will help lay the foundation for future earthquake prediction.

References

BRACE W. F., BYERLEE J. D. (1966), *Stick-slip as a mechanism for earthquakes*, Science, *153*, p990-992

DIETERICH, J. H. (1979), *Modelling of Rock Friction 1. Experimental Results and Constitutive Equations*, J. Geophys. Res. *84*, 2161–2168.

KING, G. and NABELEK, J. (1985), *Role of Fault Bends in the Initiation and Termination of Earthquake Rupture*, Science *228*, 984–987.

KATO, N., SATOH, T., LEI, X., and HIRASAWA, T. (1999), *Effects of Fault Bend on the Rupture Propagation Process of Stick-Slip*, Tectonophysics *310*, 81–99.

OGLESBY, D. D. and DAY, S. M. (2001), *Fault Geometry and the Dynamics of the 1999 Chi-Chi Earthquake*, Bull. Seismol. Soc. Am. *91*, 1099–1111.

RUINA, A. L. (1983), *Slip Instability and State Variable Friction Laws*, J. Geophys. Res. *88*, 10,359—10,370.

SCHOLZ, C. H. (1998), *Earthquakes and Friction Laws*, Nature *391*, 37–42.

XING H. L. and MAKINOUCHI, A. (1998a), *FE Modelling of 3-D Multi-Elasto-Plastic-Body Contact in Finite Deformation, Part 1. A Node-to-Point Contact Element Strategy, Part 2. Applications*. In Proceedings of Japanese Spring Conference for Technology of Plasticity (JSCTP), (JSTP, Osaka) Nos.322 and 323.

XING H. L., FUJIMOTO, T., and MAKINOUCHI, A. (1998b), *Static-Explicit FE Modelling of 3-D Large Deformation Multibody Contact Problems on Parallel Computer*. In Proceedings of NUMIFORM'98, (Eds. Huetink, J. and Baaijens, F. P. T.) (A.A. Balkema, Rotterdam), pp. 207–212.

XING, H. L. and MAKINOUCHI, A. (2000), *A Node-to-point Contact Element Strategy and its Applications*, RIKEN Review: High Performance Computing in RIKEN *30*, 35–39.

XING, H. L. and MAKINOUCHI, A. (2002a), *Finite-element Analysis of a Sandwich Friction Experiment Model of Rocks,* pure Appl. Geophys. *159*, 1985–2009.

XING, H. L. and MAKINOUCHI, A. (2002b), *Finite-element Modeling of Multiboby Contact and its Application to Active Faults,* Concurrency and Computation: Practice and Experience *14*, 431–450.

XING, H. L. and MAKINOUCHI, A. (2002c), *Three-dimensional Finite-element Modeling of Thermomechanical Frictional Contact Between Finite Deformation Bodies Using R-Minimum Strategy*, Computer Methods in Appl. Mech. Engin. *191:* 4193–4214.

YAMADA, Y., YOSHIMURA, N., and SAKURAI, T. (1968), *Plastic Stress-strain Matrix and its Application for the Solution of Elastic-Plastic Problems by Finite-element Method,* Int. J. Mech. Sci. *10*, 343–354.

(Received September 27, 2003, revised March 20, 2004, accepted March 30, 2004)

To access this journal online:
http://www.birkhauser.ch

Pure appl. geophys. 161 (2004) 2103–2118
0033–4553/04/102103–16
DOI 10.1007/s00024-004-2551-0

© Birkhäuser Verlag, Basel, 2004

▎Pure and Applied Geophysics

Quasi-static and Quasi-dynamic Modeling of Earthquake Failure at Intermediate Scales

GERT ZÖLLER[1], MATTHIAS HOLSCHNEIDER[2], and YEHUDA BEN-ZION[3]

Abstract—We present a model for earthquake failure at intermediate scales (space: 100 m–100 km, time: 100 m/v_{shear}- 1000's of years). The model consists of a segmented strike–slip fault embedded in a 3-D elastic solid as in the framework of BEN-ZION and RICE (1993). The model dynamics is governed by realistic boundary conditions consisting of constant velocity motion of the regions around the fault, static/ kinetic friction laws with possible gradual healing, and stress transfer based on the solution of CHINNERY (1963) for static dislocations in an elastic half-space. As a new ingredient, we approximate the dynamic rupture on a continuous time scale using a finite stress propagation velocity (quasi–dynamic model) instead of instantaneous stress transfer (quasi–static model). We compare the quasi–dynamic model with the quasi–static version and its mean field approximation, and discuss the conditions for the occurrence of frequency-size statistics of the Gutenberg–Richter type, the characteristic earthquake type, and the possibility of a spontaneous mode switching from one distribution to the other. We find that the ability of the system to undergo a spontaneous mode switching depends on the range of stress transfer interaction, the cell size, and the level of strength heterogeneities. We also introduce time-dependent log (t) healing and show that the results can be interpreted in the phase diagram framework. To have a flexible computational environment, we have implemented the model in a modular C++ class library.

Key words: Earthquakes, fault models, dynamic properties, seismicity.

1. Introduction

In recent years various models of earthquake sequences have been developed. Although the verification of such models is difficult due to limited data, considerable progress has been made with respect to the generation and understanding of various seismicity patterns. An important goal is the development of conceptual models, which are simple enough to allow some analytical understanding of the relevant processes, but also produce seismic dynamics that is to some degree realistic. Such models have in general a set of tuning parameters that should not be too large. For

[1] Institute of Physics and Institute of Mathematics, University of Potsdam, Potsdam, Germany. E-mail: gert@agnld.uni-potsdam.de

[2] Institute of Mathematics, University of Potsdam, Potsdam, Germany, on leave from CNRS Rennes, France. E-mail: hols@math.uni-potsdam.de

[3] Department of Earth Sciences, University of Southern California, Los Angeles, U.S.A. E-mail: benzion@terra.usc.edu

some parameters empirical values are available, while others can be used to tune the model dynamics towards an expected behavior. The model of BEN-ZION and RICE (1993) of a fault in a 3-D elastic half-space appears to meet these criteria. Using this model, several observed frequency-size and temporal statistics could be explained in terms of structural properties of a given fault.

The most striking feature of seismicity is the frequency-size distribution, which follows in regional domains the Gutenberg-Richter power-law relation. On an individual fault, the situation can be different: although power-law scaling is always observed for a certain range of magnitudes, deviations are found for large magnitudes. On one hand, the Gutenberg-Richter scaling can break down; that is, large events may be suppressed. On the other hand, the large magnitude range can be dominated by a frequently occurring "characteristic" earthquake. Using a mean field approximation of the model of BEN-ZION and RICE (1993), it has been shown that different frequency-size distributions can result from different values of a dynamic weakening coefficient controlling the brittle properties of the fault, and a conservation parameter that determines the amount of stress transfer remaining on the fault (DAHMEN *et al.* 1998; FISHER *et al.* 1997). For certain parameter values, the system can spontaneously switch from one state to the other. In HAINZL and ZÖLLER (2001) a cellular automaton model of Burridge-Knopoff type (BURRIDGE and KNOPOFF, 1967; HAINZL *et al.*, 1999) has been analyzed with the result that the degree of disorder and stress concentration can tune the frequency-size statistics. The question whether or not a small earthquake can grow into a system wide event is connected to the degree of smoothness (or roughness) of the stress field. In BEN-ZION (1996) and BEN-ZION *et al.* (2003) it is argued that the smoothing of long wavelength components of stress can be interpreted as a development of long-range correlations or, in terms of critical phenomena, as an increase of the spatial correlation length, which prepares the fault for a large event (SORNETTE and SAMMIS, 1995; ZÖLLER *et al.* 2001; ZÖLLER and HAINZL, 2002). Consequently, all processes that influence the smoothness of the stress field, e.g., the type of stress transfer during a rupture and quenched heterogeneities, are relevant for the dynamics.

Most existing models for earthquake failure are governed by instantaneous stress transfer (quasi-static) during the rupture process, and in part by unrealistic stress transfer functions like nearest-neighbor interaction or homogeneous stress transfer independent of the position on the fault and the rupture dimension (see BEN-ZION, 2001, for a recent summary of fully dynamic earthquake models). As in the present work, the fault is usually discretized into uniform cells. To account for a more realistic rupture process, the quasi-static approach is extended here to a quasi-dynamic one by introducing a finite communication speed (GABRIELOV *et al.*, 1994). Using Chinnery's solution for a strike-slip fault in a 3-D elastic half-space (CHINNERY, 1963), realistic boundary conditions, dynamic weakening, and optionally a gradual time-dependent healing, provides a realistic, but still relatively simple earthquake model on the intra- and the inter-event time scale.

The results indicate that the quasi-dynamic model favors for certain parameter values the occurrence of large events. In particular, the truncation of the Gutenberg-Richter law in the quasi-static approach vanishes. We further show that the ability of a system to undergo a mode switching between the Gutenberg-Richter law and a characteristic earthquake distribution depends on the range of the stress transfer interaction, the cell size, and the strength heterogeneities. If $\log(t)$ healing is included, the time-dependence can be absorbed in an effective dynamic friction and stress loss in the model without healing.

2. Model Framework

We assume conceptually a hierarchical model that consists of three hierarchies: the system (top level) as a whole contains a set of faults (middle level); each fault is composed of an array of cells (bottom level). The system is embedded in a three-dimensional elastic half-space. At the system level, the interaction between the faults is accounted for. The fault controls the interaction of the cells during an event, while the accumulation and the release of stress takes place on the cell level.

At the present state of model development, only a single rectangular fault is considered. Unless stated otherwise, a fault of 70-km length and 17.5-km depth is covered by a computational grid, divided into 128×32 uniform cells, where deformational processes are calculated. As discussed in BEN-ZION and RICE (1993), this geometry corresponds approximately to the San Andreas fault near Parkfield, CA. Tectonic loading is imposed by a motion with constant velocity $v_{pl} = 35$ mm/year of the regions around the computational grid. The space-dependent loading rate provides realistic boundary conditions. Using the static stress transfer function $K(i, j; k, l)$ from CHINNERY (1963), the tectonic loading for each cell (i, j) is a linear function of time t and plate velocity $v_{p/1}$:

$$\Delta\tau(i, j; t) = (-v_{pl} \cdot t) \cdot \sum_{k,l \in \text{fault}} K(i, j; k, l), \tag{1}$$

where the minus sign stems from the fact that forward (right-lateral) slip of regions around a locked fault segment is equivalent to back (left-lateral) slip of the locked fault segment. The grid of cells is governed by a static/kinetic friction law, i.e., a cell slips initially if the static friction τ_s is exceeded. The threshold decreases instantaneously to the dynamic friction $\tau_d < \tau_s$ and remains there until the earthquake is terminated (model without healing during events). The stress itself drops to the arrest stress $\tau_a < \tau_d$. This process of dynamic weakening can be parameterized by the dynamic overshoot coefficient $D = (\tau_s - \tau_a)/(\tau_s - \tau_d)$. Following the description in DAHMEN et al. (1998), we set $\tau_s = 1$ and $\tau_a = 0$ and use the dynamic weakening coefficient $\varepsilon = (\tau_s - \tau_d)/\tau_s = 1 - \tau_d$ to connect static/kinetic friction and arrest stress. Consequently, the dynamic weakening parameter

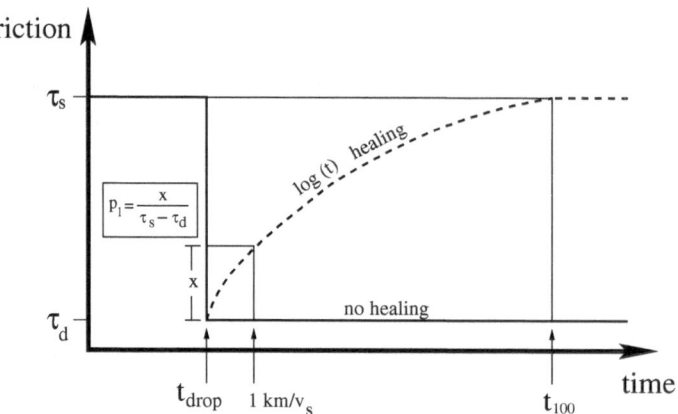

Figure 1

Sketch of frictional processes: in the model without healing during events (solid line), the friction drops from τ_s to τ_d and remains there until the earthquake is terminated, at which time there is instantaneous healing. In the model with healing during events (dashed line) the friction increases as $\log(t)$ after the stress drop. The shape of the healing curve is parameterized by the values of p_1 and t_{100}.

$\varepsilon = 1 - \tau_d$ is bounded between 0 and 1. The case $\varepsilon = 0$ represents the unrealistic case of static friction equals kinetic friction, or instantaneous healing. To account for heterogeneities in the brittle properties, some uniformly distributed noise $\in [-0.5; 0.5]$ is added to τ_d.

As yet, the frictional properties of the fault are strongly simplified; the strength envelope, which describes the change of the friction coefficient by means of static and dynamic frictions, is a piecewise constant function (see solid line in Fig. 1). It is, however, known from laboratory experiments that ruptures are governed by more complicated friction laws (DIETERICH, 1972; 1978). Most of the experimental results are described by the rate- and state-dependent constitutive law (RUINA, 1983; SCHOLZ, 1998). Within this law, the response to a sudden change of the sliding velocity is an instantaneous effect on the friction coefficient followed by a more gradual evolution.

As a first step towards a more realistic friction law, we introduce a gradual time-dependent healing of the form (see dashed line in Fig. 1)

$$\tau_d(t) = \tau_{d,0} + C \log(1 + (t/t_0)), \tag{2}$$

where $\tau_{d,0}$ is the initial dynamic friction coefficient, t_0 is a reference time and C is a free parameter. The healing begins after the instantaneous drop of the friction from τ_s to $\tau_{d,0}$. For our numerical simulations, we parameterize the healing with two parameters p_1 and t_{100}, instead of C and t_0. The value of p_1 gives the fraction of $\tau_s - \tau_{d,0}$ that has healed after a time of $1\ km/v_s$, and the value t_{100} gives the time interval required for complete healing, $\tau_d(t_{100}) = \tau_s$.

The stress transfer during an earthquake is calculated by means of the three-dimensional solution of CHINNERY (1963) for static dislocations on rectangular patches in an elastic Poisson solid with rigidity $\mu = 30$GPa. In particular, we approximate the $3 + 1$ dimensional space-time stress transfer by

$$\Delta\tau(i,j;t) = (1 - \gamma) \cdot \sum_{k,l \in \text{fault}} K(i,j;k,l)\Delta u(k,l;t - r/v_s), \tag{3}$$

where Δu is the slip, r is the spatial distance between the cells (i,j) and (k,l), and v_s is the shear-wave velocity. The factor $1 - \gamma \in (0;1]$ corresponds to a given ratio of rigidities governing during instabilities the self-stiffness of a slipping cell (diagonal elements of the stiffness matrix) and the stress transfer to the surrounding domain (off-diagonal elements). A ratio smaller than 1 represents stress loss during rapid slip on the fault to internal free surfaces in the solid associated with porosity and cracks. We refer to γ and $1 - \gamma$ the stress loss parameter and stress conservation parameter, respectively. The slip $\Delta u(i,j)$ of a cell at a position (i,j) is related to the stress drop $\Delta\tau(i,j)$ at the same position through the self-stiffness: $\Delta u(i,j) = \Delta\tau(i,j)/K(i,j;i,j)$. The size of an earthquake is measured with two quantities: 1. the area, which is proportional to the number of cells that participated in the earthquake failure, and the potency, which is the integral of the slip over the rupture area. This approach extends the model of BEN-ZION and RICE (1993) to a quasi-dynamic procedure with a finite communication speed v_s for stress transfer and a related causal rupture process. In one version of the model, the quasi-dynamic rupture process is calculated on a continuous time scale. This may, however, lead to a numerical explosion during the simulation, for certain parameter values. Therefore, we also study simplifications of the continuous time process, e.g., by discretizing the time scale for the stress transfer during an earthquake.

At the present state, the model is characterized by two separate time scales: the inter-event time scale during which the fault is loaded between two events, and the intra-event time scale defined by the travel time of a shear wave along the fault, where coseismic stress redistribution takes place. During the event, the tectonic loading is neglected. The stress conservation parameter $1 - \gamma$, which controls the amount of stress remaining on the computational grid, is varied between $\gamma \approx 1$ (no stress redistribution) and the conservative case $\gamma = 0$ (stress drop $\tau - \tau_a$ is completely redistributed). In general, γ controls the size of the generated earthquakes: for high values of γ, small amounts of stress are redistributed and the evolution of cascading failure events stops earlier than in corresponding cases of small γ, for which large earthquakes can develop. In the latter case, numerical problems may occur, because large runaway events that cover the entire fault and have multiple slip episodes of each cell, result in a memory exhaustion on the continuous intra-event time scale. Therefore, we also study a simplification of the quasi-dynamic rupture: the intra-event time scale is discretized into N time intervals. Each slip of a cell is assigned to one of the time intervals. Note that the case $N = 1$ represents the quasi-static model

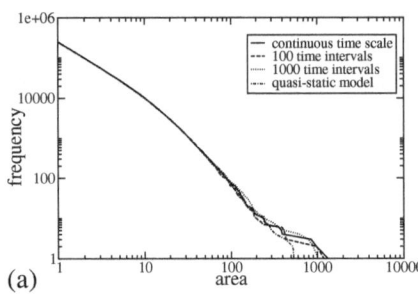
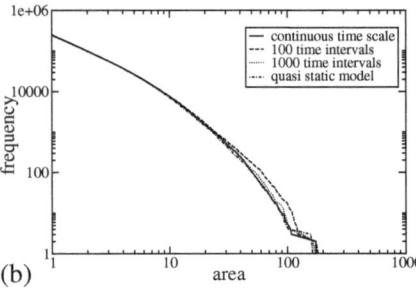

(a)   (b)

Figure 2

Frequency-size distribution for different model versions (without healing during events) employing $\gamma = 0.3$ with $\varepsilon = 0.8$ (Figure (a)) and $\varepsilon = 0.0$ (Figure (b)). The solid line refers to the continuous intra-event time scale; the dashed and the dotted line refer to discretized time scales with the 100 and 1000 time intervals, respectively. The dash-dotted line refers to the quasi-static model of BEN-ZION and RICE (1993). The area is given in units of the used cell size.

used in BEN-ZION and RICE (1993). Although it is mathematically not clear that the limit $N \to \infty$ converges to the continuous time scale, we assume that a value of N exists, which approximates the continuous case with a reasonable accuracy. In our simulations we used $N = 1000$. Given the dimensions of our grid, a stress signal may travel ten times along the fault before the time error becomes comparable to the travel time between neighboring cells. The model simulations are started with a random distribution of initial stress. To account for transient effects in the dynamics, the first 50,000 earthquakes are neglected in each simulation.

3. Model Simulations

3.1. Influence of Intra-event Dynamic

Figure 2 gives the frequency-size distribution for different versions of the intra-event time scale. Figure 2(a) shows results calculated with $\gamma = 0.3$ and a realistic value of the dynamic weakening coefficient $\varepsilon = 0.8$ corresponding to a dynamic overshoot coefficient $D = 1.25$ (MADARIAGA, 1975). Figure 2(b) shows the same calculation for the unrealistic case of $\varepsilon = 0$ and instantaneous healing ($\tau_d = \tau_s$), where the frequency-size statistics is a truncated power law. The results indicate that the continuous time scale (solid line) can be approximated quite well by the discretized time scale with 1000 intervals (dotted line). In the case without large earthquakes, there is no significant difference between the quasi-dynamic model and the quasi-static model (dash-dotted line). In contrast, Figure 2(a) shows a clear fall-off for large events in the quasi-static case. This difference is a stable feature; that is, it is also present for a broad range of parameter values ε and γ that allow large earthquakes to occur. Consequently, the quasi-static approximation seems to

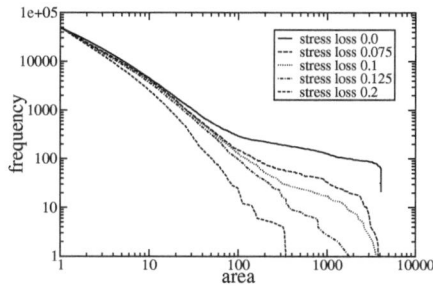

Figure 3

Frequency-size statistics for different values of the stress loss factor γ with $\varepsilon = 0.6$ (no healing during events). The transition from the Gutenberg-Richter distribution to the characteristic earthquake distribution takes place at $\gamma \approx 0.125$.

suppress large earthquakes leading to a truncation in the frequency-size statistics. For large earthquakes, the propagation of stress with finite communications speed v_s is obviously a relevant feature.

3.2. The Phase Diagram Framework

In DAHMEN *et al.* (1998) a simplified two parameter mean field version of the quasi-static model of BEN-ZION and RICE (1993) has been analyzed. A "conservation parameter" c (similar to our $1 - \gamma$) has been introduced, and an infinite-range mean-field stress transfer function $G \sim c/N$ (N = total number of cells) was used instead of Chinnery's solution for the 3-D elastic stress transfer. For this model, a phase diagram spanned by c and the dynamic weakening coefficient ε contains two distinct phases: one phase where the frequency-size distribution follows the Gutenberg-Richter law, and a second phase governed by a spontaneous mode switching between Gutenberg-Richter statistics and the characteristic earthquake distribution. In particular, it is hypothesized that these phases are generic for more realistic stress transfer functions. In the first part of this study, we analyze this claim for various types of interactions. We observe in agreement with WEATHERLEY *et al.* (2002) that with coarse cell size there is a threshold of the interaction range, below which no mode-switching occurs. However, with small enough cells and strong heterogeneities, mode-switching also occurs for the realistic case of elastic stress transfer. In the second part, we consider the influence of time-dependent $\log(t)$ healing. The results show that the model can be mapped onto the same phase diagram if the net effect of healing is absorbed in effective dynamic threshold and stress loss.

3.2.1. The phase diagram for the quasi-dynamic and elastic model.

Figure 3 shows frequency-size statistics for a fixed dynamic weakening coefficient $\varepsilon = 0.6$ as a function of the stress loss factor γ. High values of γ prevent the occurrence of large earthquakes, in that stress that is needed to bring cells to failure is

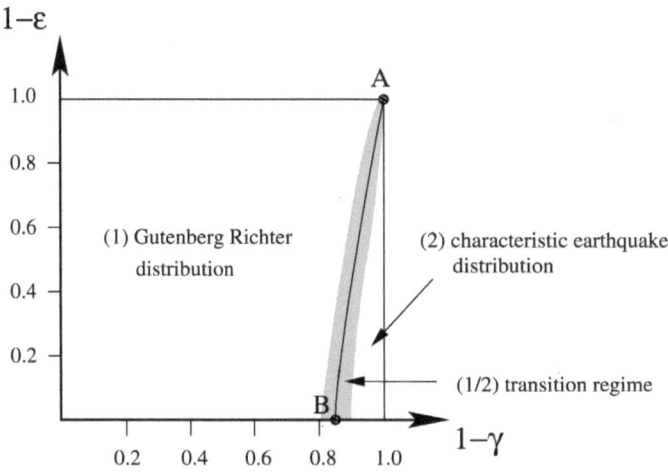

Figure 4

Schematic phase diagram for a fault with $N = 128 \times 32$ cells as a function of the dynamic weakening parameter ε and the stress loss factor γ (model without healing during events). The line AB separates the Gutenberg-Richter phase and the phase where large characteristic earthquakes occur.

lost and earthquakes are stopped earlier. Furthermore, the figure shows a transition from a truncated Gutenberg-Richter distribution ($\gamma = 0.2$) to a characteristic earthquake distribution ($\gamma = 0$). For $\gamma \approx 0.125$, the system bifurcates to the two regimes.

Our calculations with the quasi-dynamic elastic model lead (Fig. 4) to a large Gutenberg-Richter regime with small earthquake sizes, and a regime governed by a characteristic earthquake distribution with some short time fluctuations. Note that the factor c in DAHMEN *et al.* (1998) is now replaced by $1 - \gamma$. The characteristic earthquake regime is smaller than the mode switching phase in DAHMEN *et al.* (1998), because in the elastic model a small fraction of the stress drop is lost even for $\gamma = 0$ due to the finite fault size. We have estimated that on average 80% of the stress drop remains on the fault for the employed 70 km and 17.5 km dimensions with $\gamma = 0$. Consequently, the case $\gamma = 0$ for the elastic model corresponds to $c = 0.8$ of the mean field model. This is approximately confirmed by a comparison of Figure 4 with the phase diagram in DAHMEN *et al.* (1998).

3.2.2. *Mode-switching between Gutenberg-Richter and characteristic earthquake statistics.*

To address the question of a possible mode switching, long simulations are required. Therefore, we use for the following analysis a fault of 10.9-km length and 2.7-km depth divided into 20×5 uniform cells. Figure 5 shows a typical earthquake sequence (rupture area vs. time) from a catalog representing the characteristic earthquake regime. The quasi-periodically occurring characteristic earthquakes are

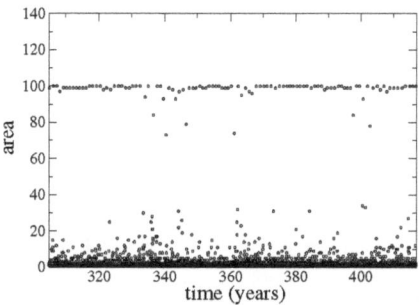

Figure 5
Earthquake area as a function of time for the quasi-static model ($N = 20 \times 5$, $\varepsilon = 0.6$, $\gamma = 0.03$, no healing during events). The sequence follows the characteristic earthquake distribution, although the system attempts to switch to a Gutenberg-Richter phase, e.g., at $t \approx 361$.

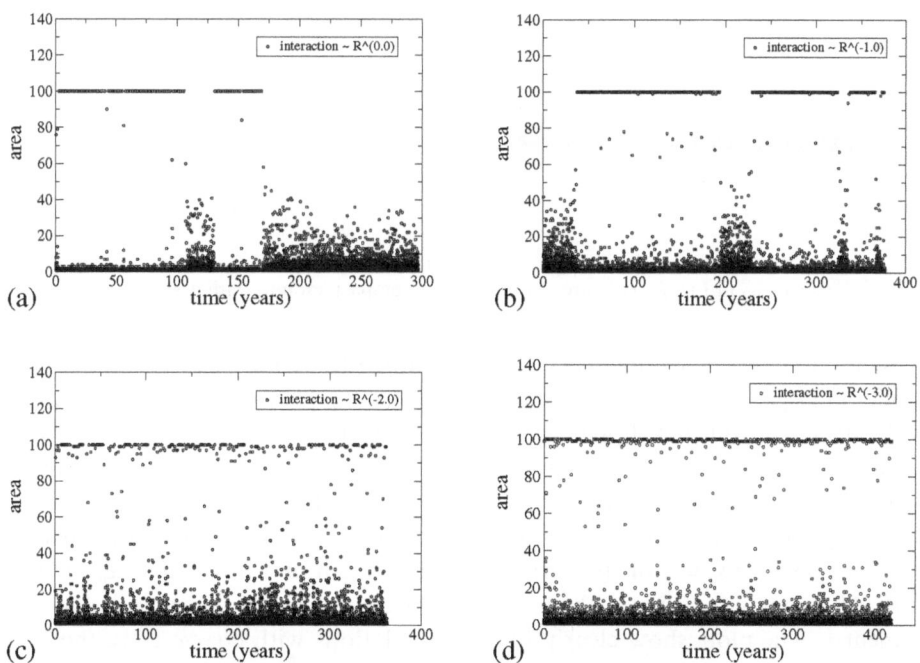

Figure 6
Earthquake area as a function of time for the quasi-static model ($N = 20 \times 5$, $\varepsilon = 0.6$, no healing during events) and different types of interaction $\sim r^{-x}$. The stress loss factor γ has been adjusted in order to bring the system into the transition regime between Gutenberg-Richter and characteristic earthquake distributions. The case $x = 0$ represents the mean field approximation [DAHMEN et al., 1998], while $x = 3$ corresponds to the elastic solution of CHINNERY [1963].

interrupted for short time by clusters of small and intermediate earthquakes. It seems as if the system attempts to switch into the other regime and flips back after a short time, e.g., for $t = 361$ in Figure 5. In Figure 6 we show that the ability of the system

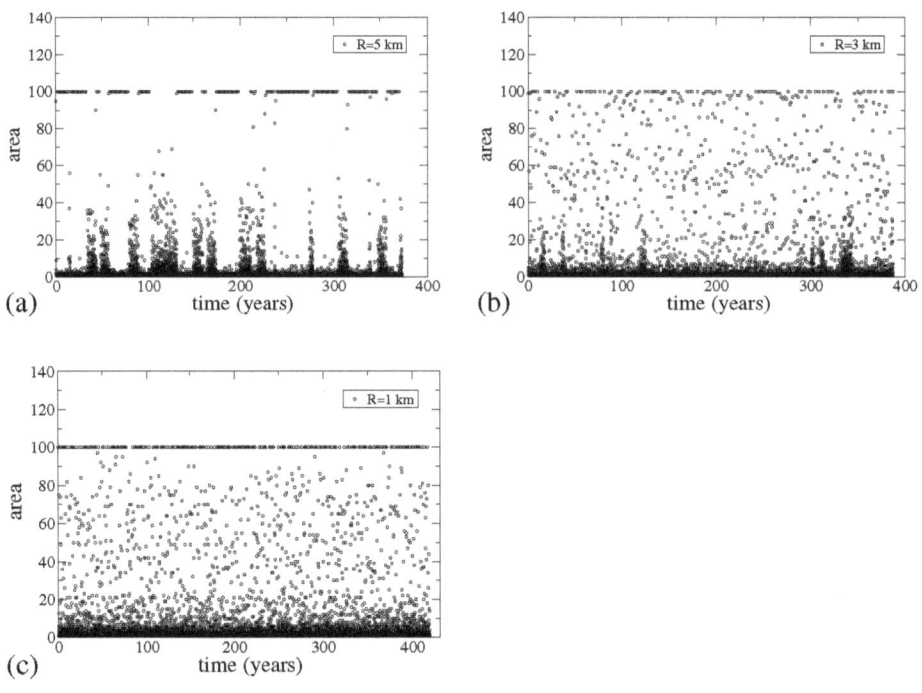

Figure 7

Earthquake area as a function of time for the quasi-static model ($N = 20 \times 5$, $\varepsilon = 0.6$, no healing during events) as in Fig. 6, with stress transfer on a compact support with a radius R.

to undergo a mode-switching depends significantly on the kind of stress transfer. We consider interactions between the limit cases of mean field interaction ($\Delta\tau \sim r^0$, where r is the distance from the slipped cell) and the elastic solution of CHINNERY (1963), $\Delta\tau \sim r^{-3}$. These simulations are performed on a 20×5 grid with dynamic weakening $\varepsilon = 0.6$. Since mode-switching most likely occurs in the transition regime between Gutenberg-Richter and characteristic event distribution, we adjust the stress loss factor γ to be in this part of the parameter space. Each plot in Figure 6 gives results for 10,000 simulated events with stress transfer $\Delta\tau \sim r^{-x}$ and a certain value of the exponent x. The plots show clearly that in simulations with coarse cells, the mode-switching in the mean field approximation degenerates with increasing exponent x to the short-term fluctuations described above. We also performed similar calculations for a uniform stress transfer on a compact support (circle with radius R), $\Delta\tau = \tau_0 \cdot \Theta(r - R)$. Figure 7 shows earthquake sequences as in Figure 6 for this kind of interaction. Again, the mode-switching behavior, which is clearly visible for large radii, degenerates to short-term fluctuations for small radii. Although we did not observe mode-switching of the type described in DAHMEN *et al.* (1998) in earthquake catalogs up to 20,000,000 events, it cannot be ruled out that this behavior may occur with a very large persistence time in one mode.

In sum, we have found that the ability of the system to undergo a mode-switching from the Gutenberg-Richter distribution to the characteristic earthquake distribution with coarse cell size depends on the range of the stress transfer interaction. Similar results were found in (WEATHERLEY et al., 2002). While mode-switching occurs quite frequently for the infinite range mean field interaction, this behavior could not be observed for the more realistic long-range 3-D elastic interaction, although the system attempts to switch from time to time. In this case, the transition from the Gutenberg-Richter regime to the characteristic earthquake regime with coarse grid is continuous; that is, a decrease of γ leads to the growth of the average magnitude. The limit case ($\gamma \rightarrow 0$) is the characteristic earthquake distribution with quasi-periodically occurring earthquakes that rupture the entire fault. It is also visible that in contrast to the mean field approximation, intermediate size earthquakes are also present in the characteristic earthquake regime. These observations also hold for the quasi-dynamic model.

Until now we have kept the cell size fixed and have varied only the range of stress transfer interaction. It is, however, possible that smaller numerical cells will produce mode-switching, since such will lead to larger stress concentrations near failure areas and larger stress fluctuations. To investigate this hypothesis we have to consider a grid with numerous cells. Note that a change of the fault dimension in terms of total length and depth produces the same earthquake sequences with a rescaled time axis. Such an analysis requires very time-consuming numerical simulations due to two reasons: 1. A new phase diagram corresponding to Figure 4 must be calculated; 2. the transition between both phases must be scanned with a very high resolution in order to extract the range of parameters within which mode-switching is expected in reasonable short simulations. In this context it is important to note that the persistence time in a certain mode depends not only sensitively on γ and ε, but also on the number of cells N. In the mean field model, the persistence time increases according to $\exp(N)$ (DAHMEN et al., 1998).

In Figure 8 we show two simulations using the elastic stress transfer on a 128×50 grid and values for τ_d which are uniformly distributed between 0.15 and 0.35 ($\langle \varepsilon \rangle = 1 - \langle \tau_d \rangle = 0.75$), resulting in a more heterogeneous distribution of the brittle properties. The two plots refer to different values of the stress loss γ. The results clearly show a tendency towards mode-switching behavior, e.g., for $t \in [6.4; 9.0]$ in Figure 8(a). We can thus conclude that mode-switching also depends on the degree of heterogeneity in a system, determined by the cell size and the distribution of the brittle properties. While in the constant mean-field interaction, a large fraction of the parameter-space is governed by mode-switching. This phenomenon seems to occur with the more realistic $1/r^3$ elastic interaction in considerably smaller ranges of parameters. However, the region in parameter space producing mode-switching in the case of $1/r^3$ interaction is expected to increase with further decreasing of cell size and increasing of heterogeneities.

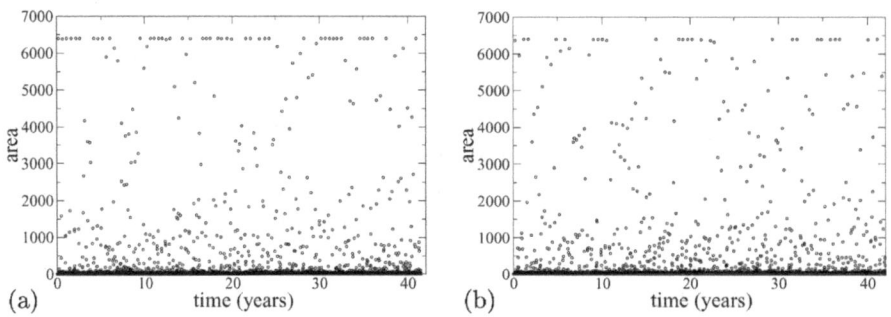

Figure 8
Earthquake area as a function of time for simulations of the quasi-static model (without healing during
events) on a fault with $N = 128 \times 50$ cells. The stress loss parameter is $\gamma = 0.01$ in (a) and $\gamma = 0.015$ in (b).
The dynamic weakening coefficients for the cells are drawn from a uniform distribution $\varepsilon \in [0.65; 0.85]$.

3.2.3 Time-dependent healing.

The introduction of time-dependent healing according to Eq. (2) replaces the piecewise constant strength envelope by a function consisting of a strength drop followed by a $\log(t)$ increase of the strength. This kind of healing requires three parameters: the value of p_1 (healing rate after 1 km/v_s stress propagation, the value t_{100} (time when healing is complete) and the values $\tau_{d,0}$ to which the strength drops when a cells begins to slide. For the present analysis we keep t_{100} fixed $(t_{100} = 100 km/v_s)$ and tune the shape of the healing curve by means of p_1 and $\tau_{d,0}$. The stress loss γ is an additional parameter independent of the healing.

It is a reasonable assumption that in the case of healing, effective values of $\tau_{d,0}$ and γ exist that correspond to τ_d in the case without healing. To investigate this hypothesis we determine for different values of p_1 and γ the dynamic threshold $\tau_{d,0}^*$, where the transition between the Gutenberg-Richter law and the characteristic earthquake distribution occurs. The value $\tau_{d,0}^*$ as a function of p_1 and γ is then compared with the corresponding dynamic threshold τ_d^* in the system without healing.

Figure 9 shows for the 20×5 grid the dependence of $\tau_{d,0}^*$ on p_1 for different values of γ (Fig. 9(a)) and the dependence of $\tau_{d,0}^*$ on γ for different values of p_1 (Fig. 9(b)). The curves for the model with healing $(p_1 > 0)$ are normalized to the model without healing $(p_1 = 0)$. In particular, the vertical axis gives $\tau_{d,0}^*(p_1)/\tau_{d,0}^*(p_1 = 0)$. Note that $\tau_{d,0}^*(p_1 = 0) = \tau_d^*$ describes the model without healing. In Figure 9(b) the curves are normalized to the horizontal line $\tau_{d,0}^*(p_1)/\tau_{d,0}^*(p_1 = 0) \equiv 1$. Figure 9(a) shows clearly a systematic decrease of $\tau_{d,0}^*$ for growing values of p_1. Moreover, $\tau_{d,0}^*$ also decreases as a function of p_1, although this decrease is less significant. The dependence of $\tau_{d,0}^*$ on p_1 and γ can be described by the formula

$$\tau_{d,0}^*(p_1, \gamma) = \tau_d^* \cdot (1 - s(p_1, \gamma)), \tag{4}$$

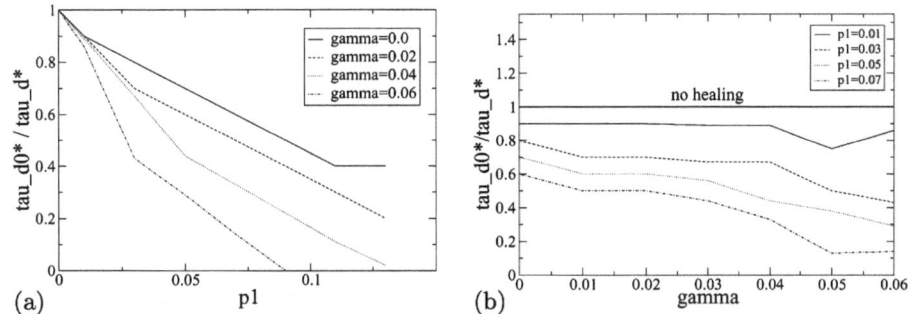

Figure 9

Model with healing: dynamical threshold $\tau_{d,0}^*$ at the transition between Gutenberg-Richter law and characteristic earthquake distribution (a) as a function of p_1 for different values of the stress loss γ and (b) as a function of γ for different values of p_1. The vertical axis in (a) is normalized to 1 at $p_1 = 0$; in (b) the curves for $p_1 > 0$ are normalized to the model without healing ($p_1 = 0$, horizontal line).

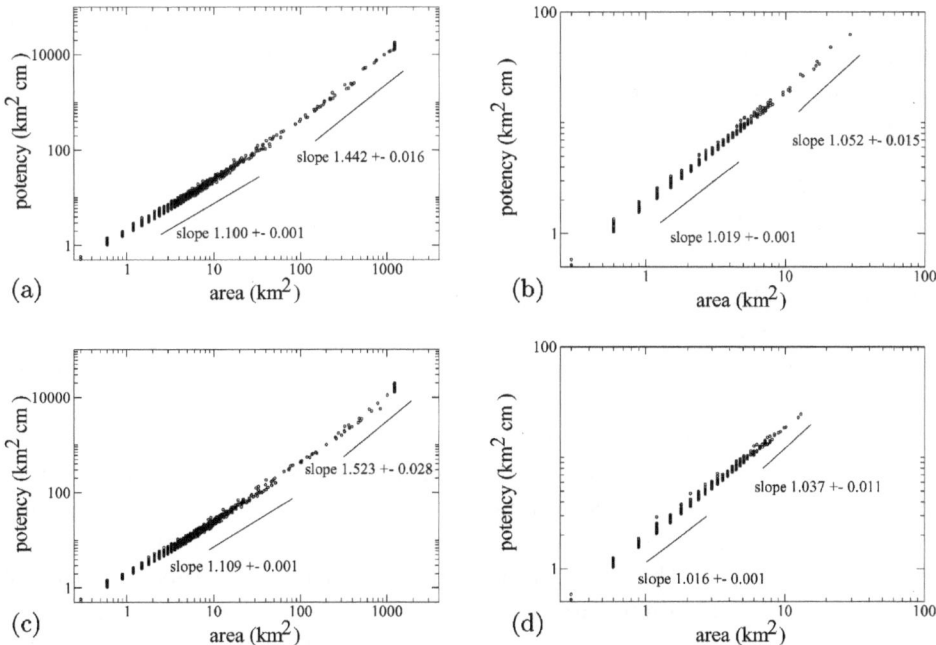

Figure 10

Relation between potency and rupture area in a model realization with $\varepsilon = 0.8$ and (a) $\gamma = 0.0$ or (b) $\gamma = 0.3$ (model without healing during events). Panels (c) and (d) show the corresponding results for the quasi-static model.

where $s(p_1, \gamma)$ is given by the curves in Figure 9. This rule connects the model with healing ($p_1 > 0$) and the model without healing ($p_1 = 0$). Consequently, p_1 is not an independent parameter like τ_d and γ. Rather, the model with healing can be obtained from the model without healing by a shift in the phase diagram according to Eq. (4). The observation that the dependence of $s(p_1, \gamma)$ on p_1 is more pronounced than the dependence on γ arises from the origin of p_1: this parameter defines the rate of healing; in other words, the rate of increase of τ_d. It is, therefore, reasonable that an effective dynamic threshold (in the model without healing) exists that can to some degree absorb the time-dependence in the model with healing.

In general, it is an interesting task to also map other models systematically onto the present one with respect to the phase diagram. This would, however, require two independent state variables that allow positioning a given catalog in the phase diagram without pre-knowledge of the parameters. This work is left for the future.

3.3. Area-potency Relations

Finally, we investigate the area-potency relation. Figure 10 shows calculations for $\varepsilon = 0.8$ and $\gamma = 0.0$ and $\gamma = 0.3$, for both the quasi-dynamic and the quasi-static models. The change of slope for the transition from small to large earthquakes is in agreement with the quasi-static simulations in BEN-ZION and RICE (1993) and the data analysis in BEN-ZION and ZHU (2002). The slope for the large earthquakes is close to the exponent in the classical crack relation $a \sim p^{2/3}$ (KANAMORI and ANDERSON, 1975).

4. Conclusions

We have developed a model for earthquake failure at intermediate scales that is based on the framework of BEN-ZION and RICE (1993). The model is implemented in a flexible C++ class library and therefore allows the incorporation of additional processes and structural fault properties in a straightforward manner. In this study we extended the model of BEN-ZION and RICE (1993) by two important features. First, the stress propagation is modeled on a continuous intra-event time scale using a finite communication speed. This quasi-dynamic model results in more realistic behavior of the stress propagation. Second, we have introduced a gradual time-dependent healing after a cell has slipped. This mechanism modifies the frictional properties from the simple static/kinetic friction towards the rate- and state-dependent constitutive law.

The results indicate that the finite stress propagation velocity is a relevant feature even for the frequency-size statistics. The quasi-dynamic model produces larger events compared to the quasi-static model in which the occurrence of large events is suppressed. This is in agreement with the fully dynamic 2-D calculations of BEN-ZION and RICE (1997). Future activities with the quasi-dynamic version of the model include analysis of slip histories and possible calculations of synthetic seismograms. In another research direction we examined the ability of the model to undergo a spontaneous mode-switching between the Gutenberg-Richter law and the characteristic earthquake distribution. While the mean field model, which is characterized by a constant stress transfer function, exhibits clear mode-switching behavior, this is not the case in the model with the coarse grid and 3-D elastic stress transfer function. Instead, it is found that in the characteristic earthquake regime, the system attempts to switch into the Gutenberg-Richter regime, but flips back after a very short time. However, if the cell size is decreased and the degree of strength heterogeneity is increased, mode-switching behavior is observed, even for the $1/r^3$ elastic stress transfer function. This is in agreement with similar studies in more complex models (BEN-ZION et al., 1999; LYAKHOVSKY et al., 2001). Remaining important questions are which mechanisms are responsible for this behavior and whether these mechanisms are relevant for real faults.

The incorporation of time-dependent healing leads to a modification of the phase diagram shown in Figure 4. As we have shown, the frequency-size event distribution for a model with a $\log(t)$ healing can be reproduced by a model without healing and with different values of γ and τ_d. It is desirable to find two independent state variables that will allow quantification of a catalog and assignment of unique values of γ and τ_d. With such state variables one could map other model classes onto the present model.

Due to the modular design of our model code, it is easy to include more structural features and mechanisms into the model, e.g., a more refined frictional behavior and heterogeneities in the distribution of the arrest stress. A detailed analysis of statistical properties and seismicity patterns with various model versions and ranges of parameters will result in a deeper understanding of natural seismicity.

Acknowledgments

We are grateful to Karin Dahmen and Sebastian Hainzl for stimulating discussions. The reviews of Yaolin Shi and Yongxian Zhang aided our improvement of the manuscript. This work was supported by the "Sonderforschungsbereich 555" of the "Deutsche Forschungsgemeinschaft" and the USGS grant 02HQGR0047.

REFERENCES

BEN-ZION, Y. (1996), *Stress, Slip, and Earthquakes in Models of Complex Single-fault Systems Incorporating Brittle and Creep Deformations*, J. Geophys. Res. *101*, 5677–5706.

BEN-ZION, Y. (2001), *Dynamic Ruptures in Recent Models of Earthquake Faults*, J. Mech Phys. Sol. *49*, 2209–2244.

BEN-ZION, Y., DAHMEN, K., LYAKHOVSKY, V., ERTAS, D., and AGNON, A. (1999), *Self-driven Mode-switching of Earthquake Activity on a Fault System*, Earth and Plan. Sci. Lett. *172*, 11–21.

BEN-ZION, Y., M. ENEVA, and LIU, Y. (2003), *Large Earthquake Cycles and Intermittent Criticality on Heterogeneous Faults due to Evolving Stress and Seismicity*, 108, 10.1029/2002JB002121 J. Geophys. Res.

BEN-ZION, Y. and RICE, J. R. (1993), *Earthquake Failure Sequences along a Cellular Fault Zone in a Three-dimensional Elastic Solid Containing Asperity and Nonasperity Regions*, J. Geophys. Res. *98*, 14, 109–14, 131.

BEN-ZION, Y. and RICE J. R. (1997), *Dynamic Simulations of Slip on a Smooth Fault in an Elastic Solid*, J. Geophys. Res. *102*, 17, 771–17, 784.

BEN-ZION, Y. and ZHU, L. (2002), *Potency-magnitude Scaling Relations for Southern California Earthquakes with* $1.0 \leq M_L \leq 7.0$, Geophys. J. Int. *148*, F1–F5.

BURRIDGE, R. and KNOPOFF, L. (1967), *Model and Theoretical Seismicity*, Bull. Seimol. Soc. Am. *57*, 341–371.

CHINNERY, M. (1963), *The Stress Changes that Accompany Strike-slip Faulting*, Bull. Seimol. Soc. Am. *53*, 921–932.

DAHMEN, K., ERTAS, D., and BEN-ZION, R. (1998), *Gutenberg-Richter and Characteristic Earthquake Behavior in Simple Mean-field Models of Heterogeneous Faults*, Phys. Rev. E *58*, 1494–1501.

DIETERICH, J. (1972), *Time-dependent Friction in Rocks*, J. Geophys. Res. *77*, 3690–3697.

DIETERICH, J. (1978), *Time-dependent Friction and the Mechanics of Stick-slip*, Pure Appl. Geophys. *116*, 790–806.

FISHER, D. S., DAHMEN, K., RAMANATHAN, S., and BEN-ZION, Y. (1997), *Statistics of Earthquakes in Simple Models of Heterogeneous Faults*, Phys. Rev. Lett. *78*, 4885–4888.

GABRIELOV, A., NEWMAN, W. I., and KNOPOFF, L. (1994), *Lattice Models of Failure: Sensitivity to the Local Dynamics*, Phys. Rev. E *50*, 188–197.

HAINZL, S. and ZÖLLER, G. (2001), *The Role of Disorder and Stress Concentration in Nonconservative Fault Systems*, Physica A *294*, 67–84.

HAINZL, S., ZÖLLER, G., and KURTHS, J. (1999), *Similar Power Laws for Fore- and Aftershock Sequences in a Spring-block Model for Earthquakes*, J. Geophys. Res. *104*, 7243–7253.

KANAMORI, H. and ANDERSON, D. L. (1975), *Theoretical Basis of Some Empirical Relations in Seismology*, Bull. Seimol. Soc. Am. *65*, 1073–1095.

LYAKHOVSKY, V., BEN-ZION, Y., and AGNON, A. (2001), *Earthquake Cycle, Fault Zones, and Seismicity Patterns in a Rheologically Layered Lithosphere*, J. Geophys. Res. *106*, 4103–4120.

MADARIAGA, R. (1975), *Dynamics of an Expanding Circular Fault*, Bull. Seimol. Soc. Am. *66*, 639–666.

RUINA, A.L. (1983), *Slip Instability and State Variable Friction Laws*, J. Geophys. Res. *88*, 10,359–10,370.

SCHOLZ, C. H. (1998), *Earthquakes and Friction Laws*, Nature *391*, 37–42.

SORNETTE, D. and SAMMIS, C. G. (1995), *Complex Critical Exponents from Renormalization Theory Group of Earthquakes: Implications for Earthquake Predictions*, J. Phys. I France *5*, 607–619.

WEATHERLEY, D., MORA, P., and XIA, M. F. (2002), *Long-range Automaton Models of Earthquakes: Power-law Accelerations, Correlation Evolution, and Mode-switching*, Pure Appl. Geophys. *159*, 2469–2490.

ZÖLLER, G. and HAINZL, S. (2002), *A Systematic Spatio-temporal Test of the Critical Point Hypothesis for Large Earthquakes*, Geophys. Res. Lett. *29*, 10.1029/2002GL014856.

ZÖLLER, G., HAINZL, S., and KURTHS, J. (2001), *Observation of Growing Correlation Length as an Indicator for Critical Point Behavior prior to Large Earthquakes*, J. Geophys. Res. *106*, 2167–2176.

(Received September 27, 2003, revised March 20, 2004, accepted March 30, 2004)

 To access this journal online:
http://www.birkhauser.ch